T0176911

The General Theory of Relativity

The general theory of relativity, Einstein's theory of gravitation, has been included as a compulsory subject in undergraduate and graduate courses in Physics and Applied Mathematics all over the world. However, the physics-first approach that is taken by many textbooks is not universally used, as the approach often depends on the instructors' or students' background. Conceived from the lecture notes made by the author over a teaching career spanning 18 years, this book introduces the general theory of relativity for advanced students with a strong mathematical background.

The proposed book takes a 'math-first approach', for which the mathematical formalism comes first and is then applied to physics. It presents a concise yet comprehensive and structured understanding of the general theory of relativity. The book discusses the mathematical foundation of the general theory of relativity and focuses heavily on topics such as tensor calculus, geodesics, Einstein field equations, linearized gravity, Lie derivatives and their applications, the causal structure of spacetime, rotating black holes, and basic knowledge of cosmology and astrophysics. All of these are explained through a large number of worked examples and exercises.

Farook Rahaman is a Professor of Mathematics at Jadavpur University, Kolkata. Besides writing a book, *The Special Theory of Relativity*, he has published numerous research papers on galactic dark matter, wormhole geometry, charged fluid model, topological defects in the early universe, gravastars, black hole physics, star modeling, and the cosmological model of the universe.

The General Theory of Relativity

A Mathematical Approach

Farook Rahaman

CAMBRIDGE
UNIVERSITY PRESS

CAMBRIDGE
UNIVERSITY PRESS

University Printing House, Cambridge CB2 8BS, United Kingdom

One Liberty Plaza, 20th Floor, New York, NY 10006, USA

477 Williamstown Road, Port Melbourne, VIC 3207, Australia

314 to 321, 3rd Floor, Plot No.3, Splendor Forum, Jasola District Centre, New Delhi 110025, India

79 Anson Road, #06 04/06, Singapore 079906

Cambridge University Press is part of the University of Cambridge.

It furthers the University's mission by disseminating knowledge in the pursuit of education, learning and research at the highest international levels of excellence.

www.cambridge.org
Information on this title: www.cambridge.org/9781108837996

© Farook Rahaman 2021

This publication is in copyright. Subject to statutory exception
and to the provisions of relevant collective licensing agreements,
no reproduction of any part may take place without the written
permission of Cambridge University Press.

First published 2021

Printed in India by Thomson Press India Ltd.

A catalogue record for this publication is available from the British Library

Library of Congress Cataloging-in-Publication Data
Names: Rahaman, Farook, author.
Title: The general theory of relativity : a mathematical approach / Farook Rahaman.
Description: Cambridge, United Kingdom ; New York, NY : Cambridge University Press, 2021. |
    Includes bibliographical references and index.
Identifiers: LCCN 2020037664 (print) | LCCN 2020037665 (ebook) |
    ISBN 9781108837996 (hardback) | ISBN 9781108936903 (ebook)
Subjects: LCSH: Relativity (Physics)
Classification: LCC QC173.55 .R334 2021 (print) | LCC QC173.55 (ebook) |
    DDC 530.11–dc23 LC record available at https://lccn.loc.gov/2020037664 LC ebook record
    available at https://lccn.loc.gov/2020037665

ISBN 978-1-108-83799-6 Hardback

Cambridge University Press has no responsibility for the persistence or accuracy
of URLs for external or third party internet websites referred to in this publication,
and does not guarantee that any content on such websites is, or will remain,
accurate or appropriate.

To
my parents
Majeda Rahaman and Late Obaidur Rahaman
and
my son and wife
Md Rahil Miraj and Pakizah Yasmin

Contents

Figures

Tables

Preface

At the beginning of the twentieth century, Einstein spent many years developing a new theory in physics. The newly developed theory is known as the theory of relativity. This is basically a combination of two theories: the first one is known as the special theory of relativity and latter one is dubbed as the general theory of relativity. The special theory of relativity is based on two postulates, namely the principle of relativity or equivalence, that is, the laws of physics are the same in all inertial systems, which means no preferred inertial system exists, while the second postulate is the principle of the constancy of the speed of light. The general theory of relativity asserts that there is no difference between the local effects of a gravitational field and that of acceleration of an inertial system. In other words, spacetime is warped or distorted by the matter and energy in it as an effect of gravity. According to the general theory of relativity, massive objects cause the outer space to twist due to gravity like a heavy ball bending a thin rubber sheet that is holding the ball. Heavier balls bend spacetime far more than lighter ones. Like the special theory of relativity, the general theory of relativity attracted scientists a lot, immediately after its discovery by Einstein. As a result, it has been included as a compulsory subject in graduate and postgraduate courses of physics and applied mathematics all over the globe. Einstein proposed the field equations for the general theory of relativity by applying his own intuition. Later, many other methods were developed to construct Einstein's field equations.

This book on the general theory of relativity is an outcome of a series of lectures delivered by me, over several years, to postgraduate students of mathematics at Jadavpur University. I should mention that it is not a fundamental book. This book has been written, from a mathematical point of view, after consulting several books existing in the literature. I have provided the list of the reference books. During my lectures, many students asked questions that helped me know their needs as well as the shortcomings in their understanding. Therefore, it is a well-planned textbook that has been organized in a logical order and every topic has been dealt with in a simple and lucid manner. A number of problems with hints, taken from the question papers of different universities, are included in each chapter.

The book is organized as follows:

In Chapter One a brief overview of tensor calculus, including the different types of tensors as well as operations on tensors, is given. Generalized Kronecker delta, Christoffel symbols, affine connection, covariant derivatives, geodesic coordinate, and various forms of tensors are described, with examples, as a foreground to understand the basics of general relativity. Chapter Two starts with a discussion of the geodesic equation in curved spacetime. In addition, several problems for different spacetimes are provided on geodesics. Chapter Three begins with the statement of three basic principles, namely Mach's principle, equivalence principle, and the principle of covariance. Next, the Einstein gravitational field equations are derived from the variational principle.

Also, in this chapter, the outline of some modified theories of gravity, such as f(R) theory of gravity, Gauss–Bonnet gravity, f(G) theory of gravity or modified Gauss–Bonnet gravity, f(T) theory of gravity, f(R,T) theory of gravity, Brans–Dicke theory of gravity, and Weyl gravity, are provided. A discussion on linearized gravity is given in Chapter Four. Newtonian limit of Einstein field equations or weak field approximation of Einstein field equations is derived. It is shown that Poisson's equation can be viewed as an approximation of Einstein field equations. A short mathematical description of gravitational wave is also provided. Chapter Five is dedicated to a short discussion on Lie derivatives and their applications. Killing equations and Killing vectors are also discussed with several examples. A short note on conformal Killing vector is also provided. Chapter Six is devoted to discussions on spacetimes of spherically symmetric distributions of matter. The exact exterior and interior solutions of Einstein field equations in spherically symmetric spacetimes are discussed. The proof of Birkoff's theory is provided. It states that a spherically symmetric gravitational field in vacuum is necessarily static and must have Schwarzschild form. The Tolman–Oppenheimer–Volkov (TOV) equation is discussed. Isotropic coordinate system is a new coordinate system whose spatial distance is proportional to the Euclidean square of the distances. Some static spherically symmetric spacetimes are rewritten in an isotropic coordinate system. A short discussion on interaction between the gravitational and electromagnetic fields are provided. Reissner–Nordström solution is a static solution of the gravitational field outside of a spherically symmetric charged body. Particle and photon orbits in the Schwarzschild spacetime are discussed in Chapter Seven. Also, in this chapter, using the trajectory in the gravitational field of sun (i.e., in the Schwarzschild spacetime), several tests of the theory of general relativity, namely the precession of the perihelion motion of mercury, bending of light, radar echo delay, and gravitational redshift, are explained. A discussion on the stable circular orbits in the Schwarzschild spacetime is given. A general treatment is provided for the experimental test of general theory of relativity for a general static and spherically symmetric configuration. Causal structure in the special theory of relativity, i.e., in Minkowski spacetime or flat spacetime, is characterized so that no massive particle can travel faster than light. In general relativity, locally there is no difference of the causality relation with Minkowski spacetime. However, globally, the causality relation is significantly different due to various spacetime topologies. A short discussion on causal structure of spacetimes is given in Chapter Eight. Several basic definitions and some standard theorems related to causality are explained. Chapter Nine deals with discussions on causal structures of specific spacetimes, which are the standard exact solutions of Einstein field equations such as Minkowski spacetime, de Sitter and anti-de Sitter spacetimes, Robertson–Walker spacetime, Bianchi-I spacetime, Schwarzschild spacetime, and Reissner–Nordström black hole. A short elementary discussion on rotating black holes is given in Chapter Ten. After introducing the tetrad, an outline of the derivation of the Kerr and Kerr–Newman solutions is illustrated through the complex transformation algorithm for both in four and higher dimensions. Some of the different forms of the Kerr solution are mentioned. Some elementary properties of the Kerr solution including the maximal extension of Kerr spacetime are discussed. Finally, brief discussions on Hawking radiation, Penrose process of extraction of energy from a Kerr black hole, and laws of black hole thermodynamics are given. Chapters Eleven and Twelve provide some simple applications of general theory of relativity in astrophysics and cosmology, respectively. Some preliminary concepts of extrinsic curvature, Lagrangian formalism of the general theory of relativity, and 3 + 1 decomposition of spacetime are given as appendices.

Acknowledgments

This book has been made possible through the support, contributions, and assistance of many people and various organizations. I take this opportunity to express my sincere gratitude to all of them. I would like to deeply and sincerely thank my mother (Majeda), wife (Pakizah), and son (Rahil), without whose loving support and encouragement this book could not have been completed. I also express my sincere gratitude to my father-in-law (Abdul Hannan), mother-in-law (Begum Nurjahan), brother-in-law (Dr. Ruhul Amin), younger brother (Mafrook Rahaman), and niece (Ayat Nazifa) for their patience and support during the entire period of the preparation of the manuscript. It is a pleasure to thank Dr. Nupur Paul, Dr. Sayeedul Islam, Dr. Banashree Sen, Dr. Mosiur Rahaman, Dr. Indrani Karar, Monsur Rahaman, Dr. Shyam Das, Lipi Baskey, Nayan Sarkar, Md Rahil Miraj, Dr. Arkopriya Mallick, Sabiruddin Molla, Dr. Ayan Banerjee, Dr. Tuhina Manna, Dr. Amna Ali, Dr. Nasarul Islam, Ksh. Newton Singh, Somi Aktar, Bidisha Samanta, Dr. Sourav Roychowdhury, Dr. Debabrata Deb, Dr. Amit Das, Dr. Abdul Aziz, Dr. Anil Kumar Yadav, Monimala Mandal, Antara Mapdar, Dr. Saibal Ray, Dr. Mehedi Kalam, Susmita Sarkar, Dr. Piyali Bhar, Dr. Gopal Chandra Shit, Dr. Ranjan Sharma, Dr. Shounak Ghosh, and Dr. Iftikar Hossain Sardar for their technical assistance in the preparation of the book. I remain thankful to all the professors and non-teaching staff members of the Department of Mathematics, Jadavpur University for providing me with all the available facilities and services whenever needed. Particularly I would like to mention the library staff for their excellent support. Finally, I am also thankful to the authority of the Inter-University Centre for Astronomy and Astrophysics (IUCAA), Pune, India for providing all kinds of working facility and hospitality under the Associateship Scheme.

Tensor Calculus — A Brief Overview

1.1 Introduction

The principal target of tensor calculus is to investigate the relations that remain the same when we change from one coordinate system to any other. The laws of physics are independent of the frame of references in which physicists describe physical phenomena by means of laws. Therefore, it is useful to exploit tensor calculus as the mathematical tool in which such laws can be formulated.

1.2 Transformation of Coordinates

Let there be two reference systems, S with coordinates $(x^1, x^2, \ldots, x^n)$ and $\overline{S}$ with coordinates $(\overline{x}^1, \overline{x}^2, \ldots, \overline{x}^n)$ (Fig. 1). The new system $\overline{S}$ depends on the old system S as

$$\overline{x}^i = \phi^i(x^1, x^2, \ldots, x^n); \quad i = 1, 2, \ldots, n. \tag{1.1}$$

Here ϕ^i are single-valued continuous differentiable functions of $x^1, x^2, \ldots, x^n$ and further the Jacobian

$$\left| \frac{\partial \phi^i}{\partial x^j} \right| = \begin{vmatrix} \frac{\partial \phi^1}{\partial x^1} & \frac{\partial \phi^1}{\partial x^2} & \frac{\partial \phi^1}{\partial x^3} & \cdots & \frac{\partial \phi^1}{\partial x^n} \\ \frac{\partial \phi^2}{\partial x^1} & \frac{\partial \phi^2}{\partial x^2} & \frac{\partial \phi^2}{\partial x^3} & \cdots & \frac{\partial \phi^2}{\partial x^n} \\ \cdots & \cdots & \cdots & \cdots & \cdots \\ \frac{\partial \phi^n}{\partial x^1} & \frac{\partial \phi^n}{\partial x^2} & \frac{\partial \phi^n}{\partial x^3} & \cdots & \frac{\partial \phi^n}{\partial x^n} \end{vmatrix} \neq 0.$$

Differentiation of Eq. (1.1) yields

$$d\overline{x}^i = \sum_{r=1}^n \frac{\partial \phi^i}{\partial x^r} dx^r = \sum_{r=1}^n \frac{\partial \overline{x}^i}{\partial x^r} dx^r = \sum_{r=1}^n \overline{a}^i_r dx^r.$$

Now and onward, we use the Einstein summation convention, i.e., omit the summation symbol $\sum$ and write the above equations as

$$d\overline{x}^i = \frac{\partial \overline{x}^i}{\partial x^r} dx^r = \overline{a}^i_r dx^r, \tag{1.2}$$

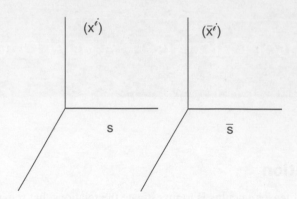

Figure 1 S and $\bar{S}$ frames.

or

$$dx^i = \frac{\partial x^i}{\partial \bar{x}^m} d\bar{x}^m = a^i_m d\bar{x}^m. \tag{1.3}$$

The repeated index r or m is known as **dummy index**. The index i is not dummy and is known as **free index**.

The transformation matrices are inverse to each other

$$\bar{a}^i_r a^m_i = \delta^m_r. \tag{1.4}$$

The symbol δ^m_r is Kronecker delta, is defined as

$$\delta^m_r = 1 \ \ if \ m = r$$
$$= 0 \ \ if \ m \neq r$$

Obviously vectors in $(\bar{S})$ system are linked with (S) system.

1.3 Covariant and Contravariant Vector and Tensor

Usually one can describe the tensors by means of their properties of transformation under coordinate transformation. There are two possible ways of transformations from one coordinate system (x^i) to the other coordinate system $(\bar{x}^i)$.

Let us consider a set of n functions A_i of the coordinates x^i. The functions A_i are said to be the components of **covariant vector** if these components transform according to the following rule

$$\bar{A}_i = \frac{\partial x^j}{\partial \bar{x}^i} A_j. \tag{1.5}$$

Also, one can find by multiplying $\frac{\partial \bar{x}^i}{\partial x^k}$ and using $\frac{\partial \bar{x}^i}{\partial x^k} \frac{\partial x^j}{\partial \bar{x}^i} = \delta^j_k$ and $\delta^j_k A_j = A_k$

$$A_k = \frac{\partial \bar{x}^i}{\partial x^k} \bar{A}_i.$$

Gradient of a scalar B, i.e., $B_i = \frac{\partial B}{\partial x_i}$ is a covariant vector.

Here, A_i is known as the **covariant tensor of first order or of the type** $(0, 1)$.

The functions A^i are said to be the components of the **contravariant vector** if these components transform according to the following rule

$$\bar{A}^i = \frac{\partial \bar{x}^i}{\partial x^j} A^j \tag{1.6}$$

Also, one can find by multiplying both sides with $\frac{\partial x^k}{\partial \bar{x}^i}$ and using $\delta^k_j A^j = A^k$

$$A^k = \frac{\partial x^k}{\partial \bar{x}^i} \bar{A}^i.$$

Here, A^i is known as the **contravariant tensor of first order or of the type** $(1, 0)$.

Exercise 1.2

Tangent vector $\frac{dx^i}{du}$ of the curve $x^i = x^i(u)$ is a contravariant vector.

Exercise 1.3

Let components of velocity vector in Cartesian coordinates are $\dot{x}$ and $\dot{y}$. Find corresponding components in polar coordinates.

Hint: Here, $x^1 = x$, $x^2 = y$, and $\bar{x}^1 = r$, $\bar{x}^2 = \theta$ with $x = r\cos\theta$, $y = r\sin\theta$, i.e., $r = \sqrt{x^2 + y^2}$, $\theta = \tan^{-1}(\frac{y}{x})$.

Let $A^1 = \dot{x}$, $A^2 = \dot{y}$. We will have to find $\bar{A}^1, \bar{A}^2$.

("dot" denotes differentiation with respect to t.)

Using the definition $\bar{A}^i = \frac{\partial \bar{x}^i}{\partial x^j} A^j$, we have

$$\bar{A}^1 = \frac{\partial \bar{x}^1}{\partial x^1} A^1 + \frac{\partial \bar{x}^1}{\partial x^2} A^2 \text{ or, } \bar{A}^1 = \frac{\partial r}{\partial x}\dot{x} + \frac{\partial r}{\partial y}\dot{y} = \dot{r}.$$

Similarly,

$$\bar{A}^2 = \frac{\partial \theta}{\partial x}\dot{x} + \frac{\partial \theta}{\partial y}\dot{y} = \dot{\theta}.$$

Exercise 1.4

Let components of acceleration vector in Cartesian coordinates be $\ddot{x}$ and $\ddot{y}$. Find corresponding components in polar coordinates.

Hint: Let $A^1 = \ddot{x}, A^2 = \ddot{y}$. We will have to find $\overline{A}^1, \overline{A}^2$.
Here,

$$\overline{A}^1 = \frac{\partial r}{\partial x}\ddot{x} + \frac{\partial r}{\partial y}\ddot{y} = \ddot{r} - r\dot{\theta}, \quad \overline{A}^2 = \frac{\partial \theta}{\partial x}\ddot{x} + \frac{\partial \theta}{\partial y}\ddot{y} = \ddot{\theta} + \frac{2}{r}\dot{\theta}\dot{r}.$$

1.3.1 Invariant

Let ϕ be a function of coordinate system (x^i) and $\overline{\phi}$ be its transform in another coordinate system $(\overline{x}^i)$. Then, ϕ is said to be **invariant** if $\overline{\phi} = \phi$.

Exercise 1.5

The expression $A^i B_i$ is an invariant or scalar, i.e.,

$$\overline{A}^i \overline{B}_i = A^i B_i. \tag{1.7}$$

Hint: Use definitions given in Eqs. (1.5) and (1.6).

An **invariant or scalar** is known as the **tensor of the type** $(0, 0)$.

1.3.2 Contravariant and covariant tensors of rank two

Let C^i and B^j be two contravariant vectors with n components, then $C^i B^j = A^{ij}$ has n^2 quantities, i.e., A^{ij} are the set of n^2 functions of the coordinates x^i. If the transformation of A^{ij} is like

$$\overline{A}^{ij} = \frac{\partial \overline{x}^i}{\partial x^k}\frac{\partial \overline{x}^j}{\partial x^l}A^{kl}, \tag{1.8}$$

then A^{ij} is known as **contravariant tensor of rank two**. Here, A^{ij} is also known as the **contravariant tensor of order two or of the type** $(2, 0)$.

If we multiply both sides of (1.8) by $\frac{\partial x^r}{\partial \overline{x}^i}\frac{\partial x^s}{\partial \overline{x}^j}$, then

$$A^{rs} = \frac{\partial x^r}{\partial \overline{x}^i}\frac{\partial x^s}{\partial \overline{x}^j}\overline{A}^{ij}.$$

Again, if C_i and B_j are two covariant vectors with n components, then $C_i B_j = A_{ij}$ form n^2 quantities, i.e., A_{ij} are the set of n^2 functions of the coordinates x^i.
If the transformation of A_{ij} is like

$$\overline{A}_{ij} = \frac{\partial x^k}{\partial \overline{x}^i}\frac{\partial x^l}{\partial \overline{x}^j}A_{kl}, \tag{1.9}$$

then A_{ij} is known as **covariant tensor of rank two**.
Here, A_{ij} is also known as the **covariant tensor of order two or of the type** $(0, 2)$.

If we multiply both sides of (1.9) by $\frac{\partial \bar{x}^i}{\partial x^r} \frac{\partial \bar{x}^j}{\partial x^s}$, then

$$A_{rs} = \frac{\partial \bar{x}^i}{\partial x^r} \frac{\partial \bar{x}^j}{\partial x^s} \bar{A}_{ij}.$$

1.3.3 Mixed tensor of order two A_j^i

Suppose A_j^i is a set of n^2 functions of n coordinates. If the transformation obeys the following rule

$$\bar{A}_j^i = \frac{\partial \bar{x}^i}{\partial x^k} \frac{\partial x^l}{\partial \bar{x}^j} A_l^k,$$

then A_l^k is known as the **mixed tensor of order two or of the type** $(1, 1)$.

Thus, mixed tensor of order two can be obtained by taking a covariant vector A_i and a contravariant vector B^j, i.e., $C_i^j = A_i B^j$.

Exercise 1.6

Kronecker delta δ_i^j is a mixed tensor of order two.

Hint: If δ_i^j can be combined with components of two vectors to form a scalar, then δ_i^j will be a tensor. Now

$$A^i B_j \delta_i^j = A^i B_i = scalar.$$

If the transformation obeys the following rule

$$\bar{A}_{j_1 j_2 \ldots j_q}^{i_1 i_2 \ldots i_p} = \frac{\partial \bar{x}^{i_1}}{\partial x^{k_1}} \frac{\partial \bar{x}^{i_2}}{\partial x^{k_2}} \cdots \frac{\partial \bar{x}^{i_p}}{\partial x^{k_p}} \frac{\partial x^{l_1}}{\partial \bar{x}^{j_1}} \frac{\partial x^{l_2}}{\partial \bar{x}^{j_2}} \cdots \frac{\partial x^{l_q}}{\partial \bar{x}^{j_q}} A_{l_1 l_2 \ldots l_q}^{k_1 k_2 \ldots k_p},$$

then $A_{l_1 l_2 \ldots l_q}^{k_1 k_2 \ldots k_p}$ is known as **mixed tensor of the type** (p, q).

1.3.4 Symmetric and skew-symmetric tensors

If a tensor is unaltered after changing every pair of contravariant or covariant indices, then it is said to be a symmetric tensor. Let $T_{\alpha\beta}$ be a covariant tensor of rank two.

If $T_{\alpha\beta} = T_{\beta\alpha}$, *then it is known as* **symmetric tensor**.

If a tensor is altered in its sign but not in magnitude after changing every pair of contravariant or covariant indices, then it is said to be a skew-symmetric tensor.

If $T_{\alpha\beta} = -T_{\beta\alpha}$, *then it is known as* **antisymmetric or skew-symmetric tensor**.

Exercise 1.7

Kronecker delta δ_{ij} is a symmetric tensor.

Exercise 1.8

If A_i is covariant vector, then $curlA_i = \frac{\partial A_i}{\partial x_j} - \frac{\partial A_j}{\partial x_i}$ is a skew-symmetric tensor.

Hint: Use $curlA_i = \frac{\partial A_i}{\partial x_j} - \frac{\partial A_j}{\partial x_i} = B_{ij}$ and show that $B_{ij} = -B_{ji}$.

Note 1.1

Symmetry property of a tensor is independent of the coordinate system.

Note 1.2

A symmetric tensor of order two in n-dimensional space has at most $\frac{n(n+1)}{2}$ independent components whereas an antisymmetric tensor of order two has at most $\frac{n(n-1)}{2}$ independent components.

1.4   Operations on Tensors

i. The addition and subtraction of two tensors of the same type is a tensor of same type.

Exercise 1.9

$$A_{ij} \pm B_{ij} = C_{ij}, \ A^{ij} \pm B^{ij} = C^{ij}, \ A_i^j \pm B_i^j = C_i^j$$

Exercise 1.10

Any covariant or contravariant tensor of second order can be expressed as a sum of a symmetric and a skew-symmetric tensor of order two.
Hint:

$$a_{ij} = \frac{1}{2}(a_{ij} + a_{ji}) + \frac{1}{2}(a_{ij} - a_{ji}), \ \text{etc.}$$

ii. The type of the tensor remains invariant by multiplication of a scalar α.

Exercise 1.11

$$\alpha A_{ij} = C_{ij}, \ \alpha A^{ij} = C^{ij}, \ \alpha A_i^j = C_i^j$$

iii. **Outer product:** The outer product of two tensors is a new tensor whose order is the sum of the orders of the given tensors.

Exercise 1.12

Let two tensors of types (2,3) and (1,2) be respectively, A^{ij}_{klm} and B^{a}_{bc}, then the outer product of these tensors has type (3,5), i.e.,

$$A^{ij}_{klm}B^{a}_{bc} = T^{ija}_{klmbc}$$

iv. **Contraction:** The particular type of operation by which the order (r) of a mixed tensor is lowered by order $(r-2)$ is known as contraction.

Exercise 1.13

Let A^{ij}_{klm} be a mixed tensor of order five. The new tensor A^{ij}_{kim} can be obtained by replacing lower index l by the upper index i and taking summation over i, one gets the tensor of order three.

$$A^{ij}_{kim} = B^{j}_{km}$$

v. **Inner product:** The outer product of two tensors followed by contraction with respect to an upper index and a lower index of the other results in a new tensor which is called an inner product.

Exercise 1.14

$$A^{ij}_{k}\,B^{k}_{mn} \equiv C^{ijk}_{kmn} = D^{ij}_{mn}, \quad A^{ij}_{k}\,B^{m}_{ij} = D^{m}_{k}$$

1.4.1 Test for tensor character: Quotient Law

An entity whose inner product by an arbitrary tensor (covariant or contravariant) always gives a tensor is itself a tensor.

Exercise 1.15

If $C(i,j)A^{i}B^{j}$ is an invariant, then $C(i,j) = C_{ij}$ is a tensor of the type (0,2).

Exercise 1.16

If $C(p,q,r)B^{qs}_{r} = A^{s}_{p}$, then $C(p,q,r) = C^{r}_{pq}$ is a tensor of the type (1,2).

Exercise 1.17

Let λ^{i}, μ^{i} be the components of two arbitrary vectors with $a_{hijk}\lambda^{h}\mu^{i}\lambda^{j}\mu^{k} = 0$, then prove that

$$a_{hijk} + a_{hkji} + a_{jihk} + a_{jkhi} = 0.$$

Hint: Given that

$$A = a_{hijk}\lambda^h \mu^i \lambda^j \mu^k = 0.$$

Differentiating with respect to λ^h, we get

$$\frac{\partial A}{\partial \lambda^h} = a_{hijk}\mu^i \lambda^j \mu^k + a_{pihk}\lambda^p \mu^i \mu^k = 0.$$

Again, differentiating with respect to λ^j, we get

$$\frac{\partial^2 A}{\partial \lambda^h \partial \lambda^j} = a_{hijk}\mu^i \mu^k + a_{jihk}\mu^i \mu^k = 0.$$

Now, differentiating with respect to μ^i and μ^k, one will find, respectively,

$$\frac{\partial^3 A}{\partial \lambda^h \partial \lambda^j \partial \mu^i} = a_{hijk}\mu^k + a_{hkji}\mu^k + a_{jihk}\mu^k + a_{jkhi}\mu^k = 0,$$

$$\frac{\partial^4 A}{\partial \lambda^h \partial \lambda^j \partial \mu^i \partial \mu^k} = a_{hijk} + a_{hkji} + a_{jihk} + a_{jkhi} = 0.$$

Exercise 1.18

If A^i is an arbitrary contravariant vector and $C_{ij}A^i A^j$ is an invariant, then show that $C_{ij} + C_{ji}$ is a covariant tensors of the second order.

Hint: Given $C_{ij}A^i A^j$ is an invariant for arbitrary contravariant vector A^i, therefore,

$$C_{ij}A^i A^j = C'_{ij}A'^i A'^j.$$

Tensor law of transformation yields

$$C_{ij}A^i A^j = C'_{ij}\frac{\partial x'^i}{\partial x^\alpha}A^\alpha \frac{\partial x'^j}{\partial x^\beta}A^\beta.$$

Now interchanging the suffix i and j

$$C_{ji}A^j A^i = C'_{ji}\frac{\partial x'^j}{\partial x^\alpha}\frac{\partial x'^i}{\partial x^\beta}A^\alpha A^\beta = C'_{ji}\frac{\partial x'^i}{\partial x^\alpha}\frac{\partial x'^j}{\partial x^\beta}A^\alpha A^\beta.$$

(interchanging the dummy suffixes α and β)
Thus,

$$(C_{ji} + C_{ij})A^i A^j = (C'_{ji} + C'_{ij})\frac{\partial x'^i}{\partial x^\alpha}\frac{\partial x'^j}{\partial x^\beta}A^\alpha A^\beta,$$

$$\Rightarrow (C_{\alpha\beta} + C_{\beta\alpha})A^\alpha A^\beta = (C'_{ji} + C'_{ij})\frac{\partial x'^i}{\partial x^\alpha}\frac{\partial x'^j}{\partial x^\beta}A^\alpha A^\beta,$$

$$\Rightarrow \left[(C_{\alpha\beta} + C_{\beta\alpha}) - (C'_{ij} + C'_{ji})\frac{\partial x'^i}{\partial x^\alpha}\frac{\partial x'^j}{\partial x^\beta}\right]A^\alpha A^\beta = 0.$$

Since A^α is arbitrary, therefore, the expression within the square bracket vanishes. Hence, $C_{\alpha\beta} + C_{\beta\alpha}$ is a $(0, 2)$-tensor.

1.4.2 Conjugate or reciprocal tensor of a tensor

Consider a symmetric covariant tensor of second order a_{ij}, i.e., of the type $(0,2)$ whose determinant, $|a_{ij}|$ is nonzero; then

$$b^{ij} = \frac{cofactor\ of\ a_{ij}\ in\ |a_{ij}|}{|a_{ij}|}$$

is known as reciprocal tensor of a_{ij}. It is of the type $(2,0)$.

Note 1.3

Reciprocal tensor exists for any tensor. Only condition being its determinant is nonzero. Here, $a_{ij}b^{ik} = \delta_j^k$ and $|a_{ij}||b^{ik}| = |\delta_j^k| = 1$. Usually, conjugate of a_{ij} is written as a^{ij} and $a_{ij}a^{ij} = \delta_j^j = n$.

Note 1.4

Tensor equations in one system (x^i) remain valid in all other coordinate systems $(\bar{x}^i)$, e.g., if $T^i_{jkl} = 2T^i_{ljk}$, then $\bar{T}^i_{jkl} = 2\bar{T}^i_{ljk}$.

1.5 Generalized Kronecker Delta

The generalized Kronecker Delta $\delta^{\alpha\beta}_{\mu\nu}$ is defined as follows:

$$\delta^{\alpha\beta}_{\mu\nu} = \begin{vmatrix} \delta^\alpha_\mu & \delta^\beta_\mu \\ \delta^\alpha_\nu & \delta^\beta_\nu \end{vmatrix}$$

$$= +1, \ \alpha \neq \beta, \ \alpha = \mu, \ \beta = \nu$$

$$= -1, \ \alpha \neq \beta, \ \alpha = \nu, \ \beta = \mu$$

$$= 0, \ otherwise.$$

We can define $\delta^{\alpha\beta\gamma}_{\mu\nu\xi}$ and $\delta^{\alpha\beta\gamma\rho}_{\mu\nu\xi\omega}$ as follows:

$$\delta^{\alpha\beta\gamma}_{\mu\nu\xi} = \begin{vmatrix} \delta^\alpha_\mu & \delta^\beta_\mu & \delta^\gamma_\mu \\ \delta^\alpha_\nu & \delta^\beta_\nu & \delta^\gamma_\nu \\ \delta^\alpha_\xi & \delta^\beta_\xi & \delta^\gamma_\xi \end{vmatrix},$$

$$\delta^{\alpha\beta\gamma\rho}_{\mu\nu\xi\omega} = \begin{vmatrix} \delta^\alpha_\mu & \delta^\beta_\mu & \delta^\gamma_\mu & \delta^\rho_\mu \\ \delta^\alpha_\nu & \delta^\beta_\nu & \delta^\gamma_\nu & \delta^\rho_\nu \\ \delta^\alpha_\xi & \delta^\beta_\xi & \delta^\gamma_\xi & \delta^\rho_\xi \\ \delta^\alpha_\omega & \delta^\beta_\omega & \delta^\gamma_\omega & \delta^\rho_\omega \end{vmatrix}.$$

Exercise 1.19

$$\delta_{123}^{123} = \delta_{231}^{123} = 1,$$
$$\delta_{213}^{123} = \delta_{132}^{123} = -1.$$

Exercise 1.20

Show that

$$\delta_{\mu\beta}^{\alpha\beta} = 3\delta_\mu^\alpha.$$

Exercise 1.21

Show that

$$\delta_\alpha^\alpha = 4.$$

Exercise 1.22

Show that

$$\delta_{\mu\gamma\tau}^{\alpha\beta\tau} = 2\delta_{\mu\gamma}^{\alpha\beta}.$$

Hint:

$$\delta_{\mu\gamma\tau}^{\alpha\beta\tau} = \begin{vmatrix} \delta_\mu^\alpha & \delta_\mu^\beta & \delta_\mu^\tau \\ \delta_\gamma^\alpha & \delta_\gamma^\beta & \delta_\gamma^\tau \\ \delta_\tau^\alpha & \delta_\tau^\beta & \delta_\tau^\tau \end{vmatrix}.$$

Now, expand along third row and use $\delta_\tau^\tau = 4$

Exercise 1.23

Show that

$$\delta_{\mu\nu\gamma\rho}^{\alpha\beta\tau\rho} = -\begin{vmatrix} \delta_\mu^\alpha & \delta_\mu^\beta & \delta_\mu^\tau \\ \delta_\nu^\alpha & \delta_\nu^\beta & \delta_\nu^\tau \\ \delta_\gamma^\alpha & \delta_\gamma^\beta & \delta_\gamma^\tau \end{vmatrix}.$$

Exercise 1.24

Show that

$$\delta^{\alpha\beta\tau\rho}_{\mu\nu\tau\rho} = -2(\delta^{\alpha}_{\mu}\delta^{\beta}_{\nu} - \delta^{\alpha}_{\nu}\delta^{\beta}_{\mu}).$$

Exercise 1.25

Show that

$$\delta^{\alpha\beta\tau\rho}_{\mu\beta\tau\rho} = -6\delta^{\alpha}_{\mu}.$$

Exercise 1.26

Show that

$$\delta^{\alpha\beta\tau\rho}_{\alpha\beta\tau\rho} = -24.$$

Symbols: Symmetric and skew-symmetric tensors of second order:

$$T_{(ab)} = \frac{1}{2}(T_{ab} + T_{ba}), \quad T_{[ab]} = \frac{1}{2}(T_{ab} - T_{ba}).$$

For the tensors of third order, we can construct symmetric and skew-symmetric tensors as

$$T_{(abc)} = \frac{1}{3!}(T_{abc} + T_{bca} + T_{cab} + T_{bac} + T_{acb} + T_{cba}),$$

$$T_{[abc]} = \frac{1}{3!}(T_{abc} + T_{bca} + T_{cab} - T_{bac} - T_{acb} - T_{cba}).$$

We can express skew-symmetry symbols by means of generalized Kronecker delta as

$$T_{[ab]} = \frac{1}{2!}T_{cd}\delta^{cd}_{ab},$$

$$T_{[abc]} = \frac{1}{2!}T_{cde}\delta^{cde}_{abc}.$$

1.6 The Line Element

The distance between two neighboring points $P(\vec{r}(x^i))$ and $F(\vec{r}(x^i) + d\vec{r}(x^i))$ (x^i are the coordinates of the space) in an n-dimensional space is given by (see Fig. 2)

$$ds^2 = d\vec{r} \cdot d\vec{r} = g_{ab}dx^a dx^b \tag{1.10}$$

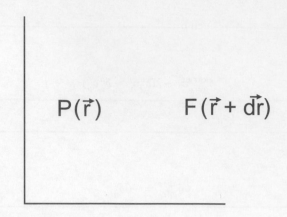

Figure 2 Two neighboring points in a space.

Here,

$$d\vec{r}(x^i) = \frac{\partial \vec{r}}{\partial x^1}dx^1 + \frac{\partial \vec{r}}{\partial x^2}dx^2 + \ldots\ldots\ldots + \frac{\partial \vec{r}}{\partial x^n}dx^n = \alpha_1 dx^1 + \alpha_2 dx^2 + \ldots + \alpha_n dx^n$$

with

$$\alpha_i = \frac{\partial \vec{r}}{\partial x^i} \ and \ g_{ab} = \alpha_a \cdot \alpha_b.$$

The distance between two neighboring points is referred as **line element** and is given by Eq. (1.10).

Here, g_{ab} are known as metric tensor, which are functions of x^a. If $g = |g_{ab}| \neq 0$ and ds is adopted to be invariant, then the space is called **Riemannian space**.

In mathematics, Riemannian space is used for a positive-definite metric tensor, whereas in theoretical physics, spacetime is modeled by a pseudo-Riemannian space in which the metric tensor is indefinite.

The metric tensor g_{ab} is also called **fundamental tensor** (covariant tensor of order two).

In Euclidean space:

$$ds^2 = dx^2 + dy^2 + dz^2.$$

In Minkowski flat spacetime, the line element

$$ds^2 = dx^{0^2} - dx^{1^2} - dx^{2^2} - dx^{3^2}.$$

Since the distance ds between two neighboring points is real, the Eq. (1.10) will be amended to

$$ds^2 = eg_{ij}dx^i dx^j,$$

where e is known as the indicator and assumes the value $+1$ or -1 in order that ds^2 be always positive.

The contravariant tensor g^{ij} is defined by

$$g^{ij} = \frac{\Delta^{ij}}{g},$$

here Δ^{ij} is the cofactor of g_{ij} and g is the determinant of g_{ij}.

Obviously

$$g_{ab} \, g^{bc} = g_a^c = \delta_a^c.$$

With the help of g^{ab} and g_{ab}, one can raise or lower the indices of any tensor as

$$g_{ac} T^{ab} = T_c^b$$

$$T_{ab} \, g^{ac} = T_b^c$$

$$g_{ab} A^b = A_a$$

$$g^{ab} A_a = A^b$$

Here, A_a and A^a are known as **associated vectors.**

$$g^{ab} \, g^{cd} \, g_{bd} = g^{ac}.$$

$$A_a B^b = g^{ab} A_a B_b = g_{ab} A^a B^b.$$

Exercise 1.27

Show that the determinant of the metric tensor is not a scalar. Also, prove that the expression $\sqrt{-g} \, d^4x$ where $d^4x = dx^1 dx^2 dx^3 dx^4$ is an invariant volume element.

Hints: We know

$$g'_{ab} = \frac{\partial x^c}{\partial x'^a} \frac{\partial x^d}{\partial x'^b} g_{cd}$$

$$\Rightarrow \quad det(g'_{ab}) = \left| \frac{\partial x}{\partial x'} \right|^2 det(g_{cd})$$

$$\Rightarrow \quad g' = \left| \frac{\partial x}{\partial x'} \right|^2 g \qquad (1.11)$$

This indicates that the determinant of the metric tensor is not a scalar.

Also the volume element d^4x transform into d^4x' as

$$d^4x' = \left| \frac{\partial x'}{\partial x} \right| d^4x \qquad (1.12)$$

From (1.11) and (1.12), we get

$$\sqrt{-g'}\, d^4x' = \sqrt{-g}d^4x.$$

Exercise 1.28

Find out the metric tensor of a three-dimensional Euclidean space in cylindrical and polar coordinates.

Hint: Here, for cylindrical coordinates,

$$y^1 = x^1 \cos x^2, \; y^2 = x^1 \sin x^2, \; y^3 = x^3$$

The metric tensor in three-dimensional Euclidean space is

$$ds^2 = dy^{1^2} + dy^{2^2} + dy^{3^2}$$

Now,

$$dy^1 = dx^1 \cos x^2 - x^1 \sin x^2 dx^2, \; dy^2 = dx^1 \sin x^2 + x^1 \cos x^2 dx^2, \; dy^3 = dx^3$$

Substituting these we get

$$ds^2 = dx^{1^2} + x^{1^2} dx^{2^2} + dx^{3^2}$$

For polar coordinates,

$$y^1 = x^1 \sin x^2 \cos x^3, \; y^2 = x^1 \sin x^2 \sin x^3, \; y^3 = x^1 \cos x^2$$

Using the same procedure, one can find

$$ds^2 = dx^{1^2} + x^{1^2} dx^{2^2} + x^{1^2} \sin^2 x^2 dx^{3^2}$$

Exercise 1.29

A curve in spherical coordinates x^i is given by

$$x^1 = t, \quad x^2 = \sin^{-1}\left(\frac{1}{t}\right), \quad x^3 = 2\sqrt{t^2 - 1}.$$

Find the length of arc for $1 \le t \le 2$.

Hint: In a spherical coordinate, the metric is given by

$$ds^2 = (dx^1)^2 + (x^1)^2(dx^2)^2 + (x^1 sinx^2)^2(dx^3)^2$$

$$= (dt)^2 + t^2 \left(-\frac{dt}{t\sqrt{t^2-1}} \right)^2 + \left(t.\frac{1}{t} \right)^2 \left(\frac{2t}{\sqrt{t^2-1}} dt \right)^2$$

$$= \frac{5t^2}{t^2-1}(dt)^2$$

Therefore, the required length of the arc $1 \leq t \leq 2$ is given by

$$\int_{t_1}^{t_2} ds = \sqrt{5} \int_1^2 \frac{t}{\sqrt{t^2-1}} dt = \sqrt{15} \text{ units.}$$

1.6.1 Norm

Let A^μ (A_μ) be any contravariant (covariant) vector. Then **norm or magnitude or length l of the vector A^μ (A_μ)** is defined as

$$l^2 = A^\mu A_\mu = g_{\mu\nu}A^\mu A^\nu = g^{\mu\nu}A_\mu A_\nu.$$

Exercise 1.30

Magnitude l of a vector is an invariant.
Hint: Try to show

$$A^\mu A_\mu = \overline{A}^\mu \overline{A}_\mu$$

1.6.2 Unit vector

A vector is said to be unit vector (unit covariant or unit contravariant) if

$$g^{ij}A_i A_j = 1 = g_{ij}A^i A^j.$$

1.6.3 Null vector

A vector is said to be null vector (covariant or contravariant) if

$$g^{ij}A_i A_j = 0 = g_{ij}A^i A^j.$$

1.6.4 Time-like vector

A vector is said to be time-like vector (covariant or contravariant) if

$$g^{ij}A_i A_j = g_{ij}A^i A^j > 0 \text{ with signature } (+,-,-,-).$$

Alternatively,

A vector is said to be time-like vector (covariant or contravariant) if

$$g^{ij}A_iA_j = g_{ij}A^iA^j < 0 \text{ with signature } (-,+,+,+).$$

1.6.5 Space-like vector

A vector is said to be space-like vector (covariant or contravariant) if

$$g^{ij}A_iA_j = g_{ij}A^iA^j < 0 \text{ with signature } (+,-,-,-).$$

Alternatively, a vector is said to be space-like vector (covariant or contravariant) if

$$g^{ij}A_iA_j = g_{ij}A^iA^j > 0 \text{ with signature } (-,+,+,+).$$

The time-like, space-like, and null vectors have important physical relevance as follows: Two events are causally connected by a time-like vector when they lie within a light cone, whereas a space-like vector connects two events that lie outside the light cone, i.e., the events are causally disconnected. Two events that lie on the light cone are connected by a null vector. Actually, collection of all null vectors in a Lorentzian space forms a light cone.

Note 1.5

The signature (p, q) of a metric tensor g is defined as the number of positive and negative eigenvalues of the real symmetric matrix g_{ab} of the metric tensor, with respect to a certain basis. However, in practice, the signature of a nondegenerate metric tensor is denoted by a single number $s = p - q$, e.g., $s = 1 - 3 = -2$ for $(+,-,-,-)$ and $s = 3 - 1 = +2$ for $(-,+,+,+)$. A metric with a positive definite signature $(p, 0)$ is known as a Riemannian metric, whereas a metric with signature $(p, 1)$ or $(1, q)$ is called a Lorentzian metric.

A light cone in special and general relativity is the surface describing the temporal evolution of a blaze of light originating from a sole event and roving in all directions in spacetime.

Two events are causally connected if one event in spacetime can influence the other event; in other words, one can join one event to the other event with a time-like or null vector.

Exercise 1.31

$(1, 0, 0, -1)$ is a null vector, whereas $(1, 0, 0, \sqrt{2})$ is a unit vector in Minkowski space

$$ds^2 = dt^2 - dx^2 - dy^2 - dz^2.$$

Hint: Here in the first case,

$$g_{00} = 1, \ g_{11} = -1, \ g_{22} = -1, \ g_{33} = -1,$$

and

$$A^0 = 1, \quad A^1 = 0, \quad A^2 = 0, \quad A^3 = -1.$$

Now,

$$l^2 = g_{ij}A^iA^j = g_{00}A^0A^0 + g_{11}A^1A^1 + g_{22}A^2A^2 + g_{33}A^3A^3 = 0, \text{ etc.}$$

1.6.6 Angle between two vectors A^μ and B^μ

In ordinary vector algebra, we know angle between two vectors $\vec{A}$ and $\vec{B}$ is defined as

$$\cos\theta = \frac{\vec{A} \cdot \vec{B}}{|\vec{A}| |\vec{B}|}.$$

Similarly, one can define the angle between two vectors A^μ and B^μ as

$$\cos\theta = \frac{\text{scalar product of } A^\mu \text{ and } B^\mu}{\text{length of } A^\mu \times \text{length of } B^\mu}$$

$$= \frac{A^\mu B_\mu}{\sqrt{(A^\mu A_\mu)(B^\mu B_\mu)}}$$

$$= \frac{g^{\mu\nu}A_\mu B_\nu}{\sqrt{(g^{\alpha\beta}A_\alpha A_\beta)(g^{\rho\sigma}B_\rho B_\sigma)}}.$$

1.6.7 Orthogonal vectors

Two covariant vectors A_i, B_j or contravariant vectors A^i, B^j are said to be orthogonal if

$$g^{ij}A_iB_j = 0 = g_{ij}A^iB^j.$$

Exercise 1.32

If θ be the angle between two non-null vectors A^i and B^i at a point, show that

$$\sin^2\theta = \frac{\left(g_{ij}g_{pq} - g_{ip}g_{jq}\right)A^iB^pA^jB^q}{\left(g_{ij}A^iA^j\right)\left(g_{pq}B^pB^q\right)}.$$

Hint: Let θ be the angle between two non-null vectors A^i and B^i at a point; then from the above definition

$$\cos\theta = \frac{g_{ij}A^iB^j}{\sqrt{g_{ij}A^iA^j}\sqrt{g_{pq}B^pB^q}}.$$

Now,

$$\sin^2 \theta = 1 - \cos^2 \theta = 1 - \frac{g_{ij}A^iB^j \; g_{pq}A^pB^q}{(g_{ij}A^iA^j)(g_{pq}B^pB^q)}$$

$$= \frac{g_{ij}g_{pq}A^iA^jB^pB^q - g_{ij}g_{pq}A^iA^pB^jB^q}{(g_{ij}A^iA^j)(g_{pq}B^pB^q)}$$

$$= \frac{g_{ij}g_{pq}A^iB^pA^jB^q - g_{ip}g_{jq}A^iB^pA^jB^q}{(g_{ij}A^iA^j)(g_{pq}B^pB^q)}$$

(Replacing the dummy indices j and p by p and j)

$$= \frac{\left(g_{ij}g_{pq} - g_{ip}g_{jq}\right)A^iB^pA^jB^q}{\left(g_{ij}A^iA^j\right)\left(g_{pq}B^pB^q\right)}.$$

Exercise 1.33

$(1, 1, 0, -1)$ and $(1, 0, 1, -1)$ are orthogonal vectors in Minkowski space

$$ds^2 = dt^2 - dx^2 - dy^2 - dz^2.$$

Hint: Here

$$g_{ij}A^iB^j = g_{00}A^0B^0 + g_{11}A^1B^1 + g_{22}A^2B^2 + g_{33}A^3B^3 = 0, \text{ etc.}$$

1.7 Levi-Civita Tensor or Alternating Tensor

Levi-Civita tensor is a tensor of order three in three dimensions and is denoted by ϵ_{abc} and defined as

$$\epsilon_{abc} = +1,$$

if a,b,c is an even permutation of $1, 2, 3$, i.e., in cyclic order.

$$= -1,$$

if a,b,c is odd permutation of $1, 2, 3$, i.e., not in cyclic order.

$$= 0$$

if any two indices are equal.

Levi-Civita tensor is a tensor of order four in four dimensions and denoted by ϵ^{abcd}.

$$\epsilon^{abcd} = +1,$$

if a,b,c,d is an even permutation of $0, 1, 2, 3$, i.e., in cyclic order.

$$= -1,$$

if a,b,c,d is odd permutation of $0, 1, 2, 3$, i.e., not in cyclic order.

$$= 0$$

if any two indices are equal.

The components of ϵ_{abcd} can be found from ϵ^{abcd} by lowering the indices in a typical way, just multiplying it by $(-g)^{-1}$:

$$\epsilon_{abcd} = g_{a\mu}\, g_{b\nu}\, g_{c\gamma}\, g_{d\sigma}(-g)^{-1}\epsilon^{\mu\nu\gamma\sigma}.$$

For example,

$$\epsilon_{0123} = g_{0\mu}\, g_{1\nu}\, g_{2\gamma}\, g_{3\sigma}\, (-g)^{-1}\epsilon^{\mu\nu\gamma\sigma}$$
$$= (-g)^{-1}det\, g_{\mu\nu} = -1.$$

In general,

$$\epsilon_{abcd} = 1,$$

if a,b,c,d is an even permutation of $0, 1, 2, 3$.

$$= -1,$$

if a,b,c,d is odd permutation of $0, 1, 2, 3$.

$$= 0 \ \ otherwise.$$

Here,

$$\epsilon_{abcd}\epsilon^{abcd} = -24.$$

Hints: The explicit form of $\epsilon_{abcd}\epsilon^{pqnm}$ is

$$\epsilon_{abcd}\epsilon^{pqnm} = -g_a^p g_b^q g_c^n g_d^m + g_a^q g_b^n g_c^m g_d^p - g_a^n g_b^m g_c^p g_d^q + g_a^m g_b^p g_c^q g_d^n + g_a^q g_b^p g_c^n g_d^m - g_a^p g_b^n g_c^m g_d^q$$
$$+ g_a^n g_b^m g_c^q g_d^p - g_a^m g_b^q g_c^p g_d^n + g_a^n g_b^q g_c^p g_d^m - g_a^q g_b^p g_c^m g_d^n + g_a^p g_b^m g_c^n g_d^q - g_a^m g_b^n g_c^q g_d^p$$
$$+ g_a^m g_b^q g_c^n g_d^p - g_a^q g_b^n g_c^p g_d^m + g_a^n g_b^p g_c^m g_d^q - g_a^p g_b^m g_c^q g_d^n + g_a^p g_b^n g_c^q g_d^m - g_a^n g_b^q g_c^m g_d^p$$
$$+ g_a^q g_b^m g_c^p g_d^n - g_a^m g_b^p g_c^n g_d^q + g_a^p g_b^q g_c^m g_d^n - g_a^q g_b^m g_c^n g_d^p + g_a^n g_b^p g_c^q g_d^m - g_a^p g_b^n g_c^q g_d^m.$$

The new tensor can be obtained by replacing upper index p by the lower index a as

$$\epsilon_{abcd}\epsilon^{aqnm} = -g_b^q g_c^n g_d^m - g_b^n g_c^m g_d^q - g_b^m g_c^q g_d^n + g_b^q g_c^m g_d^n + g_b^m g_c^n g_d^q + g_b^n g_c^q g_d^m.$$

This implies

$$\epsilon_{abcd}\epsilon^{abnm} = -2(g_c^n g_d^m + g_c^m g_d^n).$$

Finally one obtains

$$\epsilon_{abcd}\epsilon^{abcm} = -6g_d^m \quad and \quad \epsilon_{abcd}\epsilon^{abcd} = -24.$$

1.8 Christoffel Symbols

Partial derivative of a tensor is not, in general, a tensor. However, to build up expressions involving partial derivatives of a tensor, which are the components of a tensor, the fundamental tensor plays a significant role. In order to achieve this goal, in 1869 E.B. Christoffel introduced two notations, which are formed in terms of partial derivatives of the fundamental tensors g_{ij}. These are the Christoffel symbols of the first and second kinds, defined as follows:

First kind:

$$[ij,k] = \Gamma_{ijk} = \frac{1}{2}\left[\frac{\partial g_{ik}}{\partial x_j} + \frac{\partial g_{jk}}{\partial x_i} - \frac{\partial g_{ij}}{\partial x_k}\right].$$

Second kind:

$$\{^l_{ij}\} = \Gamma^l_{ij} = g^{lk}\Gamma_{ijk}.$$

Note 1.6

These contain n^3 components. For Minkowski space, i.e., in flat space all Christoffel Symbols vanish.

Properties:

1. $[ij,k] = [ji,k]$

2. $\{^l_{ij}\} = \{^l_{ji}\}$

3. $[ij,k] = g_{lk}\{^l_{ij}\}$

4. $[ij,k] + [kj,i] = \frac{\partial g_{ik}}{\partial x^j}$

5. $\{^i_{il}\} = \frac{\partial(\log\sqrt{g})}{\partial x^l}$ *where* $g = |g_{ij}|$, *if* $g < 0$, *then* $\{^i_{il}\} = \frac{\partial(\log\sqrt{-g})}{\partial x^l}$

Hint: We have

$$g^{ij} = \frac{G_{ij}}{g}, \ where \ G_{ij} \ are \ cofactor \ of \ g_{ij}$$

Differentiating g with respect to x^l, we get

$$\frac{\partial g}{\partial x^l} = \frac{\partial g_{ij}}{\partial x^l} G_{ij} = g g^{ij} \frac{\partial g_{ij}}{\partial x^l} = g g^{ij}([il,j] + [jl,i])$$

$$= g\{^i_{il}\} + g\{^j_{jl}\} = 2g\{^i_{il}\}$$

Therefore,

$$\{^i_{il}\} = \frac{\partial(\log \sqrt{g})}{\partial x^l}$$

Exercise 1.34

For diagonal metric, i.e., $g_{ij} = 0$ *for* $i \neq j$, show that

$$\{^i_{ii}\} = \frac{1}{2}\frac{\partial(\ln |g_{ii}|)}{\partial x^i} \ ; \ \{^i_{ij}\} = \frac{1}{2}\frac{\partial(\ln |g_{ii}|)}{\partial x^j} \ ; \ \{^i_{jj}\} = -\frac{1}{2g_{ii}}\frac{\partial(g_{jj})}{\partial x^i} \ , \ \{^i_{jk}\} = 0$$

Hint: Use the definition of $\{^i_{jk}\}$.

Exercise 1.35

Find $\{^i_{jk}\}$ for the metric

$$ds^2 = dr^2 + r^2 d\theta^2 + r^2 \sin^2\theta d\phi^2.$$

1.8.1 Transformation of Christoffel symbols
We will try to find

$$\overline{[ij,k]} = \frac{1}{2}\left[\frac{\partial \bar{g}_{ik}}{\partial \bar{x}_j} + \frac{\partial \bar{g}_{jk}}{\partial \bar{x}_i} - \frac{\partial \bar{g}_{ij}}{\partial \bar{x}_k}\right]$$

Since g_{ij} is tensor of the type (0,2), therefore,

$$\bar{g}_{ij} = \frac{\partial x_a}{\partial \bar{x}_i}\frac{\partial x_b}{\partial \bar{x}_j} g_{ab}$$

differentiating both sides with respect to $\bar{x}_k$, we obtain

$$\frac{\partial \bar{g}_{ij}}{\partial \bar{x}_k} = \frac{\partial g_{ij}}{\partial x_c}\frac{\partial x_c}{\partial \bar{x}_k}\frac{\partial x_a}{\partial \bar{x}_i}\frac{\partial x_b}{\partial \bar{x}_j} + \frac{\partial^2 x_a}{\partial \bar{x}_k \partial \bar{x}_i}\frac{\partial x_b}{\partial \bar{x}_j}g_{ab} + \frac{\partial x_a}{\partial \bar{x}_i}\frac{\partial^2 x_b}{\partial \bar{x}_k \partial \bar{x}_j}g_{ab}$$

Similarly, one can find $\frac{\partial \bar{g}_{ik}}{\partial \bar{x}_j}$ and $\frac{\partial \bar{g}_{jk}}{\partial \bar{x}_i}$ and adding these two and subtract $\frac{\partial \bar{g}_{ij}}{\partial \bar{x}_k}$, we obtain after some mathematical manipulation

$$\overline{[ij,k]} = [ab,c]\frac{\partial x_a}{\partial \bar{x}_i}\frac{\partial x_b}{\partial \bar{x}_j}\frac{\partial x_c}{\partial \bar{x}_k} + g_{ab}\frac{\partial x_a}{\partial \bar{x}_k}\frac{\partial^2 x_b}{\partial \bar{x}_i \partial \bar{x}_j}$$

The presence of second term indicates that $[ab,c]$ do not transform like a tensor.

1.9 Affine Connection

At first, we are providing some basic concepts:

A set M, which is locally Euclidean of dimension n is called a **manifold** of dimension n. **Locally Euclidean** means that every x that belongs to M possesses a neighborhood, which is homeomorphic to an open subset of R^n (see Fig. 3).

A mapping $f : X \to Y$ is **homeomorphic**, if f is bijective mapping, continuous and f^{-1} exists.

A manifold M is said to be **Hausdorff**, if for any two distinct points x and y in M, there exist disjoint neighborhoods of x and y.

A manifold M is said to be **compact** if each open cover of M has a finite subcover.

A manifold M is said to be **paracompact** if every open cover of M has an open refinement that is locally finite.

In the absence of paracompact, a manifold does not admit a real analytic differentiable structure.

A manifold M is said to be **connected** space if it cannot be represented as the union of two or more disjoint nonempty open subsets.

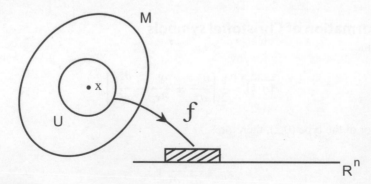

Figure 3 Locally Euclidean space.

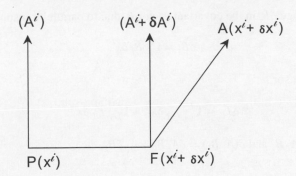

Figure 4 Parallel transport.

A manifold is said to be **differentiable manifold** if it is continuous and differentiable. Usually in physics, one can describe the spacetime by a differentiable manifold.

Suppose M is a differential manifold of dimension m. The **tangent space** T_pM is a collection of tangent vectors v_p to M at the point $p \in M$. The **tangent bundle** TM of a manifold M is defined by $TM = \cup_{p \in M} T_p M = (p, v) \mid p \in M, v \in T_pM$.

Let a contravariant vector A^i at point P (x^i). F $(x^i + \delta x^i)$ be a neighborhood point of P. Now we shift the vector A^i from P to F such that its magnitude and direction do not change. This scheme is known as **parallel transport** (see Fig. 4). In this scheme, the tangent vector is propagated parallel to itself. The changes δA^i of the components of A^i in going from P to F under such a parallel transport will be proportional to the original components of A^i and also to the displacement δx^l, i.e., the changes δA^i will be linear functions of δx^l and A^k. Thus

$$\delta A^i = -{}^a\Gamma^i_{kl} A^k \delta x^l.$$

The notations ${}^a\Gamma^i_{kl}$ are known as the **affine connection** of the spacetime region, which contains $4^3 = 64$ components entity. This connection is torsion-free. Here, the notion of local parallelism, i.e., parallelism over infinitesimal distances or the parallel transport of connecting two nearby vectors is affine connection of the spacetime. These symbols are mentioned as **Christoffel symbols**, i.e., ${}^a\Gamma^i_{kl} = \Gamma^i_{kl}$.

In the above, we have seen the change δA^i in the contravariant vector A^i due to parallel transport. Now we try to find the similar law for covariant vector A_i.

We know scalars are invariant under parallel transport; therefore,

$$\delta \left(A^i B_i \right) = 0,$$

$$or, \ \delta A^i B_i + A^i \delta B_i = 0,$$

$$or, \ -\Gamma^i_{kl} A^k \delta x^l B_i + A^k \delta B_k = 0,$$

$$or, \ A^k [\delta B_k - \Gamma^i_{kl} \delta x^l B_i] = 0.$$

Thus, we find the change δB_i in the covariant vector B_i due to parallel transport as

$$\delta B_k = \Gamma^i_{kl}\, B_i\, \delta x^l.$$

For second rank tensor,

$$\delta T_{ik} = \Gamma^l_{im}\, T_{lk}\, \delta x^m + \Gamma^l_{km}\, T_{il}\, \delta x^m.$$

Proof: Assume $T_{ik} = A_i\, B_k$ and $\delta(A_i\, B_k) = \delta A_i\, B_k + A_i\, \delta B_k$, etc.

1.10   Covariant Derivative

As the partial derivative of a tensor is not, in general, a tensor, therefore, it is demanded to introduce a new kind of differentiation, which gives rise to a tensor when applied to a tensor. This new type of derivative is actually covariant derivative. It is independent of the choice of coordinates.

Consider a contravariant vector A^i at the point P (x^i) and then displace the vector to a point F $(x^i + \delta x^i)$. The actual physical change in A^i from P to F is given by $dA^i - \delta A^i$, where dA^i is due to point differences and δA^i due to parallel transport.

We know

$$dA^i = A^i(x^i + \delta x^i) - A^i(x^i)$$

$$\cong A^i(x^i) + \delta x^l \frac{\partial A^i}{\partial x^l} - A^i(x^i) = \frac{\partial A^i}{\partial x^l} \delta x^l.$$

The rate of change with respect to x^i is

$$\frac{dA^i - \delta A^i}{\delta x^l} = A^i_{\;;\,l}.$$

This rate of change is called **covariant derivative** of A^i for $\delta x^l \longrightarrow 0$.

Now putting the values of dA^i and δA^i, we get the covariant derivative of a contravariant vector as

$$A^i_{\;;\,l} = \frac{\partial A^i}{\partial x^l} + \Gamma^i_{kl} A^k. \tag{1.13}$$

Remember that $A_{\mu;\lambda} = g_{\mu\nu} A^\nu_{\;;\lambda}.$

Hint: Differentiating both side of $A_\mu = g_{\mu\nu} A^\nu$ with respect to x^λ, we obtain

$$\frac{\partial A_\mu}{\partial x^\lambda} = \frac{\partial g_{\mu\nu}}{\partial x^\lambda} A^\nu + g_{\mu\nu} \frac{\partial A^\nu}{\partial x^\lambda} = (\Gamma^\delta_{\mu\lambda} g_{\nu\delta} + \Gamma^\delta_{\lambda\nu} g_{\mu\delta}) A^\nu + g_{\mu\nu} \frac{\partial A^\nu}{\partial x^\lambda}$$

or

$$\frac{\partial A_\mu}{\partial x^\lambda} - \Gamma^\delta_{\lambda\mu} g_{\nu\delta} A^\nu = g_{\mu\nu} \frac{\partial A^\nu}{\partial x^\lambda} + \Gamma^\nu_{\lambda\delta} g_{\mu\nu} A^\delta$$

(replacing the dummy index ν by δ)

or

$$\frac{\partial A_\mu}{\partial x^\lambda} - \Gamma^\delta_{\lambda\mu} A_\delta = g_{\mu\nu} \left(\frac{\partial A^\nu}{\partial x^\lambda} + \Gamma^\nu_{\lambda\delta} A^\delta \right)$$

Hence we have

$$A_{\mu;\lambda} = g_{\mu\nu} A^\nu_{;\lambda}.$$

1.10.1 Covariant differentiation of scalar, vectors, and tensors

(i)

$$Scalar\ function\ \phi : \quad \phi_{;l} = \phi_{,l}. \tag{1.14}$$

(ii)

$$Covariant\ vector\ B_i : \quad B_{i;l} = B_{i,l} - \Gamma^k_{il} B_k. \tag{1.15}$$

[comma $\rightarrow$ partial derivative]

Hint: Use $\delta B_k = \Gamma^i_{kl} B_i\ \delta x^l$

(iii) Covariant tensor of rank two T_{ik}:

$$T_{ik;l} = T_{ik,l} - \Gamma^m_{il} T_{mk} - \Gamma^m_{kl} T_{im}. \tag{1.16}$$

Hint: Use $T_{ik} = A_i B_k$ and $\delta(A_i B_k) = \delta A_i B_k + A_i \delta B_k$, etc.

(iv) Contravariant tensor of rank two T^{ik}:

$$T^{ik}_{;l} = T^{ik}_{,l} + \Gamma^i_{lm} T^{mk} + \Gamma^k_{lm} T^{im}. \tag{1.17}$$

(v) Mixed tensor T^i_k:

$$T^i_{k;j} = T^i_{k,j} - \Gamma^l_{jk} T^i_l + \Gamma^i_{jl} T^l_k. \tag{1.18}$$

(vi) Fundamental tensor g_{ik}:

$$g_{ik;l} = 0. \tag{1.19}$$

$$g^{ik}_{;l} = 0. \tag{1.20}$$

(vii) Gradient of an invariant ϕ:

$$grad\ \phi = \phi_{;l} = \phi_{,l}. \tag{1.21}$$

(viii) Divergence of a contravariant vector A^i:

$$div \, A^i = A^i_{\;;\,i} = \frac{1}{\sqrt{-g}} \frac{\partial(\sqrt{-g} A^k)}{\partial x^k}. \tag{1.22}$$

(ix) Divergence of a covariant vector A_i:

$$div \, A_i = g^{jk} A_{j;\,k}. \tag{1.23}$$

(x) Laplacian of an invariant ϕ:

$$\nabla^2 \phi = div \, grad \, \phi = \frac{1}{\sqrt{-g}} \frac{\partial \left[\sqrt{-g} \, g^{kj} \frac{\partial \phi}{\partial x^j} \right]}{\partial x^k}. \tag{1.24}$$

(xi) Curl of a covariant vector A_i:

$$curl \, A_i = A_{i;\,j} - A_{j;\,i} = \frac{\partial A^i}{\partial x^j} - \frac{\partial A^j}{\partial x^i}. \quad [as \; \Gamma^r_{ij} = \Gamma^r_{ji}] \tag{1.25}$$

(xii) Curl grad $\phi = 0$:

$$Let \; A_i = \phi_{;\,i} = \phi_{,\,i} = \frac{\partial \phi}{\partial x^i}, \tag{1.26}$$

$$then, curl \, A_i = A_{i;\,j} - A_{j;\,i} = \frac{\partial^2 \phi}{\partial x^i \partial x^j} - \frac{\partial^2 \phi}{\partial x^j \partial x^i} = 0.$$

Note that

$$\phi_{;\,ij} = \phi_{;\,ji}. \tag{1.27}$$

(xiii) Covariant derivative of a tensor of the type (p, q):

$$
A^{i_1 \ldots i_p}_{j_1 \ldots j_q \; :k} = \frac{\partial A^{i_1 \ldots i_p}_{j_1 \ldots j_q}}{\partial x^k} + A^{ai_2 \ldots i_p}_{j_1 \ldots j_q} \Gamma^{i_1}_{ak} + \ldots \ldots \ldots + A^{i_1 i_2 \ldots i_{p-1} a}_{j_1 \ldots j_q} \Gamma^{i_p}_{ak}
$$

$$
- A^{i_1 i_2 \ldots i_p}_{bj_2 \ldots j_q} \Gamma^b_{j_1 k} - \ldots \ldots \ldots - A^{i_1 i_2 \ldots i_p}_{j_1 \ldots j_{q-1} b} \Gamma^b_{j_q k} \tag{1.28}
$$

(xiv) The covariant derivative of a contravariant vector is a tensor.

Hint: Suppose A^l is a contravariant vector, so we have

$$A^a = A'^i \frac{\partial x^a}{\partial x'^i}.$$

Differentiating it with respect to x^b, we obtain

$$\frac{\partial A^a}{\partial x^b} = \frac{\partial A'^i}{\partial x'^j}\frac{\partial x^a}{\partial x'^i}\frac{\partial x'^j}{\partial x^b} + A'^i \frac{\partial^2 x^a}{\partial x'^i \partial x'^j}\frac{\partial x'^j}{\partial x^b}.$$

But from the above, we know

$$\Gamma'^p_{ij}\frac{\partial x^c}{\partial x'^p} = \Gamma^c_{ab}\frac{\partial x^a}{\partial x'^i}\frac{\partial x^b}{\partial x'^j} + \frac{\partial^2 x^c}{\partial x'^i \partial x'^j}.$$

Therefore, we have

$$\frac{\partial A^a}{\partial x^b} = \frac{\partial A'^i}{\partial x'^j}\frac{\partial x^a}{\partial x'^i}\frac{\partial x'^j}{\partial x^b} + A'^i \frac{\partial x'^j}{\partial x^b}\left[\Gamma'^p_{ij}\frac{\partial x^a}{\partial x'^p} - \Gamma^a_{mc}\frac{\partial x^m}{\partial x'^i}\frac{\partial x^c}{\partial x'^j}\right]$$

or

$$\frac{\partial A^a}{\partial x^b} + A'^i \frac{\partial x'^j}{\partial x^b}\frac{\partial x^m}{\partial x'^i}\frac{\partial x^c}{\partial x'^j}\Gamma^a_{mc} = \frac{\partial A'^i}{\partial x'^j}\frac{\partial x^a}{\partial x'^i}\frac{\partial x'^j}{\partial x^b} + A'^p \frac{\partial x'^j}{\partial x^b}\frac{\partial x^a}{\partial x'^i}\Gamma'^i_{pj}$$

or

$$\frac{\partial A^a}{\partial x^b} + A^m \delta^c_b \Gamma^a_{mc} = \left[\frac{\partial A'^i}{\partial x'^j} + A'^p\,\Gamma'^i_{pj}\right]\frac{\partial x^a}{\partial x'^i}\frac{\partial x'^j}{\partial x^b}$$

$$\frac{\partial A^a}{\partial x^b} + A^m\,\Gamma^a_{mb} = \left[\frac{\partial A'^i}{\partial x'^j} + A'^p\,\Gamma'^i_{pj}\right]\frac{\partial x^a}{\partial x'^i}\frac{\partial x'^j}{\partial x^b}$$

Hence,

$$A^a_{;b} = A'^i_{;j}\frac{\partial x^a}{\partial x'^i}\frac{\partial x'^j}{\partial x^b}.$$

This confirms that $A^a_{;b}$ is a mixed tensor of rank two.

Note 1.7

The notation semi colon ";" or "∇" is used to denote covariant derivative. Covariant differentiation for products, sums, and differences obeys the same rule as in the case of ordinary differentiation.

1.11   Curvature Tensor

It is known that ordinary differentiation with respect to coordinates is commutative. However, in general, covariant differentiation is not commutative. This has happened due to use of Riemannian space in which the operation is undertaken. The distinctiveness features of such a space consist in a certain tensor known as curvature tensor whose components consist of first-order derivatives of the fundamental tensor.

Let us start with a covariant vector, V_i, then

$$V_{i;\,j} = a_{ij} = \frac{\partial V_i}{\partial x^j} - \Gamma^l_{ij}\, V_l.$$

Now,

$$a_{ij;\,k} = \frac{\partial a_{ij}}{\partial x^k} - \Gamma^l_{jk}\, a_{il} - \Gamma^l_{ki}\, a_{lj} \equiv V_{i;\,jk}.$$

Hence,

$$V_{i;\,jk} - V_{i;\,kj} = \left[\frac{\partial \Gamma^l_{ik}}{\partial x^j} - \frac{\partial \Gamma^l_{ij}}{\partial x^k} + \Gamma^r_{ki}\,\Gamma^l_{rj} - \Gamma^r_{ji}\,\Gamma^l_{rk}\right] V_l = R^l_{ijk}\, V_l.$$

Therefore,

$$R^l_{ijk} = \frac{\partial \Gamma^l_{ik}}{\partial x^j} - \frac{\partial \Gamma^l_{ij}}{\partial x^k} + \Gamma^r_{ki}\,\Gamma^l_{rj} - \Gamma^r_{ji}\,\Gamma^l_{rk},$$

is a tensor of type $(1,3)$ and is called **Riemann–Christoffel curvature tensor** or simply **curvature tensor**. This tensor is made with fundamental tensor and its derivatives. Note that this tensor is independent of the choice of the vector V_i. Also in general, $V_{i;\,jk} \neq V_{i;\,kj}$. However, if $R^l_{ijk} = 0$, then $V_{i;\,jk} = V_{i;\,kj}$. Therefore, one can infer that the covariant differentiation of a vector is commutative if and only if Riemann–Christoffel curvature tensor vanishes identically. The curvature tensor has some algebraic properties.

(a) $R^l_{ijk} = -R^l_{ikj}$

(b) $R^l_{ijk} + R^l_{jki} + R^l_{kij} = 0$

By using inner product one can lower the contravariant index in R^l_{ijk}, i.e.,

$$R_{nijk} = g_{nl} R^l_{ijk}.$$

R_{nijk} is antisymmetric in first two indices as well as in last two indices, i.e.,

$$R_{nijk} = -R_{injk} \quad ; \quad R_{nijk} = -R_{nikj}.$$

R_{nijk} keeps the same value in interchanging the first and last pair of indices, i.e.,

$$R_{nijk} = R_{jkni}.$$

Also, in general, covariant differentiation of any tensor is not commutative.

Hint: Let us take covariant derivative of a mixed tensor as

$$\nabla_j A_q^p = \partial_j A_q^p + \Gamma_{sj}^p A_q^s - \Gamma_{qj}^s A_s^p.$$

Now,

$$\nabla_k \nabla_j A_q^p = \partial_k(\nabla_j A_q^p) - \Gamma_{jk}^\sigma \nabla_\sigma A_q^p + \Gamma_{\sigma k}^p \nabla_j A_q^\sigma - \Gamma_{qk}^\sigma \nabla_j A_\sigma^p$$

$$= \partial_k \partial_j A_q^p - (\partial_k \Gamma_{sj}^p) A_q^s + \Gamma_{sj}^p(\partial_k A_q^s) - (\partial_k \Gamma_{qj}^s) A_s^p - \Gamma_{qj}^s \partial_k A_s^p - \Gamma_{jk}^\sigma(\partial_\sigma A_q^p + \Gamma_{s\sigma}^p A_q^s - \Gamma_{q\sigma}^s A_s^p)$$

$$+ \Gamma_{\sigma k}^p(\partial_j A_q^\sigma + \Gamma_{sj}^\sigma A_q^s - \Gamma_{qj}^s A_s^\sigma) - \Gamma_{qk}^\sigma(\partial_j A_\sigma^p + \Gamma_{sj}^p A_\sigma^s - \Gamma_{\sigma j}^s A_s^p).$$

Interchanging k and j we obtain

$$(\nabla_k \nabla_j - \nabla_j \nabla_k) A_q^p$$

$$= (\partial_k \Gamma_{js}^p - \partial_j \Gamma_{ks}^p + \Gamma_{\sigma k}^p \Gamma_{js}^\sigma - \Gamma_{\sigma j}^p \Gamma_{ks}^\sigma) A_q^s - (\partial_k \Gamma_{jq}^s - \partial_j \Gamma_{kq}^s + \Gamma_{\sigma k}^s \Gamma_{jq}^\sigma - \Gamma_{\sigma j}^s \Gamma_{kq}^\sigma) A_s^p.$$

Thus, we have

$$(\nabla_k \nabla_j - \nabla_j \nabla_k) A_q^p = -R_{sjk}^p A_q^s + R_{qjk}^s A_s^p.$$

In a similar manner

$$(\nabla_k \nabla_j - \nabla_j \nabla_k) A_{pq} = A_{sq} R_{pjk}^s + A_{ps} R_{qjk}^s,$$

and

$$(\nabla_k \nabla_j - \nabla_j \nabla_k) A^{pq} = -R_{sjk}^p A^{sq} - R_{sjk}^q A^{ps}.$$

1.12 Ricci Tensor

The contracted Riemann–Christoffel curvature tensor, which is not identically zero, is known as Ricci Tensor and its components are denoted by R_{ij}. Thus

$$R_{ij} = R_{ijl}^l.$$

A Riemannian space whose curvature tensor is identically zero is called **flat space** (e.g., Minkowski space).

It is obvious that $R_{ij} = R_{ji}$.

If the Ricci tensor of space takes the following form

$$R_{ij} = \lambda g_{ij},$$

where λ is a constant, then it is called **Einstein space.**

Note 1.8

The Ricci tensor

$$R_{\mu\nu} = R^{\sigma}_{\mu\nu\sigma} = \frac{\partial \Gamma^{\sigma}_{\mu\sigma}}{\partial x^{\nu}} - \frac{\partial \Gamma^{\sigma}_{\mu\nu}}{\partial x^{\sigma}} + \Gamma^{\alpha}_{\mu\sigma}\Gamma^{\sigma}_{\alpha\nu} - \Gamma^{\alpha}_{\mu\nu}\Gamma^{\sigma}_{\alpha\sigma},$$

can be written in the form by using $\Gamma^{\sigma}_{\mu\sigma} = \frac{\partial}{\partial x^{\mu}} \ln \sqrt{-g}$ as

$$R_{\mu\nu} = \frac{\partial^2}{\partial x^{\mu}\partial x^{\nu}} \ln \sqrt{-g} - \frac{\partial \Gamma^{\alpha}_{\mu\nu}}{\partial x^{\alpha}} + \Gamma^{\beta}_{\mu\alpha}\Gamma^{\alpha}_{\nu\beta} - \Gamma^{\alpha}_{\mu\nu}\frac{\partial}{\partial x^{\alpha}} \ln \sqrt{-g}.$$

1.13 Ricci Scalar

Contracting Ricci tensor R_{ij}, we obtain

$$R = g^{ij} R_{ij}.$$

Here R is called Ricci scalar or the curvature invariant.

Note 1.9

$R_{ijkl} = 0$ implies $R_{ij} = 0$ and $R = 0$, but its converse is not true.

Exercise 1.36

Show that if

$$R_{ij} - \frac{1}{2}g_{ij}R = 0,$$

then $R_{ij} = 0$.
Hint: Multiplying both sides by g^{ij}, one gets

$$g^{ij}R_{ij} - \frac{1}{2}g^{ij}g_{ij}R = 0,$$

or

$$R - \frac{1}{2}nR = 0, \quad or \quad R = 0, \; i.e., \; R_{ij} = 0.$$

Exercise 1.37

Let in a Riemannian space the following relation hold.
Prove that this is Einstein space.

$$g_{ij}R_{kl} - g_{il}R_{jk} + g_{jk}R_{il} - g_{kl}R_{ij} = 0.$$

Hint: Multiplying both sides by g^{ij}, one gets

$$nR_{kl} - \delta_l^j R_{jk} + \delta_k^i R_{il} - Rg_{kl} = 0.$$

$$or, \quad nR_{kl} - R_{lk} + R_{kl} - Rg_{kl} = 0,$$

$$or, \quad R_{kl} = \frac{R}{n} g_{kl}.$$

Exercise 1.38

Show that for the spacetime with metric

$$ds^2 = e^{2\phi(r)} dt^2 - e^{2\theta(r)} dr^2 - dx^2 - dy^2,$$

Ricci tensor vanishes if

$$\phi'' - \theta'\phi' + \phi'^2 = 0.$$

Hint: Here the nonzero Christoffel symbols are

$$\Gamma_{11}^1 = \theta', \quad \Gamma_{10}^0 = \phi' = \Gamma_{01}^0, \quad \Gamma_{00}^1 = e^{2(\phi-\theta)}\phi'.$$

$R_{\mu\nu} = 0$ implies $R_{11} = 0 = R_{22} = R_{33} = R_{00}$. Here, $x_0 = t, x_1 = r, x_2 = x, x_3 = z$.
Now,

$$R_{11} = \frac{\partial^2}{\partial x^1 \partial x^1} \ln \sqrt{-g} - \frac{\partial \Gamma_{11}^\alpha}{\partial x^\alpha} + \Gamma_{1\alpha}^\beta \Gamma_{1\beta}^\alpha - \Gamma_{11}^\alpha \frac{\partial}{\partial x^\alpha} \ln \sqrt{-g} = 0.$$

Using $-g = -g_{00}g_{11}g_{22}g_{33} = e^{2(\theta+\phi)}$, we get

$$-\theta'' + \theta'^2 + \phi'^2 + (\theta'' + \phi'') - \theta'(\theta' + \phi') = 0,$$

or

$$\phi'' - \theta'\phi' + \phi'^2 = 0.$$

$R_{22} = 0 = R_{33}$ identically.
Finally,

$$R_{00} = \frac{\partial^2}{\partial x^0 \partial x^0} \ln \sqrt{-g} - \frac{\partial \Gamma_{00}^\alpha}{\partial x^\alpha} + \Gamma_{0\alpha}^\beta \Gamma_{0\beta}^\alpha - \Gamma_{00}^\alpha \frac{\partial}{\partial x^\alpha} \ln \sqrt{-g} = 0,$$

implies

$$-\frac{\partial}{\partial x^1}[e^{2(\phi-\theta)}\phi'] + \phi'e^{2(\phi-\theta)}\phi' + \phi'e^{2(\phi-\theta)}\phi' - e^{2(\phi-\theta)}\phi'(\theta'+\phi') = 0,$$

or

$$\phi'' - \theta'\phi' + \phi'^2 = 0.$$

Exercise 1.39

Prove that for an Einstein space of dimension $n \geq 2$,

$$R_{ij} = \frac{R}{n}g_{ij}.$$

Hint: Use

$$g^{ij}R_{ij} = \lambda g^{ij}g_{ij} = n\lambda, \text{ etc.}$$

1.14   Space of Constant Curvature

If a Riemannian space is such that its curvature tensor is of the form

$$R_{ijkl} = b(g_{ik}\,g_{jl} - g_{il}\,g_{jk}),$$

where b is a constant, the space is said to be of **constant curvature** b. Note that for $b = 0$, $R_{ijkl} = 0$ and the space becomes a flat space. In this sense, one can infer that flat space is a particular type of a space of constant curvature. A space of constant curvature is also an Einstein space.

Exercise 1.40

Show that a space of constant curvature is an Einstein space.
Hint: Let M be a space of constant curvature of dimension n. Then by definition

$$R_{hijk} = k(g_{hj}g_{ik} - g_{hk}g_{ij}).$$

Now multiplying both sides by g^{hk} and contracting on h and k we get

$$R_{ij} = k(\delta_j^k g_{ik} - ng_{ij}) = k(1-n)g_{ij}.$$

This confirms that the space is an Einstein space. Note that

$$k(1-n) = \frac{R}{n} \Rightarrow R = kn(1-n).$$

Exercise 1.41

Prove that in an Einstein space with dimension $n > 2$, the scalar curvature R is always a constant.

Hint: It is known that Einstein space of dimension n is defined by the relation

$$R_{ij} = \frac{R}{n} g_{ij} \ \ or, \ \ g^{ik} R_{ij} = \frac{R}{n} g^{ik} g_{ij} \ \ or, \ \ R_j^k = \frac{R}{n} \delta_j^k.$$

$$or, \ \ \nabla_k R_j^k = \nabla_k R \frac{1}{n} \delta_j^k = \frac{1}{n} \frac{\partial R}{\partial x^j}.$$

However, we know

$$\nabla_k R_j^k = \frac{1}{2} \frac{\partial R}{\partial x^j}.$$

Hence, we have

$$(n-2) \frac{\partial R}{\partial x^j} = 0.$$

Thus, if $n > 2$ then scalar curvature R is a constant.

Exercise 1.42

$$ds^2 = a^2 \, d\phi^2 + a^2 \sin^2\phi \, d\theta^2,$$

is a line element of a surface. The surface under consideration is a space of constant curvature $\frac{1}{a^2}$.

Exercise 1.43

If A_{ij} are components of the curl of a covariant vector B_i then

$$A_{ij;\,k} + A_{jk;\,i} + A_{ki,\,j} = 0.$$

Hint: Here,

$$A_{ij} = B_{i;j} - B_{j;i} = \frac{\partial B_i}{\partial x^j} - \frac{\partial B_j}{\partial x^i}$$

Now,

$$\frac{\partial A_{ij}}{\partial x^k} = \frac{\partial^2 B_i}{\partial x^k \partial x^j} - \frac{\partial^2 B_j}{\partial x^k \partial x^i}$$

and

$$A_{ij;\,k} = \frac{\partial A_{ij}}{\partial x^k} - A_{pj}\Gamma^p_{ik} - A_{ip}\Gamma^p_{jk}, \quad \text{etc.}$$

Exercise 1.44

If A_{jk} be a symmetric tensor field, then show the following formula

$$A^k_{j\,;k} = \frac{1}{\sqrt{-g}}\frac{\partial}{\partial x^k}(A^k_j\sqrt{-g}) - \frac{1}{2}A^{lk}\frac{\partial g_{lk}}{\partial x^j}$$

$$= \frac{1}{\sqrt{-g}}\frac{\partial}{\partial x^k}(A^k_j\sqrt{-g}) + \frac{1}{2}A_{lk}\frac{\partial g^{lk}}{\partial x^j}.$$

Hint: We know

$$A^k_{j\,;\,k} = A^k_{j,\,k} - \Gamma^l_{kj}A^k_l + \Gamma^k_{kl}A^l_j.$$

Given, A_{jk} is a symmetric tensor, i.e., $A_{jk} = A_{kj}$. We know

$$\Gamma^k_{km} = \frac{1}{\sqrt{-g}}\frac{\partial\sqrt{-g}}{\partial x^m}.$$

Therefore,

$$\Gamma^k_{kl}A^l_j = \Gamma^k_{km}A^m_j = A^m_j\frac{1}{\sqrt{-g}}\frac{\partial\sqrt{-g}}{\partial x^m} = A^k_j\frac{1}{\sqrt{-g}}\frac{\partial\sqrt{-g}}{\partial x^k}.$$

Now,

$$A^k_{j,\,k} + \Gamma^k_{kl}A^l_j = \frac{1}{\sqrt{-g}}\frac{\partial}{\partial x^k}(A^k_j\sqrt{-g}). \tag{1.29}$$

Also, we know

$$A^k_l\Gamma^l_{kj} = A^{kl}\Gamma_{kj,l} = A^l_k\Gamma^k_{lj}.$$

Now,

$$A^k_l\Gamma^l_{kj} = \frac{1}{2}A^{kl}(g_{kl,j} + g_{jl,k} - g_{kj,l}).$$

Interchanging l and k on the right-hand side of the above equation, we get

$$A_l^k \Gamma_{kj}^l = \frac{1}{2} A^{lk}(g_{lk,j} + g_{jk,l} - g_{lj,k}).$$

Adding these equations, we get

$$A_l^k \Gamma_{kj}^l = \frac{1}{2} A^{lk} g_{lk,j}. \tag{1.30}$$

The Eqs. (1.29) and (1.30) yield

$$A_{j;k}^k = \frac{1}{\sqrt{-g}} \left(A_j^k \sqrt{-g} \right)_{,k} - \frac{1}{2} A^{ik} \frac{\partial g_{ik}}{\partial x^j}.$$

Exercise 1.45

Show that the expression of wave operator for a scalar field ϕ can be written in the following form

$$\Box^2 \phi = g^{ik} \phi_{;ik} = \frac{1}{\sqrt{-g}} \frac{\partial}{\partial x^k} \left(\sqrt{-g} g^{ik} \frac{\partial \phi}{\partial x^i} \right).$$

Hint: We know

$$\Box^2 \phi = \left[\frac{\partial \phi}{\partial x_i} g^{ik} \right]_{;k} = \frac{1}{\sqrt{-g}} \left[\frac{\partial \phi}{\partial x_i} \sqrt{-g} g^{ik} \right]_{,k} \quad \text{by Eq. (1.22).}$$

Exercise 1.46

If

$$R_{ij;k} = 2A_k R_{ij} + A_i R_{kj} + A_j R_{ik},$$

show that

$$A_k = \frac{\partial}{\partial x^k}(\ln \sqrt{R}).$$

Hint: Multiplying both sides by g^{ij}, one gets

$$g^{ij} R_{ij;k} = 2g^{ij} A_k R_{ij} + g^{ij} A_i R_{kj} + g^{ij} A_j R_{ik}$$

$$or, \ R_{,k} = 2A_k R + A^j R_{kj} + A^i R_{ik}$$

$$or, \ R_{,k} = 2A_k R + 2A^i R_{ik} \quad (as \ R_{ij} = R_{ji}) \tag{i}$$

The given condition implies

$$R_{ij;k} - R_{ik;j} = A_k R_{ij} - A_j R_{ik}$$

$$or, \ g^{ij} R_{ij;k} - g^{ij} R_{ik;j} = g^{ij} A_k R_{ij} - g^{ij} A_j R_{ik}$$

$$or. \ R_{,k} - R^j_{k;j} = A_k R - A^i R_{ik}$$

Using the result given in example (1.39), we get

$$R_{,k} - \frac{1}{2} R_{,k} = A_k R - A^i R_{ik}$$

$$or, \qquad R_{,k} = 2(A_k R - A^i R_{ik}) \qquad \qquad (ii)$$

Equations (i) and (ii) imply that

$$A^i R_{ik} = 0.$$

This yields

$$R_{,k} = 2 A_k R \ or \ A_k = \frac{1}{2R} \frac{\partial R}{\partial x^k} = \frac{\partial}{\partial x^k} (\ln \sqrt{R}).$$

1.15   The Affine Connection in Riemannian Geometry

We now enforce two additional conditions on the affine connection Γ^i_{kl} as

$$\Gamma^i_{kl} = \Gamma^i_{lk}, \qquad \qquad (1.31)$$

$$g_{ik;l} = 0. \qquad \qquad (1.32)$$

The affine connection obeying these conditions is said to be **Riemannian connection** and the corresponding geometry under consideration is called the **Riemannian Geometry**.

Now, formulae (1.16) and (1.32), yield

$$\Gamma^m_{il} \, g_{mk} + \Gamma^m_{kl} \, g_{im} = g_{ik, \, l},$$

or

$$\Gamma_{ilk} + \Gamma_{kli} = g_{ik, \, l}. \qquad \qquad (1.33)$$

Using Eq. (1.31), we have

$$\Gamma_{ilk} \equiv g_{mk} \, \Gamma^m_{il} = g_{mk} \, \Gamma^m_{li} = \Gamma_{lik}. \qquad \qquad (1.34)$$

Now, making changes in a cyclic order in i, k, l in (1.33), we get two other relations as

$$\Gamma_{kil} + \Gamma_{lik} = g_{kl,i}, \tag{1.35}$$

$$\Gamma_{lki} + \Gamma_{ikl} = g_{li,k}, \tag{1.36}$$

Performing (1.35) + (1.36) − (1.33) and using (1.34), we get

$$\Gamma_{ikl} = \frac{1}{2} \left[g_{li,k} + g_{kl,i} - g_{ik,l} \right].$$

1.16 Geodesic Coordinate

In Euclidean space, the metric tensors are constant throughout the space and hence all Christoffel symbols vanish. This is not true in Riemannian space. However, there exists a coordinate system known as **Geodesic coordinate** system x^i such that all the Christoffel symbols vanish at a particular point P which is known as **pole**.

$$\Gamma^i_{jk} = 0 = \Gamma_{ijk} \ at \ point \ P.$$

In the geodesic system, all the Christoffel symbols vanish but their derivatives do not necessarily become zero.

An important property of geodesic coordinate system is given as **"the covariant derivative of a tensor at the pole equal to the corresponding partial derivative"** (since, Christoffel symbols vanish at the pole).

1.16.1 Local inertial coordinate system

According to equivalence principle, the local properties of curved spacetime and flat spacetime are indistinguishable. Therefore, it is possible to introduce new coordinates $x_P^{\mu'}$ in every point P of the spacetime such that

$$g_{\alpha\beta}(x_P^{\mu'}) = \eta_{\alpha\beta}, \tag{i}$$

where $\eta_{\alpha\beta} = diag(1, -1, -1, -1)$ is the flat Minkowski metric. In this $x_P^{\mu'}$, the first derivatives of the metric vanish, i.e.,

$$\frac{\partial g_{\alpha\beta}}{\partial x^{\mu'}} \Big|_{x = x_P^{\mu'}} = 0. \tag{ii}$$

This new coordinate system, satisfying (i) and (ii) at a point, is known as **local inertial coordinate system**. This new coordinate reference frame is very similar to the inertial frame of flat spacetime, which acts only in an infinitesimal neighborhood of a single point P.

1.17 Bianchi Identity

Curvature tensor follows another important identity property, which is not an algebraic but a differential identity and is known as **Bianchi identity**, which takes the form

$$R^l_{ijk;m} + R^l_{ikm;j} + R^l_{imj;k} = 0.$$

This identity has a crucial role in the formulation of Einstein's general theory of relativity. One has to use Bianchi identity to derive the curvature tensor of some metric.

Proof: For a geodesic coordinate system with a point P as a pole, we obtain,

$$R^l_{ijk;m} = \frac{\partial}{\partial x^m}(R^l_{ijk})$$

$$= \frac{\partial}{\partial x^m}\left[\frac{\partial}{\partial x^j}\Gamma^l_{ik} - \frac{\partial}{\partial x^k}\Gamma^l_{ij} + \Gamma^r_{ki}\Gamma^l_{rj} - \Gamma^r_{ji}\Gamma^l_{rk}\right]$$

$$= \frac{\partial^2}{\partial x^m \partial x^j}\Gamma^l_{ik} - \frac{\partial^2}{\partial x^m \partial x^k}\Gamma^l_{ij}. \tag{1.37}$$

Replacing $j = k, k = m, m = j$ in (1.37) we get at P

$$R^l_{ikm;j} = \frac{\partial^2}{\partial x^j \partial x^k}\Gamma^l_{im} - \frac{\partial^2}{\partial x^j \partial x^m}\Gamma^l_{ik}. \tag{1.38}$$

Again for $k = m, m = j, j = k$ in (1.38) we get

$$R^l_{imj;k} = \frac{\partial^2}{\partial x^k \partial x^m}\Gamma^l_{ij} - \frac{\partial^2}{\partial x^k \partial x^j}\Gamma^l_{im}. \tag{1.39}$$

Adding (1.37), (1.38), (1.39) we get the result.

Thus, the result is true for a geodesic system at the pole P. We know a tensor equation is independent of coordinate system, i.e., if it is true in one coordinate system then, it is true for all coordinate systems. In this case, the equation is a tensor equation; it holds in all coordinate systems at the point P. Since the point P can be an arbitrary point of an n-dimensional space V_n, therefore, the result is true for all points of V_n.

Exercise 1.47

If

$$g^{ik}R_{kj} = R^i_j \text{ and } g^{ij}R_{ij} = R,$$

then show that

$$R^i_{j;i} = \frac{1}{2}\frac{\partial R}{\partial x^j}.$$

Hint: Covariant derivative of $g^{ik}R_{kj} = R^i_j$ yields

$$R^i_{j;i} = g^{ik}R_{kj;i} \quad (as \ g^{ik}_{\;;i} = 0).$$

Thus

$$R^i_{j;i} = g^{ik}R^t_{kjt;i} = g^{ik}(g^{pt}R_{pkjt})_{;i} = g^{ik}g^{pt}R_{pkjt;i}$$

Now, using Bianchi identity, we have

$$R^i_{j;i} = g^{ik}g^{pt}R_{pkjt;i} = -g^{ik}g^{pt}[R_{pkti;j} + R_{pkij;t}]$$

$$= -g^{pt}[-R^i_{pti;j} + R^i_{pji;t}] \quad as \quad R_{kpti} = -R_{pkti} \ and \ R_{kpji} = R_{pkij}$$

$$= -g^{pt}[-R_{pt;j} + R_{pj;t}] = -[-(g^{pt}R_{pt})_{;j} + (g^{pt}R_{pj})_{;t}], \ etc.$$

1.18 Einstein Tensor

The following covariant tensor is known as Einstein tensor

$$G_{\mu\nu} = R_{\mu\nu} - \frac{1}{2}g_{\mu\nu}R.$$

One can also write it in mixed tensor form as

$$G^\nu_\mu = R^\nu_\mu - \frac{1}{2}\delta^\nu_\mu R.$$

The Einstein tensor is frequently used in the general theory of relativity.

1.18.1 Divergence of Einstein tensor is zero, i.e.,

$$G^{\mu\nu}_{\;;\nu} = 0.$$

Hint:

We know, Bianchi identity as

$$R^\rho_{\sigma\mu\nu;\alpha} + R^\rho_{\sigma\nu\alpha;\mu} + R^\rho_{\sigma\alpha\mu;\nu} = 0. \tag{1.40}$$

Now, we contract the indices ρ and ν to yield

$$R_{\sigma\mu;\alpha} + R^\rho_{\sigma\rho\alpha;\mu} + R^\rho_{\sigma\alpha\mu;\rho} = 0. \tag{1.41}$$

$$or, \quad R_{\sigma\mu;\alpha} - R^\rho_{\sigma\alpha\rho;\mu} + R^\rho_{\sigma\alpha\mu;\rho} = 0 \tag{1.42}$$

Here, we have used the symmetry property of the curvature tensor.

Using $R^{\rho}_{\sigma\alpha\rho;\mu} = R_{\sigma\alpha;\mu}$, we get from Eq. (1.42) as

$$R_{\sigma\mu;\alpha} - R_{\sigma\alpha;\mu} + R^{\rho}_{\sigma\alpha\mu;\rho} = 0. \qquad (1.43)$$

Since the covariant derivatives of $g_{\mu\nu}$'s are zero, therefore, multiplication of Eq. (1.43) by $g^{\sigma\lambda}$ yields

$$(g^{\sigma\lambda}R_{\sigma\mu})_{;\alpha} - (g^{\sigma\lambda}R_{\sigma\alpha})_{;\mu} + (g^{\sigma\lambda}R^{\rho}_{\sigma\alpha\mu})_{;\rho} = 0, \qquad (1.44)$$

$$or, \quad R^{\lambda}_{\mu;\alpha} - R^{\lambda}_{\alpha;\mu} + (g^{\sigma\lambda}R^{\rho}_{\sigma\alpha\mu})_{;\rho} = 0. \qquad (1.45)$$

The last term within the bracket can be rewritten as

$$g^{\sigma\lambda}R^{\rho}_{\sigma\alpha\mu} = g^{\sigma\lambda}g^{\rho\beta}R_{\beta\sigma\alpha\mu} = g^{\sigma\lambda}g^{\rho\beta}R_{\sigma\beta\mu\alpha} = g^{\rho\beta}g^{\sigma\lambda}R_{\sigma\beta\mu\alpha} = g^{\rho\beta}R^{\lambda}_{\beta\mu\alpha}. \qquad (1.46)$$

Therefore, Eq. (1.45) implies

$$R^{\lambda}_{\mu;\alpha} - R^{\lambda}_{\alpha;\mu} + (g^{\rho\beta}R^{\lambda}_{\beta\mu\alpha})_{;\rho} = 0. \qquad (1.47)$$

Now, we contract λ and α to yield

$$R^{\lambda}_{\mu;\lambda} - R^{\lambda}_{\lambda;\mu} + (g^{\rho\beta}R_{\beta\mu})_{;\rho} = 0. \qquad (1.48)$$

Here, the dummy index ρ can be changed to λ to yield

$$R^{\lambda}_{\mu;\lambda} - R^{\lambda}_{\lambda;\mu} + R^{\lambda}_{\mu;\lambda} = 0 \qquad (1.49)$$

or

$$R^{\lambda}_{\mu;\lambda} - \frac{1}{2}R^{\lambda}_{\lambda;\mu} = 0. \qquad (1.50)$$

We know

$$R^{\lambda}_{\lambda;\mu} = \frac{\partial}{\partial x^{\mu}}R^{\lambda}_{\lambda} = \frac{\partial}{\partial x^{\nu}}(g^{\nu}_{\mu}R^{\lambda}_{\lambda}). \qquad (1.51)$$

Using this result and changing the dummy index λ in the first term of Eq. (1.50) to ν we get

$$(R^{\nu}_{\mu} - \frac{1}{2}g^{\nu}_{\mu}R)_{;\nu} = 0.$$

Hence the proof.
Alternative hint:

$$G_{\mu\nu} = R_{\mu\nu} - \frac{1}{2}g_{\mu\nu}R \Rightarrow g^{\mu\sigma}G_{\mu\nu} = g^{\mu\sigma}R_{\mu\nu} - \frac{1}{2}g^{\mu\sigma}g_{\mu\nu}R$$

$$\Rightarrow G^{\sigma}_{\nu} = R^{\sigma}_{\nu} - \frac{1}{2}\delta^{\sigma}_{\nu}R \Rightarrow G^{\sigma}_{\nu;\sigma} = R^{\sigma}_{\nu;\sigma} - \frac{1}{2}\delta^{\sigma}_{\nu}R_{;\sigma} = \frac{1}{2}R_{;\nu} - \frac{1}{2}R_{;\nu} = 0.$$

If curl of a vector field (A_i) vanishes, then show that the vector field is a gradient of a scalar field.

Hint: Given

$$Curl\ A_i = \frac{\partial A_i}{\partial x^j} - \frac{\partial A_j}{\partial x^i} = 0.$$

$$\Rightarrow \frac{\partial A_i}{\partial x^j}dx^j = \frac{\partial A_j}{\partial x^i}dx^j$$

$$\Rightarrow dA_i = \frac{\partial A_j}{\partial x^i}dx^j$$

$$\Rightarrow A_i = \int \frac{\partial}{\partial x^i}(A_j dx^j) = \frac{\partial}{\partial x^i}\int A_j dx^j$$

Note that $\int A_j dx^j$ is a scalar quantity and let $\int A_j dx^j = \phi$, ϕ is a scalar. Hence

$$A_i = \frac{\partial \phi}{\partial x^i} = \nabla \phi.$$

Show that if

$$R_{iklm} = K(g_{il}\,g_{km} - g_{im}\,g_{kl}),$$

then K is a constant.

Hint: We have

$$R_{kl} = g^{im}K\,(g_{il}\,g_{km} - g_{im}\,g_{kl}) = -3Kg_{kl},\ \ therefore,\ \ R = -12K.$$

$$Now,\ Einstein\ tensor\ is\ G_{ik} = -3Kg_{ik} + 6Kg_{ik}.$$

$$Since,\ G^{ik}_{;k} = 0,\ we\ have, (3Kg^{ik})_{;k} = 0,\ i.e.,\ K_{,i} = 0\ as,\ g^{ik}_{;k} = 0.$$

Hence, K = constant.

1.19 Weyl Tensor

In 1921, H. Weyl proposed a new tensor, known as Weyl Tensor, which is a measure of the curvature of spacetime. Similar to the Riemann curvature tensor, the Weyl tensor expresses the tidal force felt by a body moving along a geodesic.

The Weyl tensor in n-dimensional space is defined as

$$C_{abcd} = R_{abcd} + \frac{1}{n-2}(g_{ad}R_{cb} + g_{bc}R_{da} - g_{ac}R_{db} - g_{bd}R_{ca}) + \frac{1}{(n-1)(n-2)}(g_{ac}g_{db} - g_{ad}g_{cb})R. \quad (1.52)$$

It is easy to show that Weyl tensor has the same symmetries as in Riemann tensor.

$$C_{abcd} = -C_{abdc} = -C_{bacd} = C_{cdab},$$
$$C_{abcd} + C_{adbc} + C_{acdb} = 0.$$

In addition, it possesses an additional symmetry that it is traceless, i.e., its value is zero for any two pair of contracted indices.

$$C^{\sigma}_{\rho\sigma v} = g^{\sigma\mu}C_{\rho\sigma\mu v} = 0.$$

Hint: By using inner product, one can lower the contravariant index in $C^{\sigma}_{\rho\sigma v}$, i.e.,

$$g^{\sigma\mu}C_{\rho\sigma\mu v} = g^{\sigma\mu}R_{\rho\sigma\mu v} - \frac{1}{n-2}\left(g^{\sigma\mu}g_{\rho\mu}R_{v\sigma} + g^{\sigma\mu}g_{\sigma v}R_{\mu\rho} - g^{\sigma\mu}g_{\rho v}R_{\mu\sigma} - g^{\sigma\mu}g_{\sigma\mu}R_{v\rho}\right)$$

$$+ \frac{R}{(n-1)(n-2)}(g^{\sigma\mu}g_{\rho\mu}g_{v\sigma} - g^{\sigma\mu}g_{\rho v}g_{\mu\sigma})$$

$$= -g^{\sigma\mu}R_{\rho\sigma\mu v} - \frac{1}{n-2}(\delta^{\sigma}_{\rho}R_{v\sigma} + \delta^{\mu}_{v}R_{\mu\rho} - g_{\rho v}R - nR_{v\rho})$$

$$+ \frac{R}{(n-1)(n-2)}(\delta^{\sigma}_{\rho}g_{v\sigma} - ng_{\rho v})$$

$$= -R_{\rho v} - \frac{1}{n-2}(R_{\rho v} + R_{\rho v} - g_{\rho v}R - nR_{v\rho}) + \frac{R}{(n-1)(n-2)}(g_{\rho v} - ng_{\rho v})$$

$$= -R_{\rho v} + R_{\rho v} + \frac{R}{(n-2)}g_{\rho v} - \frac{R}{(n-2)}g_{\rho v} = 0.$$

Hence, Weyl tensor is traceless, i.e., the value of Weyl tensor is zero for any two pairs of contracted indices, $C^{\sigma}_{\rho\sigma v} = 0$.

In other words, it is irreducible. In two and three dimensions, the Weyl curvature tensor vanishes identically. In general, the Weyl curvature is nonzero for $n > 3$.

Two Riemannian spaces are said to be **conformally related** if two matrices from two spaces are related by the equation

$$\bar{g}_{ab} = \Omega^2 g_{ab}, \quad (1.53)$$

where $\Omega(x)$ is a nonzero differentiable function. If two spaces are conformally related then Weyl tensor is preserved, i.e., two matrices possess the same Weyl tensor.

$$\bar{C}^{a}_{bcd} = C^{a}_{bcd}. \quad (1.54)$$

Thus, the Weyl tensor is invariant under conformal transformation.

The space is said to be conformally flat if $g_{ab} = \Omega^2 \eta_{ab}$, where η_{ab} is the Minkowski metric.

Note 1.10

Since Weyl tensors are same for two conformally related spaces, therefore, a metric will be conformally flat if and only if its Weyl tensor vanishes everywhere. Thus,

$$C_{abcd} = 0 \iff g_{ab} = \Omega^2 \eta_{ab}.$$

Exercise 1.50

Let two metrics be conformally related as $g_{\alpha\beta} \to f(x^\mu)g_{\alpha\beta}$ where f is an arbitrary function. Then show that all angles are preserved.

Hint: The angle between two vectors A^μ and B^μ in the metric space is defined as

$$\cos\theta = \frac{g_{\mu\nu}A^\mu B^\nu}{\sqrt{(g_{\alpha\beta}A^\alpha B^\beta)(g_{\rho\sigma}A^\rho B^\sigma)}} \to \frac{f(x^\mu)g_{\mu\nu}A^\mu B^\nu}{\sqrt{(f(x^\mu)g_{\alpha\beta}A^\alpha B^\beta)(f(x^\mu)g_{\rho\sigma}A^\rho B^\sigma)}}.$$

The cancelling of $f(x^\mu)$ gives the same value as before.

Note 1.11

In two conformally related spaces, the null curves remain null curves as the squares of their tangent vectors remain zero

$$0 = \mathbf{t} \cdot \mathbf{t} = t_\alpha t^\alpha = g_{\alpha\beta}t^\alpha t^\beta \to f(x^\mu)g_{\alpha\beta}t^\alpha t^\beta = 0.$$

Note 1.12

In general relativity, the Weyl tensor contributes curvature to the spacetime when the Ricci tensor vanishes. In general relativity, source of $R_{ij} \propto (T_{ij} - \frac{1}{2}g_{ij}T)$, where T_{ij} is the energy momentum tensor of the local matter distribution. Thus, absence of matter implies $R_{ij} = 0$. However, the absence of matter does not imply that the spacetime is flat as the Weyl tensor provides curvature to the Riemann curvature tensor. This indicates that the gravitational field is not zero in spacetime in the absence of matter. This term has an important implication that gravity may propagate in regions where there is no matter/energy source.

Note 1.13

Conformal transformation is also known as Weyl transformation, which is an important part of any general relativity development. This conformal or Weyl transformation is a transformation of the metric without modifying the coordinates so that

$$\bar{g}_{\mu\nu}(x) = e^{2\sigma(x)}g_{\mu\nu}(x),$$

where σ is a real function of the coordinates. The contravariant component of the transformed metric is

$$\bar{g}^{\mu\nu}(x) = e^{-2\sigma}g^{\mu\nu}(x).$$

Under conformal or Weyl transformation the Christoffel symbol transforms

$$\bar{\Gamma}_{\alpha\beta\gamma} = e^{2\sigma}\left[\Gamma_{\alpha\beta\gamma} + \left(g_{\alpha\beta}\frac{\partial\sigma}{\partial x^\gamma} + g_{\alpha\gamma}\frac{\partial\sigma}{\partial x^\beta} - g_{\beta\gamma}\frac{\partial\sigma}{\partial x^\alpha}\right)\right], (first\ kind) \tag{1.55}$$

$$\bar{\Gamma}^\alpha_{\beta\gamma} = \bar{g}^{\alpha\gamma}\bar{\Gamma}_{\lambda\beta\gamma} = \Gamma^\alpha_{\beta\gamma} + \left(\delta^\alpha_\beta\frac{\partial\sigma}{\partial x^\gamma} + \delta^\alpha_\gamma\frac{\partial\sigma}{\partial x^\beta} - g_{\beta\gamma}g^{\alpha\gamma}\frac{\partial\sigma}{\partial x^\lambda}\right), (second\ kind) \tag{1.56}$$

One can also find the transformation properties for the Riemann tensor, Ricci tensor and Ricci scalar. The Riemann tensor transforms as

$$\bar{R}_{\alpha\beta\gamma\delta} = e^{2\sigma}[R_{\alpha\beta\gamma\delta} + (g_{\alpha\delta}\sigma_{\beta\gamma} + g_{\beta\gamma}\sigma_{\alpha\delta} - g_{\alpha\gamma}\sigma_{\beta\delta} - g_{\beta\delta}\sigma_{\alpha\gamma})]$$
$$+ e^{2\sigma}[(g_{\alpha\delta}g_{\beta\gamma} - g_{\alpha\gamma}g_{\beta\delta})(\nabla_\mu\sigma\nabla^\mu\sigma)].$$

Here, ∇_α stands for covariant derivative and

$$\sigma_{\alpha\beta} = \sigma_{\beta\alpha} = \nabla_\alpha\nabla_\beta\sigma - (\nabla_\alpha\sigma)(\nabla_\beta\sigma),$$

$$\nabla_\mu\sigma\nabla^\mu\sigma = g^{\mu\nu}\nabla_\mu\sigma\nabla_\nu\sigma = g^{\mu\nu}\sigma_{;\mu\nu} = g^{\mu\nu}\frac{\partial\sigma}{\partial x^\mu}\frac{\partial\sigma}{\partial x^\nu}.$$

The Ricci tensor transforms as

$$\bar{R}_{\alpha\beta} = \bar{g}^{\rho\sigma}\bar{R}_{\rho\alpha\sigma\beta} = R_{\alpha\beta} - 2\sigma_{\alpha\beta}(\Box\sigma + 2\nabla_\mu\sigma\nabla^\mu\sigma)g_{\alpha\beta}.$$
$$\Box\sigma = \nabla_\mu\nabla^\mu\sigma = g^{\mu\nu}\nabla_\mu\nabla_\nu\sigma]$$

The Ricci scalar transforms as

$$\bar{R} = \bar{g}^{\alpha\beta}\bar{R}_{\alpha\beta} = e^{-2\sigma}(R - 6\Box\sigma - 6\nabla_\mu\sigma\nabla^\mu\sigma).$$

However, as in Eq. (1.54), Weyl tensor keeps the same form

$$\bar{C}^a_{bcd} = C^a_{bcd}.$$

Geodesics

2.1 Geodesics Equation

According to general theory of relativity, gravitation is not a force but a property of spacetime geometry. A test particle and light move in response to the geometry of the spacetime. Actually, curved spacetimes of general relativity are explored by reviewing the nature of the motion of freely falling particles and light through them. **Freely falling particles** are those particles that are free from any effects except curvature of spacetime. This chapter provides the derivation of the equations of motion of the test particles and light rays in a general curved spacetime.

The path or the differential equation of the curve having an external length, i.e., path of extremum distance between two points is called the **geodesic equation**.

Therefore, for a geodesic $I = \int_A^B ds$ must be extremum, where the limits of integration are taken to be two fixed points A and B.

Thus,

$$\delta I = \delta \int_A^B ds = 0.$$

For Riemannian space

$$ds^2 = g_{\mu\nu}dx^\mu dx^\nu,$$

$$\Rightarrow \delta \int \frac{ds}{dp}dp = 0,$$

$$\Rightarrow \delta \int_A^B \left(g_{\mu\nu} \frac{dx^\mu}{dp} \frac{dx^\nu}{dp} \right)^{1/2} dp = 0,$$

$$\Rightarrow \delta \int_A^B L dp = 0. \tag{2.1}$$

Here,

$$L = \left(g_{\mu\nu} \frac{dx^\mu}{dp} \frac{dx^\nu}{dp} \right)^{1/2} \tag{2.2}$$

is known as **Lagrangian**.

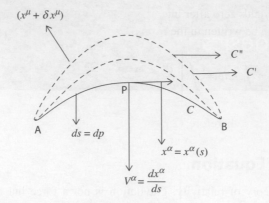

Figure 5   Curves joining two fixed points.

This implies **Euler–Lagrange** equation as

$$\frac{d}{dp}\frac{\partial L}{\partial\left(\frac{dx^\mu}{dp}\right)} - \frac{\partial L}{\partial x^\mu} = 0. \tag{2.3}$$

Here, p = affine parameter, describing the trajectory.

[In general , $d\tau$(proper time $d\tau^2 = \frac{ds^2}{c^2}$) is proportional to dp, such that for a material particle, we can normalize p so that $p = \tau$. Nevertheless, for a photon, the proportionality constant $\frac{d\tau}{dp}$ vanishes (as $ds = 0$, for photon)]

We can describe the geodesic line between two fixed points A and B. Let us consider the shortest path, i.e., the curve C is the geodesic line. The other two curves C' and C'', e.g., are different curves, other than geodesic C, with a variation δx^μ (see Fig. 5). The geodesic line is described by $x^\mu = x^\mu(s) = x^\mu(p)$, where s and p are parameters along the curve. The metric is defined as $ds^2 = g_{\mu\nu}dx^\mu dx^\nu$. The tangent vector to the curve $x^\alpha = x^\alpha(s)$ at P is the unit vector $v^\alpha = \frac{dx^\alpha}{dp}$. A and B are two fixed points.

2.2   Derivation of Euler–Lagrange Equation

We prove Euler–Lagrange equation from Eq. (2.1)

$$\delta\int L ds = \int\left[\frac{\partial L}{\partial x^\mu}\delta x^\mu + \frac{\partial L}{\partial\left(\frac{dx^\mu}{dp}\right)}\delta\left(\frac{dx^\mu}{dp}\right)\right]dp. \tag{2.4}$$

We can write the second term of (2.4) as the difference of two terms

$$\frac{d}{dp}\left[\frac{\partial L}{\partial(\frac{dx^\mu}{dp})}\delta x^\mu\right] - \frac{d}{dp}\left[\frac{\partial L}{\partial(\frac{dx^\mu}{dp})}\right]\delta x^\mu. \tag{2.5}$$

[Since $\frac{d}{dp}(\delta x^\mu) = \delta(\frac{dx^\mu}{dp})$]

The first expression yields zero after integration as the variations vanish at the end points of the curve. Thus, Eq. (2.4) can be written in the form

$$\delta \int L dp = \int \left[\frac{\partial L}{\partial x^\mu} - \frac{d}{dp} \left(\frac{\partial L}{\partial (\frac{dx^\mu}{dp})} \right) \right] \delta x^\mu dp = 0. \tag{2.6}$$

Here variation δx^μ is arbitrary, therefore, (2.6) implies the well-known **Euler–Lagrangian equation** (2.3) as

$$\frac{d}{dp} \left(\frac{\partial L}{\partial (\frac{dx^\mu}{dp})} \right) - \frac{\partial L}{\partial x^\mu} = 0.$$

2.3 Geodesic Equation in Curved Spacetime

Now we are trying to find explicit expression for the differential equation of geodesics. Now,

$$\frac{\partial L}{\partial \left(\frac{dx^\mu}{dp} \right)} = \frac{1}{2} \left[g_{\alpha\beta} \frac{dx^\alpha}{dp} \frac{dx^\beta}{dp} \right]^{-1/2} \left[g_{\alpha\beta} \delta^\alpha_\mu \frac{dx^\beta}{dp} + g_{\alpha\beta} \frac{dx^\alpha}{dp} \delta^\beta_\mu \right],$$

$$= \frac{1}{2} \left[g_{\mu\beta} \frac{dx^\beta}{dp} + g_{\alpha\mu} \frac{dx^\alpha}{dp} \right] = g_{\mu\nu} \frac{dx^\nu}{dp}.$$

[Since along the geodesic $ds^2 = g_{\alpha\beta} dx^\alpha dx^\beta$ and affine parameter p and s are equivalent. Also, $g_{\mu\nu}$ is symmetric.]

Therefore,

$$\frac{d}{dp} \left[\frac{\partial L}{\partial (\frac{dx^\mu}{dp})} \right] = g_{\mu\nu} \frac{d^2 x^\nu}{dp^2} + \frac{\partial g_{\mu\nu}}{\partial x^\beta} \frac{dx^\beta}{dp} \frac{dx^\nu}{dp}.$$

Also,

$$\frac{\partial L}{\partial x^\mu} = \frac{1}{2} \left[g_{\alpha\beta} \frac{dx^\alpha}{dp} \frac{dx^\beta}{dp} \right]^{-1/2} \frac{\partial g_{\nu\beta}}{\partial x^\mu} \frac{dx^\nu}{dp} \frac{dx^\beta}{dp} = \frac{1}{2} \frac{\partial g_{\nu\beta}}{\partial x^\mu} \frac{dx^\nu}{dp} \frac{dx^\beta}{dp}.$$

Thus, we have

$$g_{\mu\nu} \frac{d^2 x^\nu}{dp^2} + \frac{1}{2} \left[\frac{2 \partial g_{\mu\nu}}{\partial x^\beta} - \frac{\partial g_{\nu\beta}}{\partial x^\mu} \right] \frac{dx^\nu}{dp} \frac{dx^\beta}{dp} = 0.$$

Using the factor

$$\frac{2 \partial g_{\mu\nu}}{\partial x^\beta} \frac{dx^\nu}{dp} \frac{dx^\beta}{dp} = \left(\frac{\partial g_{\mu\nu}}{\partial x^\beta} + \frac{\partial g_{\mu\beta}}{\partial x^\nu} \right) \frac{dx^\nu}{dp} \frac{dx^\beta}{dp}.$$

[Fix μ and interchange dummy index β, ν]

Hence, finally we get

$$g_{\mu\nu}\frac{d^2x^\nu}{dp^2} + \Gamma_{\mu\nu\beta}\frac{dx^\nu}{dp}\frac{dx^\beta}{dp} = 0,$$

$$\Rightarrow$$

$$\frac{d^2x^\rho}{dp^2} + \Gamma^\rho_{\nu\beta}\frac{dx^\nu}{dp}\frac{dx^\beta}{dp} = 0. \qquad (2.7)$$

[Multiplying by $g^{\rho\mu}$]

This is the **geodesic equation in curved spacetime**.

Also multiplying by $2g_{\rho\sigma}\frac{dx^\sigma}{dp}$, we get

$$2g_{\rho\sigma}\frac{d^2x^\rho}{dp^2}\frac{dx^\sigma}{dp} + [g_{\sigma\nu,\beta} + g_{\sigma\beta,\nu} - g_{\beta\nu,\sigma}]\frac{dx^\nu}{dp}\frac{dx^\beta}{dp}\frac{dx^\sigma}{dp} = 0,$$

i.e.

$$2g_{\rho\sigma}\frac{d^2x^\rho}{dp^2}\frac{dx^\sigma}{dp} + g_{\sigma\nu,\beta}\frac{dx^\nu}{dp}\frac{dx^\beta}{dp}\frac{dx^\sigma}{dp} = 0.$$

[Here we have used the symmetry and antisymmetry with respect to ν and σ to simplify the second term]

Using, further, $g_{\rho\nu} = g_{\nu\rho}$ and integrating to yield

$$g_{\rho\sigma}\frac{dx^\rho}{dp}\frac{dx^\sigma}{dp} = constant. \qquad (2.8)$$

The Eq. (2.8) is known as **first integral.** If the constant is zero, the geodesic is called **null geodesic.** For positive constant ($\equiv 1$), the geodesic is **time-like** and if it is negative($\equiv -1$), the geodesic is **space-like**.

Then (2.7) and

$$g_{\rho\sigma}\frac{dx^\rho}{dp}\frac{dx^\sigma}{dp} = 1,$$

are the equations for **non-null geodesic.**

Again, equation (2.7) and

$$g_{\rho\sigma}\frac{dx^\rho}{dp}\frac{dx^\sigma}{dp} = 0,$$

are the equations for **null geodesic.**

In other words,

$$Constant = 0, \quad \text{for photon [for photon } ds = 0 \text{ on the light cone]}$$

$$= 1, \quad \text{for a massive particle.}$$

2.4 Geodesic Deviation

Let geodesic curves be parametrized by a parameter λ. Suppose at any instant s^i is the joining vector between two points with the same value of λ on two curves, i.e., s^i be a separation vector between neighboring geodesics at same parameter value (see Fig. 6). This joining vector s^i is known as the measure of geodesic deviation of the curves. In Euclidean space, two geodesics (which are straight lines) maintain the same distance if they are parallel or the separation distance increases if they intersect at any point. However, in curved space this situation varies significantly.

Let v be an affine parameter in each geodesic. Then any point x^i on the geodesic depends on two parameters λ and v, i.e., $x^i(\lambda, v)$. Here, λ and v are basis vectors of the coordinate system. Then the separation vector (s^i) and tangent vector (u^i) are defined respectively as

$$s^i = \frac{\partial x^i}{\partial \lambda} \ \text{ and } \ u^i = \frac{\partial x^i}{\partial v}.$$

The geodesic deviation equation is given by

$$\frac{\partial^2 s^i}{\partial \lambda^2} + R^i_{jkl} u^j s^k u^l = 0.$$

This indicates that if one knows the geodesic deviation, then it would be possible to determine the Riemannian curvature tensor.

For a comoving geodesic normal coordinate we can have $u^i = (1, 0, 0, 0)$. Then the above geodesic deviation equation assumes the following form

$$\frac{\partial^2 s^i}{\partial \lambda^2} + R^i_k s^k = 0.$$

2.5 Geodesics Are Auto Parallel

Proof: We know $u^i = \frac{dx^i}{dp}$ is the tangent vector of any geodesic curve

$$\frac{d^2 x^i}{dp^2} + \Gamma^i_{jk} \frac{dx^j}{dp} \frac{dx^k}{dp} = 0.$$

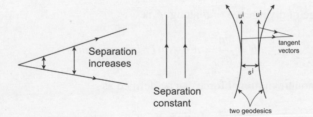

Figure 6 Geodesic deviation.

Now one can write the above geodesic equation in terms of tangent vector as

$$\frac{du^i}{dp} + \Gamma^i_{jk}u^j u^k = 0.$$

Using $\frac{du^i}{dp} = \frac{\partial u^i}{\partial x^k}\frac{dx^k}{dp} = \frac{\partial u^i}{\partial x^k}u^k$, we get from the above equation

$$\left(\frac{\partial u^i}{\partial x^k} + \Gamma^i_{jk}u^j\right)u^k = 0,$$

or

$$u^i_{;k}u^k = 0.$$

Hence, geodesics are auto parallel.

2.6   Raychaudhuri Equation

In 1955, Professor A. K. Raychaudhuri derived an important equation for the rate of change of the divergence of $u^k_{;k}$. Here, u^k is the time-like four-velocity of the particle along the geodesic and $\theta = u^k_{;k}$ is the expansion/contraction of volume, i.e., expansion/contraction of the congruence of geodesics defined by u^a. From the definition of curvature tensor, we can write

$$u^a_{;bc} - u^a_{;cb} = R^a_{dbc}u^d.$$

Now, we can deduce after contracting a and b and multiplying by u^c as

$$u^c u^a_{;ac} - u^c u^a_{;ca} = R_{dc}u^c u^d,$$

or

$$u^c\theta_{;c} - u^c u^a_{;ca} = R_{dc}u^c u^d. \tag{i}$$

We can rewrite the second term as

$$(u^a_{;c}u^c)_{;a} - u^a_{;c}u^c_{;a} = \dot{u}^a_{;a} - u_{a;b}u^{b;a}. \tag{ii}$$

The rate of change of divergence of θ along u^c is

$$\dot{\theta} = \frac{d\theta}{d\tau} = u^c\theta_{;c}. \quad (\tau \text{ is proper time}) \tag{iii}$$

Acceleration due to nongravitational force can be defined as

$$\dot{u}_a = u_{a;b}u^b,$$

which is orthogonal to u^a.

Let us now define

$$h_{ab} = g_{ab} + u_a u_b, \text{ where, } u^a h_{ab} = h_{ab} u^b = 0,$$

and

$$B_{ab} = u_{a;b}, \text{ here, } B_{ab} u^b = u_{a;b} u^b = 0.$$

Now, B_{ab} can be decomposed into its antisymmetric, symmetric-traceless, and trace part as

$$u_{a;b} = B_{ab} = \omega_{ab} + \sigma_{ab} + \frac{1}{3}\theta h_{ab}, \tag{iv}$$

where

$$\omega_{ab} = \frac{1}{2}(B_{ab} - B_{ba}),$$

$$\sigma_{ab} = \frac{1}{2}(B_{ab} + B_{ba}) - \frac{1}{3}\theta h_{ab},$$

$$\theta = h^{ab} B_{ab} = g^{ab} B_{ab} = u^a_{;a}.$$

The quantities ω_{ab} and σ_{ab} are known as **rotation tensor** and **shear tensor**.

Now combining (i), (ii), (iii), and (iv), we get the Raychaudhuri equation as

$$\frac{d\theta}{d\tau} = -\frac{1}{3}\theta^2 - \sigma_{ab}\sigma^{ab} + \omega_{ab}\omega^{ab} + \dot{u}^a_{;a} + R_{ab}u^a u^b.$$

Raychaudhuri equation has many applications in general relativity, particularly in the context of singularity theorems of Penrose and Hawking.

Exercise 2.1

Find the geodesic in three-dimensional Euclidean space.

Hint: The geodesic equation is given by

$$\frac{d^2 x^\rho}{dp^2} + \Gamma^\rho_{\nu\beta} \frac{dx^\nu}{dp} \frac{dx^\beta}{dp} = 0.$$

In Euclidean space $g_{ij} = 1$, $\forall i, j$, therefore, all Christoffel symbols vanish identically. The above geodesic equation implies

$$\frac{d^2 x^\rho}{dp^2} = 0.$$

This yields

$$x^\rho = a^\rho p + b^\rho, \text{ where } a^\rho \text{ and } b^\rho \text{ are constants.}$$

This represents a straight line.

Exercise 2.2

Calculate the radial null geodesic for the line element with the plane $\theta = constant$ and $\phi = constant$.

$$ds^2 = dt^2 - f(t)[dr^2 + r^2(d\theta^2 + \sin^2\theta \, d\phi^2)].$$

Hint:

For $\theta = constant$ and $\phi = constant$, the above geodesic equation

$$\frac{d^2x^\rho}{dp^2} + \Gamma^\rho_{\nu\beta} \frac{dx^\nu}{dp} \frac{dx^\beta}{dp} = 0,$$

yields

$$\frac{d^2t}{dp^2} + \Gamma^t_{rr} \frac{dr}{dp} \frac{dr}{dp} = 0.$$

Using

$$\Gamma^b_{aa} = -\frac{1}{2g_{bb}} g_{aa,b},$$

we obtain

$$\frac{d^2t}{dp^2} + \frac{1}{2}f'(t)\left(\frac{dr}{dp}\right)^2 = 0.$$

Also, first integral gives

$$\left(\frac{dt}{dp}\right)^2 = f(t)\left(\frac{dr}{dp}\right)^2.$$

Now, for a given $f(t)$, one can find geodesic equation in terms of r and t. For example, say, $f(t) = e^{2Ht}$ for de Sitter spacetime, the solutions of the above two equations yield

$$t = \frac{1}{H} \ln\left(1 + \frac{p}{p_0}\right), \quad r = \frac{1}{H}\left(\frac{p}{p + p_0}\right).$$

Exercise 2.3

Consider the spacetimes

$$ds^2 = x^2 dt^2 - dx^2.$$

Find the shape of $x(t)$ for all the time-like geodesics in this spacetime.

Hints:

Method 1: Write the Lagrangian $(L(\dot{x}_\alpha, x_\alpha))$ as

$$L(\dot{x}^\alpha, x^\alpha) = (g_{\alpha\beta}\dot{x}^\alpha\dot{x}^\beta)^{\frac{1}{2}} = (x^2\dot{t}^2 - \dot{x}^2)^{\frac{1}{2}}.$$

Use Euler–Lagrangian equation to find the relation between x and t.

$$\frac{d}{dp}\left(\frac{\partial L}{\partial(\frac{dx^\mu}{dp})}\right) - \frac{\partial L}{\partial x^\mu} = 0.$$

Solution:

$$t = \cosh^{-1}\left(\frac{c}{x}\right) + t_0.$$

Method 2: Use directly the geodesic equation and first integral

$$\frac{d^2x^\rho}{dp^2} + \Gamma^\rho_{\nu\beta}\frac{dx^\nu}{dp}\frac{dx^\beta}{dp} = 0,$$

$$g_{\rho\sigma}\frac{dx^\rho}{dp}\frac{dx^\sigma}{dp} = 0, \text{ for photon}$$

$$= 1, \text{ time-like particle}$$

Exercise 2.4

Consider the spacetime

$$ds^2 = y^{-2}[dx^2 + dy^2].$$

Find the shape of $y(x)$ for all geodesics in this spacetime.
Hint: Here the nonzero Christoffel symbols are

$$-\Gamma^x_{xy} = \Gamma^y_{xx} = -\Gamma^y_{yy} = \frac{1}{y}.$$

The geodesic equation for x component is

$$\frac{d^2x}{dp^2} = \frac{2}{y}\frac{dx}{dp}\frac{dy}{dp}.$$

This implies

$$y^2 \frac{d}{dp} \left(\frac{1}{y^2} \frac{dx}{dp} \right) = 0.$$

$$\textit{This yields, } \frac{dx}{dp} = \frac{y^2}{C}.$$

C is an integration constant.

The first integral yields

$$\frac{1}{y^2} \left[\left(\frac{dx}{dp} \right)^2 + \left(\frac{dy}{dp} \right)^2 \right] = 1.$$

Using the above result, we get

$$\left(\frac{dy}{dp} \right)^2 = \left[y^2 - \frac{y^4}{C^2} \right].$$

Solution of these equations yield the geodesic

$$(x - x_0)^2 + y^2 = C^2.$$

Exercise 2.5

Let us consider a flat spacetime in the frame (t, x, y, z), which is rotating with an angular velocity Ω about z axis of an inertial frame. In this frame, the line element is given by

$$ds^2 = -[1 - \Omega^2(x^2 + y^2)]dt^2 + 2\Omega(ydx - xdy) + dx^2 + dy^2 + dz^2.$$

(i) Obtain the geodesic equations for x, y, z and (ii) prove that these reduce to the usual equations of Newtonian mechanics, for a free particle, in a frame comprising the centrifugal force and the Coriolis force in the nonrelativistic limit.

Hint: Here, the affine parameter τ is the extremal proper time. In the nonrelativistic limit it is just time t.

The equations for x, y and z are

$$-\frac{d^2x}{d\tau^2} - 2\Omega \frac{dy}{d\tau} \frac{dt}{d\tau} + \Omega^2 x \left(\frac{dt}{d\tau} \right)^2 = 0.$$

$$-\frac{d^2y}{d\tau^2} + 2\Omega \frac{dx}{d\tau} \frac{dt}{d\tau} + \Omega^2 y \left(\frac{dt}{d\tau} \right)^2 = 0.$$

$$\frac{d^2z}{d\tau^2} = 0.$$

In the nonrelativistic limit, this equation becomes (using $\tau \approx t$)

$$\frac{d^2x}{dt^2} = -2\Omega\frac{dy}{dt} + \Omega^2 x.$$

We know the centrifugal force is $\vec{\Omega} \times (\vec{\Omega} \times \vec{r})$, therefore, the second term of the right-hand side is the x component of centrifugal force, where $\vec{\Omega} = \Omega\vec{e_z}$. The first term is the x component of Coriolis force, $2\vec{\Omega} \times \left(\frac{d\vec{r}}{dt}\right)$.

Exercise 2.6

Show that the norm of four velocity $\| \mathbf{u} \| = \mathbf{u} \cdot \mathbf{u}$ is a constant along a geodesic.
Hint: Suppose

$$U = \mathbf{u} \cdot \mathbf{u} = g_{ik}u^i u^k.$$

Now,

$$\frac{dU}{d\tau} = 2g_{ik}\frac{du^i}{d\tau}u^k + \frac{\partial g_{ik}}{\partial x^j}u^i u^k u^j.$$

Using $u^i = \frac{dx^i}{d\tau}$, the geodesic equation $\frac{d^2x^i}{d\tau^2} + \Gamma^i_{jl}\frac{dx^j}{d\tau}\frac{dx^l}{d\tau} = 0$, yields

$$\frac{du^i}{d\tau} = -\Gamma^i_{jl}u^j u^l.$$

Therefore,

$$\frac{dU}{d\tau} = -2g_{ik}\Gamma^i_{jl}u^k u^j u^l + \frac{\partial g_{ik}}{\partial x^j}u^i u^k u^j.$$

To get the same factor $u^i u^j u^k$, we change the dummy indices to yield

$$\frac{dU}{d\tau} = \left(-2g_{kl}\Gamma^l_{ji} + \frac{\partial g_{ik}}{\partial x^j}\right)u^i u^k u^j,$$

$$= \left(-\frac{\partial g_{kj}}{\partial x^i} - \frac{\partial g_{ki}}{\partial x^j} + \frac{\partial g_{ji}}{\partial x^k} + \frac{\partial g_{ik}}{\partial x^j}\right)u^i u^k u^j.$$

The right-hand side is zero as $g_{ik} = g_{ki}$ and adjusting the dummy indices. Hence, U is constant.

Exercise 2.7

Calculate the radial null geodesic for the line element

$$ds^2 = A(r)dt^2 - B(r)dr^2 - r^2(d\theta^2 + \sin^2\theta d\phi^2).$$

Hint:

Using the geodesic equation

$$\frac{d^2x^i}{dp^2} + \Gamma^i_{jk}\frac{dx^j}{dp}\frac{dx^k}{dp} = 0,$$

we find t component as

$$\frac{d^2t}{dp^2} + \Gamma^t_{tt}\left(\frac{dt}{dp}\right)^2 + \Gamma^t_{rr}\left(\frac{dr}{dp}\right)^2 + \Gamma^t_{\theta\theta}\left(\frac{d\theta}{dp}\right)^2 + \Gamma^t_{\phi\phi}\left(\frac{d\phi}{dp}\right)^2$$

$$+ 2\Gamma^t_{rt}\frac{dr}{dp}\frac{dt}{dp} + 2\Gamma^t_{\theta t}\frac{d\theta}{dp}\frac{dt}{dp} + 2\Gamma^t_{\phi t}\frac{d\phi}{dp}\frac{dt}{dp} = 0.$$

Using the values of Christoffel symbols, we get

$$\frac{d^2t}{dp^2} + \frac{1}{A}\frac{\partial A}{\partial r}\frac{dr}{dp}\frac{dt}{dp} = 0.$$

Similarly, we determine the other components

$$\frac{d^2r}{dp^2} + \Gamma^r_{tt}\left(\frac{dt}{dp}\right)^2 + \Gamma^r_{rr}\left(\frac{dr}{dp}\right)^2 + \Gamma^r_{\theta\theta}\left(\frac{d\theta}{dp}\right)^2 + \Gamma^r_{\phi\phi}\left(\frac{d\phi}{dp}\right)^2$$

$$+ 2\Gamma^r_{tr}\frac{dr}{dp}\frac{dt}{dp} + 2\Gamma^r_{\theta r}\frac{d\theta}{dp}\frac{dr}{dp} + 2\Gamma^r_{\phi r}\frac{d\phi}{dp}\frac{dr}{dp} = 0,$$

$$\Rightarrow \frac{d^2r}{dp^2} + \frac{1}{2B}\frac{\partial B}{\partial r}\left(\frac{\partial r}{\partial p}\right)^2 + \frac{1}{2B}\frac{\partial A}{\partial r}\left(\frac{dt}{dp}\right)^2 - \frac{r}{B}\left(\frac{d\theta}{dp}\right)^2 - \frac{r\sin^2\theta}{B}\left(\frac{d\phi}{dp}\right)^2 = 0,$$

$$\frac{d^2\theta}{dp^2} + \Gamma^\theta_{tt}\left(\frac{dt}{dp}\right)^2 + \Gamma^\theta_{rr}\left(\frac{dr}{dp}\right)^2 + \Gamma^\theta_{\theta\theta}\left(\frac{d\theta}{dp}\right)^2 + \Gamma^\theta_{\phi\phi}\left(\frac{d\phi}{dp}\right)^2$$

$$+ 2\Gamma^\theta_{tt}\frac{d\theta}{dp}\frac{dt}{dp} + 2\Gamma^\theta_{\theta r}\frac{d\theta}{dp}\frac{dr}{dp} + 2\Gamma^\theta_{\phi t}\frac{d\phi}{dp}\frac{d\theta}{dp} = 0,$$

$$\Rightarrow \frac{d^2\theta}{dp^2} - \sin\theta\cos\theta\left(\frac{d\phi}{dp}\right)^2 + \frac{2}{r}\frac{d\theta}{dp}\frac{dr}{dp} = 0.$$

$$\frac{d^2\phi}{dp^2} + \Gamma^\phi_{tt}\left(\frac{dt}{dp}\right)^2 + \Gamma^\phi_{rr}\left(\frac{dr}{dp}\right)^2 + \Gamma^\phi_{\theta\theta}\left(\frac{d\theta}{dp}\right)^2 + \Gamma^\phi_{\phi\phi}\left(\frac{d\phi}{dp}\right)^2$$

$$+ 2\Gamma^\phi_{t\phi}\frac{dt}{dp}\frac{d\phi}{dp} + 2\Gamma^\phi_{\phi r}\frac{d\phi}{dp}\frac{dr}{dp} + 2\Gamma^\phi_{\phi\theta}\frac{d\phi}{dp}\frac{d\theta}{dp} = 0$$

$$\Rightarrow \frac{d^2\phi}{dp^2} + \frac{2}{r}\frac{d\phi}{dp}\frac{dr}{dp} + \frac{2\cos\theta}{\sin\theta}\frac{d\theta}{dp}\frac{d\phi}{dp} = 0.$$

Also first integral yields

$$A \left(\frac{dt}{dp} \right)^2 - B \left(\frac{dr}{dp} \right)^2 - r^2 \left(\frac{d\theta}{dp} \right)^2 - r^2 \sin^2 \theta \left(\frac{d\phi}{dp} \right)^2 = \epsilon,$$

where $\epsilon = 0$ *or* 1 for null or time-like geodesic. For given values of A and B, one can find the exact geodesic path.

Exercise 2.8

Calculate the radial null geodesic for the line element

$$ds^2 = \left(1 - \frac{2m}{r} \right) dt^2 - \left(1 - \frac{2m}{r} \right)^{-1} dr^2 - r^2 (d\theta^2 + \sin^2 \theta d\phi^2).$$

Hint: Use

$$A = B^{-1} = \left(1 - \frac{2m}{r} \right),$$

in the above problem.

Exercise 2.9

Calculate the radial null geodesic for the line element

$$ds^2 = A(r) dt^2 - B(r) dr^2 - C(r) dy^2 - D(r) dz^2.$$

Hint:
Using the geodesic equation, we find

$$\frac{d^2 t}{dp^2} + \frac{1}{A} \frac{\partial A}{\partial r} \frac{dr}{dp} \frac{dt}{dp} = 0,$$

$$\frac{d^2 r}{dp^2} + \frac{1}{2B} \frac{\partial B}{\partial r} \left(\frac{dr}{dp} \right)^2 - \frac{1}{2B} \frac{\partial A}{\partial t} \left(\frac{dr}{dp} \right)^2 - \frac{1}{2B} \frac{\partial C}{\partial r} \left(\frac{dy}{dp} \right)^2 - \frac{1}{2B} \frac{\partial D}{\partial r} \left(\frac{dz}{dp} \right)^2 = 0,$$

$$\frac{d^2 y}{dp^2} + \frac{1}{C} \frac{\partial C}{\partial r} \frac{dr}{dp} \frac{dy}{dp} = 0,$$

$$\frac{d^2 z}{dp^2} + \frac{1}{D} \frac{\partial D}{\partial r} \frac{dr}{dp} \frac{dz}{dp} = 0.$$

Also, the first integral yields

$$A \left(\frac{dt}{dp} \right)^2 - B \left(\frac{dr}{dp} \right)^2 - C \left(\frac{dy}{dp} \right)^2 - D \left(\frac{dz}{dp} \right)^2 = \epsilon,$$

where $\epsilon = 0$ *or* 1 for null or time-like geodesic. For given values of A, B, C, and D, one can find the exact geodesic path.

Exercise 2.10

Determine the geodesic for the line element

$$ds^2 = dr^2 + r^2 d\theta^2.$$

Hint: Here,

$$g_{11} = 1, \quad , g_{22} = r^2, \quad \Gamma^1_{22} = -r, \quad \Gamma^2_{12} = \frac{1}{r}.$$

The geodesic equations read

$$\frac{d^2r}{dp^2} - r\left(\frac{d\theta}{dp}\right)^2 = 0 \quad and \quad \frac{d^2\theta}{dp^2} + \frac{2}{r}\frac{dr}{dp}\frac{d\theta}{dp} = 0.$$

Solving these equations, one can get the geodesic equation as

$$\theta = \pm \cos^{-1}\left(\frac{a}{r}\right) + b, \qquad (a,\ b\ are\ integration\ constants).$$

Exercise 2.11

Determine the geodesic for the line element

$$ds^2 = a^2(d\theta^2 + \sin^2\theta d\phi^2).$$

Hint: Here,

$$g_{11} = a^2, \quad g_{22} = a^2\sin^2\theta, \quad \Gamma^1_{22} = -\sin\theta\cos\theta, \quad \Gamma^2_{12} = \cot\theta.$$

The geodesic equations read

$$\frac{d^2\theta}{dp^2} - \sin\theta\cos\theta\left(\frac{d\phi}{dp}\right)^2 = 0 \quad and \quad \frac{d^2\phi}{dp^2} + 2\cot\theta\frac{d\phi}{dp}\frac{d\theta}{dp} = 0.$$

Solving these equations, one can get the geodesic equation as

$$\theta = b - \sin^{-1}(c\cot\theta) \qquad (c,\ b\ are\ integration\ constants).$$

Exercise 2.12

Show that a vector A^j, with constant magnitude, undergoes parallel displacement along a geodesic if

$$A^j_{;k} \frac{dx^k}{dp} = 0.$$

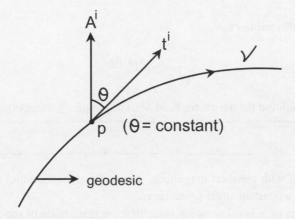

Figure 7 The angle θ between A^i and t^i is constant.

Hint: Without any loss of generality, we assume that the magnitude of the vector A^j at the point P is unity, which is not a tangent. Let $t^i = \frac{dx^i}{dp}$ be the unit tangent at P. The vector A^j is parallel along the curve, which means that the angle between A^i and t^i is constant throughout the curve γ (see Fig. 7). This implies

$$\frac{d(\cos\theta)}{dp} = \frac{d(g_{ij}t^i A^j)}{dp} = 0,$$

$$or, \quad \frac{\partial(g_{ij}t^i A^j)}{\partial x^k} \frac{dx^k}{dp} = 0.$$

Since angle is a scalar, therefore, its covariant derivative is equal to its partial derivative. As a consequence the above equation yields

$$(g_{ij}t^i A^j)_{;k} \frac{dx^k}{dp} = 0,$$

$$or, \quad (g_{ij}t^i)A^j_{;k} \frac{dx^k}{dp} + g_{ij}A^j(t^i_{;k}t^k) = 0.$$

As geodesics are auto parallel, then $t^i_{;k}t^k = 0$.

Hence,

$$t_j A^j_{;k} \frac{dx^k}{dp} = 0.$$

As the tangent vector to the geodesic is an arbitrary vector field, hence we have

$$A^j_{;k} \frac{dx^k}{dp} = 0.$$

This equation can be also written as

$$\frac{dA^j}{dp} = -A^k \Gamma^j_{ki} \frac{dx^i}{dp}.$$

This is actually the condition for the vector field A^j to be parallelly transported.

Exercise 2.13

Show that a vector A^j, with constant magnitude, which undergoes parallel displacement along a geodesic is inclined at a constant angle to the curve.

Hint: Without any loss of generality, we assume that the magnitude of the vector A^j at the point P is unity, which is not a tangent. Let $t^i = \frac{dx^i}{dp}$ be the unit tangent at P. The vector A^j is parallelly transported along the curve, therefore,

$$A^j_{;k} \frac{dx^k}{dp} = 0.$$

As geodesics are auto parallel, then $t^i_{;k} t^k = 0$. Now, if the angle between A^i and t^i is θ, then

$$\cos \theta = g_{ij} t^i A^j.$$

Taking derivative both sides, we get

$$\frac{d(\cos \theta)}{dp} = \frac{d(g_{ij} t^i A^j)}{dp} = \frac{\partial(g_{ij} t^i A^j)}{\partial x^k} \frac{dx^k}{dp}.$$

Since angle is a scalar, its covariant derivative is equal to its partial derivative. As a consequence, the above equation yields

$$\frac{d(\cos \theta)}{dp} = (g_{ij} t^i A^j)_{;k} \frac{dx^k}{dp} = (g_{ij} t^i) A^j_{;k} \frac{dx^k}{dp} + g_{ij} A^j (t^i_{;k} t^k) = 0.$$

This indicates that θ is constant.

Einstein Field Equations

3.1 Introduction

To describe how matter and energy change the geometry of spacetime, Einstein proposed a new theory of gravitation, known as the general theory of relativity. This theory is characterized by a set of dynamical equations, dubbed as Einstein field equations, which demonstrate the fundamental interaction of gravitation as a result of spacetime being curved by mass and energy. Albert Einstein published this new theory of gravitation in 1915 as a tensor equation. The motion of a body in this gravitational field is explained almost perfectly by the geodesic equations.

Einstein used three principles to develop his general theory of relativity.

1. **Mach's Principle:** Ernest Mach gave some ideas without any proof and arguments that are known as Mach's principle. There are several forms of Mach's principle. Some forms are executed in general relativity and others are not.

    (i) Geometry is determined by the matter distribution.

    (ii) If matter does not exist then the geometry will not occur.

    These statements are not absolutely true as there are vacuum solutions of the Einstein field equations (see later).

    (iii) The local inertial frame is entirely persistent by the dynamical fields in the universe.

    Actually, this form of Mach's principle is the key idea of Einstein.

2. **Equivalence Principle:** The laws of physics are the same in all inertial systems. No preferred inertial system exists.

    (The above statements, i.e., containing the words "Laws of Physics" is referred to **strong equivalence principle** and when "Law of Physics" is replaced by "Law of motion of freely falling particles" then the principle is referred to **weak principle of equivalence**.)

    This principle reveals that one can eliminate gravity locally by falling freely in a gravitational field and regain special relativity.

3. **Principle of covariance:** The law of gravitation should be independent of the coordinate system. This means the field equations of gravitation should be invariant in form with respect to arbitrary coordinate transformations. In other words, the field equations should have a tensorial form.

3.2 Three Types of Mass

1. **Inertial mass:** This mass is a measure of the body's resistance to change its state, i.e., either in a rest position or in motion. The mass appears in Newton's second law $F = m^I f$, i.e., here m^I is known as inertial mass.

2. **Passive gravitational mass:** Passive gravitational mass m^p deals with a body's response when it is placed in a gravitational field. Suppose the gravitational potential of some source at some point is φ, then if we place the body of mass m^p at this point, it will experience a force, which is given by $F = -m^p \operatorname{grad} \varphi$.

3. **Active gravitational mass:** Active gravitational mass m^A determines the strength of the gravitational field what the body produces. Suppose we place a body of mass m^A at a particular point, then the gravitational potential it produces at any point of distance r is given by $\varphi = -\frac{Gm^A}{r}$.

In Newtonian physics, these concepts are identical. These identification are valid in general relativity, too.

3.3 Einstein Tensor

$$G_{\mu\nu} = R_{\mu\nu} - \frac{1}{2} g_{\mu\nu} R.$$

Einstein tensor (symmetric tensor of rank two) specifies the geometry of the spacetime. We have already shown in Section 1.17 that $G^{\nu}_{\mu;\nu} = 0$.

The energy-momentum tensor is represented as a bulk properties of matter. This is a symmetric tensor of rank two. Also, conservation law for the energy-momentum tensor $T^{\mu\nu}$ implies that $T^{\nu}_{\mu;\nu} = 0$. Again, one of the forms of Mach's principle is that the matter distribution in the universe is responsible for inertial effects. It seems, Einstein was guided by these two results and considering equivalence principle, principle of covariance along with Mach's principle to give the final form of his gravitational field equations, which should be tensorial equations, i.e.,

$$G^{ab} \ \alpha \ T^{ab} \Rightarrow G^{ab} = -kT^{ab},$$

where $k = \frac{8\pi G}{c^4}$ = coupling constant.

Einstein gave the above equation by applying his own intuition. Later there have been many methods developed to construct the Einstein field equation.

3.4 Some Useful Variations

1. $\delta\sqrt{-g} = \frac{1}{2}\sqrt{-g}\,g^{ik}\delta g^{ik}$.

2. $\delta g_{ik} = -g_{il}g_{km}\delta g^{lm}$.

3. $\delta L = \frac{\partial L}{\partial g_{ik}}\delta g_{ik} + \frac{\partial L}{\partial g_{ik,l}}\delta(g_{ik,l})$.

4. $\delta(g_{ik,l}) = (\delta g_{ik})_{,l}$.

5. $(\sqrt{-g})_{,l} = \frac{1}{2}\sqrt{-g}\,g^{ik}g_{ik,l}$.

6. $\delta R_{\alpha\beta} = \left[\delta\Gamma^{\nu}_{\alpha\nu}\right]_{;\beta} - \left[\delta\Gamma^{\nu}_{\alpha\beta}\right]_{;\nu}$.

Hint: $\delta R_{\mu\nu} = \delta\left[\frac{\partial\Gamma^{\rho}_{\mu\nu}}{\partial x^{\rho}} - \frac{\partial\Gamma^{\rho}_{\mu\rho}}{\partial x^{\nu}} + \Gamma^{\sigma}_{\mu\nu}\Gamma^{\rho}_{\rho\sigma} - \Gamma^{\sigma}_{\mu\rho}\Gamma^{\rho}_{\nu\sigma}\right]$.

In a geodesic coordinate system, we have

$$\delta R_{\mu\nu} = \delta\left[\frac{\partial\Gamma^{\rho}_{\mu\nu}}{\partial x^{\rho}} - \frac{\partial\Gamma^{\rho}_{\mu\rho}}{\partial x^{\nu}}\right],$$

$$= \frac{\partial(\delta\Gamma^{\rho}_{\mu\nu})}{\partial x^{\rho}} - \frac{\partial(\delta\Gamma^{\rho}_{\mu\rho})}{\partial x^{\nu}},$$

$$= (\delta\Gamma^{\rho}_{\mu\nu})_{;\rho} - (\delta\Gamma^{\rho}_{\mu\rho})_{;\nu}.$$

Since this is a tensorial equation, therefore, it is true for all coordinate systems and at all points of space and time.

7. $\delta\Gamma^{\lambda}_{\mu\nu} = \frac{1}{2}g^{\lambda\rho}\left[D_{\nu}\delta g_{\rho\mu} + D_{\mu}\delta g_{\nu\rho} - D_{\rho}\delta g_{\mu\nu}\right]$.

$D \equiv$ covariant derivative.

8. $V^{\alpha}_{;\alpha} = \frac{1}{\sqrt{-g}}\frac{\partial}{\partial x^{\alpha}}(\sqrt{-g}\,V^{\alpha})$.

3.5 Action Integral for the Gravitational Field

Most of the fundamental physical equations of the classical field can be derived from a variational principle. Therefore, it is expected that Einstein's field equation can also be derived from a variational principle. To derive it, we follow the same notion of the variational principle in classical mechanics and electromagnetism.

Now, we seek a Lagrangian L_G for the gravitational field, which has a geometrical origin that contains derivatives of fundamental tensor $g_{\mu\nu}$. Since $\sqrt{-g}\,d^4x$ is an invariant, the action integral for the gravitational field will take the following form

$$\int \sqrt{-g}\,L_G\,d^4x.$$

(Here integration is done over all spacetime coordinates between two given points.)

There are several scalars apart from constant, however, the simplest scalar as Lagrangian, which we can consider is the scalar curvature $L_G = R$. Here R contains metric tensor and Christoffel symbols $\Gamma^\alpha_{\mu\nu}$ (i.e., derivatives of $g_{\mu\nu}$). One can also think other scalars like R^2, $f(R)$, $R_{ik}R^{ik}$, $R_{ijkl}R^{ijkl}$, etc., as Lagrangian, which can be treated as modified gravity. Later, we will discuss these modified gravity theories.

So the action integral for the gravitational field will be taken as $\int \sqrt{-g}R d^4x$.

This is known as the **Einstein–Hilbert action**. In addition to the gravitational field, one should add to this integral another one which takes care of different fields presented in our physical system.

3.6   Einstein's Equation from Variational Principle

The Einstein's equation can be derived from the principle of least action $\delta I = 0$ where

$$I = \int \sqrt{-g}[L_G + 2kL_F]d^4x. \tag{3.1}$$

Here, we choose $L_G = R$ as the Lagrangian for the gravitational field with $R = g_{\mu\nu}R^{\mu\nu}$. L_F is the Lagrangian for all the other fields. k = Einstein's gravitational constant = $\frac{8\pi G}{c^4}$. The integral has been taken over the whole spacetime. Here, the action is usually assumed to be a functional of the metric and matter fields. This integral contains the metric tensor, which determines the curvature of spacetime in the presence of the gravitational field either in the absence or the presence of matter. We are varying the first part of the integral (3.1), which yields

$$\delta \int \sqrt{-g}R d^4x = \delta \int \sqrt{-g}g^{\mu\nu}R_{\mu\nu}d^4x,$$

$$= \int \sqrt{-g}g^{\mu\nu}\delta R_{\mu\nu}d^4x + \int R_{\mu\nu}\delta(\sqrt{-g}g^{\mu\nu})d^4x. \tag{3.2}$$

Since

$$\delta R_{\mu\nu} = \left[(\delta\Gamma^\rho_{\mu\nu})_{;\rho} - (\delta\Gamma^\rho_{\mu\rho})_{;\nu}\right],$$

we have

$$\sqrt{-g}g^{\mu\nu}\delta R_{\mu\nu} = \sqrt{-g}\left[(g^{\mu\nu}\delta\Gamma^\rho_{\mu\nu})_{;\rho} - (g^{\mu\nu}\delta\Gamma^\rho_{\mu\rho})_{;\nu},\right]$$

$$= \sqrt{-g}\left[(g^{\mu\nu}\delta\Gamma^\alpha_{\mu\nu})_{;\alpha} - (g^{\mu\alpha}\delta\Gamma^\rho_{\mu\rho})_{;\alpha},\right]$$

$$= \sqrt{-g}(V^\alpha)_{;\alpha} = \frac{\partial}{\partial x^\alpha}(\sqrt{-g}V^\alpha),$$

(by result (8) in Section (3.4))

where

$$V^\alpha = g^{\mu\nu}\delta\Gamma^\alpha_{\mu\nu} - g^{\mu\alpha}\delta\Gamma^\rho_{\mu\rho},$$

is contravariant vector.

Thus, the first integral of Eq. (3.2) implies,

$$\int \sqrt{-g} g^{\mu\nu} \delta R_{\mu\nu} d^4x = \int \frac{\partial}{\partial x^\alpha} (\sqrt{-g} V^\alpha) d^4x = 0.$$

The above integral transforms into a surface integral on $\sqrt{-g} V^\alpha$ due to Gauss theorem. Since variations of the Christoffel symbols on the boundaries of integration vanish, therefore, the value of the integral is zero.

Now the second integral of (3.2) is

$$\int R_{\mu\nu} \delta(\sqrt{-g} g^{\mu\nu}) d^4x = \int \sqrt{-g} R_{\mu\nu} \delta g^{\mu\nu} d^4x + \int R_{\mu\nu} g^{\mu\nu} \delta \sqrt{-g} d^4x,$$

$$= \int \sqrt{-g} R_{\mu\nu} \delta g^{\mu\nu} d^4x + \int R \left(-\frac{1}{2} \sqrt{-g} g_{\mu\nu} \delta g^{\mu\nu}\right) d^4x,$$

$$= \int \sqrt{-g} \left[R_{\mu\nu} - \frac{1}{2} g_{\mu\nu} R\right] \delta g^{\mu\nu} d^4x. \tag{3.3}$$

The second part of (3.1), which defines different fields presented in our physical system in addition to the gravitational field can be calculated as follows:

$$\delta \int \sqrt{-g} L_F d^4x = \int \left[\frac{\partial}{\partial g^{\mu\nu}} (\sqrt{-g} L_F) \delta g^{\mu\nu} + \frac{\partial(\sqrt{-g} L_F)}{\partial [\frac{\partial g^{\mu\nu}}{\partial x^\alpha}]} \delta[\frac{\partial g^{\mu\nu}}{\partial x^\alpha}]\right] d^4x$$

$$= \int \left[\frac{\partial(\sqrt{-g} L_F)}{\partial g^{\mu\nu}} \delta g^{\mu\nu} + \frac{\partial}{\partial x^\alpha} \left(\frac{\partial(L_F \sqrt{-g} \delta g^{\mu\nu})}{\partial g^{\mu\nu}_{,\alpha}}\right) - \frac{\partial}{\partial x^\alpha} \left(\frac{\partial(L_F \sqrt{-g})}{\partial(g^{\mu\nu}_{,\alpha})}\right) \delta g^{\mu\nu}\right].$$

The contribution from the second term of the integrand will be zero as it can be reduced to a surface integral over the boundary where $\delta g^{\mu\nu} = 0$.

We define the energy-momentum tensor $T_{\mu\nu}$ as

$$T_{\mu\nu} = \frac{2}{\sqrt{-g}} \left[\frac{\partial(\sqrt{-g} L_F)}{\partial g^{\mu\nu}} - \frac{\partial}{\partial x^\alpha} \left(\frac{\partial(L_F \sqrt{-g})}{\partial g^{\mu\nu}_{,\alpha}}\right)\right]. \tag{3.4}$$

Thus, we obtain,

$$\delta \int \sqrt{-g} L_F d^4x = \frac{1}{2} \int \sqrt{-g} T_{\mu\nu} \delta g^{\mu\nu} d^4x.$$

Hence, finally we get,

$$\delta I = 0 \Rightarrow \int \sqrt{-g} (R_{\mu\nu} - \frac{1}{2} g_{\mu\nu} R + k T_{\mu\nu}) \delta g^{\mu\nu} d^4x = 0.$$

Hence, variation with respect to $g^{\mu\nu}$, i.e., $\frac{\delta I}{\delta g^{\mu\nu}} = 0$ gives

$$R_{\mu\nu} - \frac{1}{2}g_{\mu\nu}R = -kT_{\mu\nu}. \tag{3.5}$$

This is the **Einstein field equation** in general theory of relativity.

Note 3.1

In this section, we have derived field equation without boundary term, which is assumed to be zero at infinity. Also, we have assumed $\delta g_{\alpha\beta}$ is zero on the boundary. However, its tangential derivatives should vanish but normal derivative of $\delta g_{\alpha\beta}$ does not mandatorily need to vanish on the boundary. In the appendix, we will generalize the field equation to a general case comprising the boundary term in action.

Note 3.2

The action integral for Einstein field equation with cosmological constant Λ is

$$I = \int \sqrt{-g}[R + 2\Lambda + 2kL_F]d^4x.$$

Varying with respect to $g_{\mu\nu}$, we get Einstein field equation with cosmological constant as

$$R_{\mu\nu} - \frac{1}{2}g_{\mu\nu}R + \Lambda g_{\mu\nu} = -kT_{\mu\nu}. \tag{3.6}$$

Contracting above equation we get

$$R = kT + 4\Lambda. \tag{3.7}$$

Using this, we can rewrite the Einstein field equation with cosmological constant as

$$R_{\mu\nu} = \Lambda g_{\mu\nu} - k\left(T_{\mu\nu} - \frac{1}{2}g_{\mu\nu}T\right). \quad (T = T^\mu_\mu) \tag{3.8}$$

In the absence of matter distribution (i.e., $T_{\mu\nu} = 0$) Einstein field equation with cosmological constant reduces to

$$R_{\mu\nu} = \Lambda g_{\mu\nu}. \tag{3.9}$$

Einstein introduced a cosmological constant to obtain favorable cosmological phenomena. Actually, he introduced it to obtain a static cosmological solution for the gravitational field equations.

Note 3.3

If L_F contains other fields say ψ, we then find ψ field equation by varying action integral I with respect to ψ as

$$\partial_\mu \left(\frac{\partial L_F}{\partial \left(\frac{\partial \psi}{\partial x^\mu} \right)} \right) = \frac{\partial L_F}{\partial \psi}.$$

Hint: Let the Lagrangian depend on ψ and on its derivative, i.e., $L_F(\psi(x), \partial_\mu \psi(x))$ where $\partial_\mu = (\frac{\partial}{\partial x^\mu})$. The principle of least action implies $\delta I = 0$ where

$$I = \int \sqrt{-g}[L_F] d^4 x.$$

Here, L_F describes all fields except the gravitational field. Now,

$$\delta \int \sqrt{-g} L_F d^4 x = \int \left[\frac{\partial}{\partial \psi}(\sqrt{-g} L_F)\delta\psi + \frac{\partial(\sqrt{-g} L_F)}{\partial \left(\frac{\partial \psi}{\partial x^\mu} \right)} \delta \left(\frac{\partial \psi}{\partial x^\mu} \right) \right] d^4 x,$$

$$= \int \left[\frac{\partial(\sqrt{-g} L_F)}{\partial \psi}\delta\psi + \frac{\partial}{\partial x^\mu} \left(\frac{\partial(L_F \sqrt{-g}\delta\psi)}{\partial \left(\frac{\partial \psi}{\partial x^\mu} \right)} \right) - \frac{\partial}{\partial x^\mu} \left(\frac{\partial(L_F \sqrt{-g})}{\partial \left(\frac{\partial \psi}{\partial x^\mu} \right)} \right) \delta\psi \right].$$

The contribution from second term of the integrand will be zero as it can be reduced to a surface integral over the boundary where $\delta\psi = 0$. So, we arrive at Euler–Lagrange (E–L) equations of motion.

$$\partial_\mu \left(\frac{\partial L_F}{\partial \left(\frac{\partial \psi}{\partial x^\mu} \right)} \right) = \frac{\partial L_F}{\partial \psi}.$$

Note 3.4

If L_F does not depend on $g_{ik,1}$ explicitly, then

$$T_{\mu\nu} = \frac{2}{\sqrt{-g}} \frac{\partial}{\partial g^{\mu\nu}}(\sqrt{-g} L_F) = 2\frac{\partial L_F}{\partial g^{\mu\nu}} - L_F g_{\mu\nu}.$$

It is interesting to consider the effect of gravitation on the physical system. In the absence of gravitational force, the governing equations are involving with tensors of special theory of relativity, e.g., in the absence of gravitational force electromagnetism is governed by the Maxwell equation. According to the principle of covariance, the physical system under consideration must be written in terms of the general tensor in the presence of gravitation. Such equations, e.g., Einstein-Maxwell equations, Klein–Gordon, etc., are actually classical field theory defined in the setting of general relativity.

I will provide some examples of the electromagnetic field and massive scalar field in the presence of gravitation based on the action principle.

Exercise 3.1

Find the energy-momentum tensor of the electromagnetic field in the presence of gravitation.
Hint: For the presence of electromagnetic field in addition to gravitation in the action integral, it takes the form

$$I = \int \sqrt{-g}[L_G + kL_{EM}]d^4x.$$

[where L_{EM} = Lagrangian of electromagnetic field]

Lagrangian of electromagnetic field without current is given by

$$L_{EM} = -\frac{1}{16\pi}F_{ik}F^{ik},$$

where the field strength F_{im} is related to electromagnetic potential A_i as

$$F_{ik} = A_{k,i} - A_{i,k}.$$

From $\delta I = 0$, the second term contributes $T_{ik}(EM)$ where

$$T_{ik}(EM) = 2\frac{\partial L_{EM}}{\partial g^{ik}} - g_{ik}L_{EM}.$$

Here,

$$L_{EM} = \frac{-1}{16\pi}F_{ik}F^{ik} = \frac{-1}{16\pi}F_{ik}F_{lm}g^{il}g^{km}.$$

Therefore,

$$\frac{\partial L_{EM}}{\partial g^{ik}} = \frac{-1}{8\pi}F_{il}F_{km}g^{lm} = \frac{-1}{8\pi}F_i^m F_{km}.$$

Hence energy-momentum tensor of the electromagnetic field in presence of gravitation is given by

$$T_{ik}(EM) = \frac{-1}{4\pi}\left[F_i^m F_{km} - \frac{1}{4}F_{lm}F^{lm}g_{ik}\right].$$

The following equation

$$R_{\mu\nu} - \frac{1}{2}g_{\mu\nu}R = -k[T_{\mu\nu} + T_{\mu\nu}(EM)], \qquad (3.10)$$

is known as coupled **Einstein–Maxwell field equations**.

Exercise 3.2

Deduce Maxwell equations in presence of gravitation.

Hint: The Lagrangian density for the electromagnetic field in the presence of gravitation is

$$L = -\frac{1}{16\pi}\sqrt{-g}f_{\alpha\beta}f^{\alpha\beta} - \frac{1}{c}\sqrt{-g}j^\alpha A_\alpha,$$

where electromagnetic potential A_μ determines the Maxwell field strength tensor $f_{\mu\nu}$ as

$$f_{\mu\nu} = \frac{\partial A_\mu}{\partial x^\nu} - \frac{\partial A_\nu}{\partial x^\mu}$$

and j^α is the electric current density vector.

We know

$$f_{\alpha\beta}f^{\alpha\beta} = g^{\alpha\mu}g^{\beta\nu}f_{\alpha\beta}f_{\mu\nu}.$$

The (E–L) equation

$$\frac{\partial}{\partial x^\beta}\left[\frac{\partial L}{\partial\left(\frac{\partial A_\alpha}{\partial x^\beta}\right)}\right] - \frac{\partial L}{\partial A_\alpha} = 0.$$

For the above Lagrangian density yields

$$\frac{\partial L}{\partial A_{\alpha,\beta}} = \frac{\partial}{\partial A_{\alpha,\beta}}\left[g^{a\mu}g^{b\nu}f_{ab}f_{\mu\nu}\right]\left(-\frac{1}{16\pi}\sqrt{-g}\right),$$

$$= 2g^{a\mu}g^{b\nu}f_{ab}\frac{\partial}{\partial A_{\alpha,\beta}}(A_{\mu,\nu} - A_{\nu,\mu})\left(-\frac{1}{16\pi}\sqrt{-g},\right)$$

$$= 2g^{a\mu}g^{b\nu}f_{ab}[\delta^{\mu\nu}_{\alpha\beta} - \delta^{\nu\mu}_{\alpha\beta}]\left(-\frac{1}{16\pi}\sqrt{-g}\right),$$

$$= 2g^{a\mu}g^{b\nu}f_{ab}[\delta^{\mu\nu}_{\alpha\beta} + \delta^{\mu\nu}_{\alpha\beta}]\left(-\frac{1}{16\pi}\sqrt{-g}\right) \quad [as \; \delta^{\nu\mu}_{\alpha\beta} = -\delta^{\mu\nu}_{\alpha\beta},]$$

$$= \left(-\frac{1}{16\pi}\sqrt{-g}\right)4g^{a\mu}g^{b\nu}f_{ab}\delta^{\mu\nu}_{\alpha\beta} = \left(-\frac{1}{4\pi}\sqrt{-g}\right)f^{\alpha\beta}.$$

$$[since \; \delta^{\mu\nu}_{\eta\beta} = 1 \; for \; \alpha = \mu \; and \; \beta = \nu]$$

Also

$$\frac{\partial L}{\partial A_\alpha} = -\frac{1}{c}\sqrt{-g}j^\alpha.$$

Then from (E–L) equation, we get

$$\frac{1}{\sqrt{-g}}\frac{\partial(\sqrt{-g}f^{\alpha\beta})}{\partial x^\beta} = \frac{4\pi}{c}J^\alpha,$$

$$\Longrightarrow$$

$$\nabla_\beta f^{\alpha\beta} = \frac{4\pi}{c}J^\alpha. \quad \text{(by result (8) in Section (3.4))}$$

Definitely, this is a generalization of the flat space counterpart of the Maxwell equations.

$$\frac{\partial f^{\mu\nu}}{\partial x^\nu} = \frac{4\pi}{c}J^\mu.$$

Thus, one can see that to extend from flat spacetime to curved spacetime, one has to replace the partial derivative by a covariant derivative. Thus, to extend other Maxwell equation in flat spacetime

$$\frac{\partial f_{\alpha\beta}}{\partial x^\nu} + \frac{\partial f_{\beta\nu}}{\partial x^\alpha} + \frac{\partial f_{\nu\alpha}}{\partial x^\beta} = 0,$$

into curved spacetime, we have to replace the partial derivatives by covariant derivatives. Hence, the final form is

$$\nabla_\nu f_{\alpha\beta} + \nabla_\alpha f_{\beta\nu} + \nabla_\beta f_{\nu\alpha} = 0.$$

Exercise 3.3

Write the equation of continuity in curved space.

Hint: Equation of continuity in electrodynamics, i.e., in flat spacetime background is

$$\frac{\partial j^\mu}{\partial x^\mu} = 0.$$

Using the above notion, we can write the equation of continuity in curved space background by changing the partial derivative by covariant derivative as

$$j^\mu_{;\mu} = 0.$$

Using result (8) in Section (3.4), we have

$$\frac{1}{\sqrt{-g}}\frac{\partial}{\partial x^\mu}(\sqrt{-g}J^\mu) = 0 \ or \ \frac{\partial}{\partial x^\mu}(\sqrt{-g}J^\mu) = 0.$$

Find the energy-momentum tensor and scalar field equation for the massive scalar field ϕ with mass m.

Hint: The Lagrangian for a massive scalar field ϕ with mass m in flat spacetime, i.e., in the absence of gravitation is given by

$$L_\phi = \frac{1}{2}\left(\eta^{\mu\nu}\frac{\partial\phi}{\partial x^\mu}\frac{\partial\phi}{\partial x^\nu} - m^2\phi^2\right).$$

The E–L equation

$$\frac{\partial}{\partial x^\beta}\left[\frac{\partial L_\phi}{\partial\left(\frac{\partial\phi}{\partial x^\beta}\right)}\right] - \frac{\partial L_\phi}{\partial\phi} = 0,$$

yields the field equation of ϕ

$$\eta^{\mu\nu}\frac{\partial^2\phi}{\partial x^\mu\partial x^\nu} + m^2\phi = 0 \ or \ \Box^2\phi + m^2\phi = 0.$$

This is known as **Klein–Gordon equation**.

The Lagrangian for a massive scalar field ϕ with mass m in the presence of gravitation is given by

$$L_\phi = \frac{1}{2}\sqrt{-g}\left(g^{\mu\nu}\frac{\partial\phi}{\partial x^\mu}\frac{\partial\phi}{\partial x^\nu} - m^2\phi^2\right).$$

Using E–L equation, we get **Klein–Gordon equation in curved spacetime**

$$\frac{\partial}{\partial x^\mu}\left(\sqrt{-g}g^{\mu\nu}\frac{\partial\phi}{\partial x^\nu}\right) + m^2\sqrt{-g}\phi = 0.$$

Since L_ϕ does not depend on $g_{ik,1}$, then using

$$T_{\mu\nu} = 2\frac{\partial L_\phi}{\partial g^{\mu\nu}} - L_\phi g_{\mu\nu},$$

we get **energy-momentum tensor for massive scalar field ϕ**

$$T_{\alpha\beta} = \frac{\partial\phi}{\partial x^\alpha}\frac{\partial\phi}{\partial x^\beta} - \frac{1}{2}g_{\alpha\beta}\left(g^{\mu\nu}\frac{\partial\phi}{\partial x^\mu}\frac{\partial\phi}{\partial x^\nu} - m^2\phi^2\right).$$

$$\left[Here, \ \frac{\partial L_\phi}{\partial g^{\alpha\beta}} = \frac{1}{2}\sqrt{-g}\frac{\partial\phi}{\partial x^\alpha}\frac{\partial\phi}{\partial x^\beta} - \frac{1}{4}\sqrt{-g}g_{\alpha\beta}\left(g^{\mu\nu}\frac{\partial\phi}{\partial x^\mu}\frac{\partial\phi}{\partial x^\nu} - m^2\phi^2\right)\right]$$

Exercise 3.5

Find the energy-momentum tensor for the scalar field ϕ with a potential $V(\phi)$.
Hint: Here the Lagrangian is given by

$$L = \nabla^\mu \phi \nabla_\mu \phi - V(\phi).$$

As before, L does not depend on $g_{ik,1}$, then using

$$T_{\mu\nu} = 2\frac{\partial L}{\partial g^{\mu\nu}} - Lg_{\mu\nu},$$

we get **energy-momentum tensor for scalar field ϕ with a potential $V(\phi)$**

$$T_{\mu\nu} = \nabla_\mu \phi \nabla_\nu \phi - \frac{1}{2}g_{\mu\nu}\nabla^\sigma \phi \nabla_\sigma \phi - g_{\mu\nu}V(\phi).$$

Using E–L equation, we get **Klein–Gordon equation in curved spacetime**, i.e., ϕ field equation

$$\frac{1}{\sqrt{-g}}\frac{\partial}{\partial x^\mu}\left(\sqrt{-g}g^{\mu\nu}\frac{\partial\phi}{\partial x^\nu}\right) + \frac{\partial V}{\partial\phi} = 0.$$

Exercise 3.6

Find the energy-momentum tensor of a perfect fluid.
Hint: We can visualize perfect fluid as a mechanical medium, which has no viscosity and no heat conduction. It does not tolerate any shearing stress but its normal stresses are isotropic. For a local observer, its normal components of stress are equal to the isotropic hydrostatic pressure. The perfect fluid is described by a four velocity u^μ in a comoving coordinate system and some of the following scalar quantities: density (ρ), isotropic pressure (p), temperature (T), specific entropy (s), and specific enthalpy ($w = \frac{p+\rho}{n}$, n is the baryon number density). The baryon number can be defined in terms of a baryon number flux vector density (n^μ) as

$$n^\mu = n\sqrt{-g}u^\mu \quad or \quad n = \sqrt{\frac{g_{\mu\nu}n^\mu n^\nu}{g}}. \quad (using\ u^\mu u_\mu = -1)$$

We consider the Lagrangian in proper coordinate system in which the fluid is supposed to be at rest as (here, $\delta n^\mu = 0$)

$$L = -\rho.$$

Using thermodynamical relation $((\frac{\partial\rho}{\partial n})_s = w)$ implies

$$\delta\rho = w\delta n.$$

Now

$$\delta n = \frac{1}{2n} \left(\frac{n^\mu n^\nu}{g} \delta g_{\mu\nu} - n^\mu n^\nu g_{\mu\nu} \frac{\delta g}{g^2} \right)$$

$$= \frac{n}{2} \left(-u^\mu u^\nu \delta g_{\mu\nu} + u^\mu u_\mu \frac{\delta g}{g} \right),$$

Using the variational formula we finally get,

$$\delta L = -\delta\rho = -w\delta n = -w\frac{n}{2} \left(u_\mu u_\nu + g_{\mu\nu} \right) \delta g^{\mu\nu},$$

Using

$$T_{\mu\nu} = 2\frac{\partial L}{\partial g^{\mu\nu}} - Lg_{\mu\nu},$$

we get,

$$T_{\mu\nu} = -[pg_{\mu\nu} + (p + \rho)u_\mu u_\nu].$$

Note 3.6

Noether's theorem

In 1915, Emmy Noether has discovered something deeply fundamental about the universe, known as Noether's theorem, which is related to the conserved quantities from symmetries of the laws of nature.

Time-independent Hamiltonian $\Rightarrow$ Energy conserved

Translation-independent Hamiltonian $\Rightarrow$ Momentum conserved

Rotation-independent Hamiltonian $\Rightarrow$ Angular momentum conserved, and so on. The precise formulation of these correspondences is given by Noether's theorem. Actually, Noether's theorem gives a formal connection between continuous symmetries of a physical system and the resulting conservation laws.

Proof of Translation-Independent Hamiltonian $\Rightarrow$ Momentum Conserved:

Consider a translation

$$x^\mu \to x^\mu + a^\mu,$$

where a^μ is a constant.

Under this displacement, a field $\phi(x^\mu)$ transforms as

$$\phi(x^\mu) \to \phi(x^\mu + a^\mu).$$

The change in the field after a displacement is given by

$$\delta\phi = \phi(x^\nu + a^\nu) - \phi(x^\nu),$$
$$= \phi(x^\nu) + a^\mu \partial_\mu \phi(x^\nu) - \phi(x^\nu),$$
$$= a^\mu \partial_\mu \phi(x^\nu).$$

[The translation $\delta x^\mu = a^\mu$]

Thus, the amount of change of the field ϕ after translation as:

$$\delta\phi = a^\mu \partial_\mu \phi.$$

In general, $\delta A = a^\mu \partial_\mu A$.

Hence, we have

$$\delta \partial_\mu \phi = a^\nu \partial_\nu (\partial_\mu \phi).$$

Now,

$$\delta L = a^\mu \partial_\mu L = \frac{\partial L}{\partial \phi} \delta\phi + \frac{\partial L}{\partial(\partial_\mu \phi)} \delta(\partial_\mu \phi),$$

$$= \frac{\partial L}{\partial \phi} a^\nu \partial_\nu \phi + \frac{\partial L}{\partial(\partial_\mu \phi)} a^\nu \partial_\nu(\partial_\mu \phi),$$

$$= a^\nu \left[\partial_\mu \frac{\partial L}{\partial(\partial_\mu \phi)} \partial_\nu \phi + \frac{\partial L}{\partial(\partial_\mu \phi)} \partial_\mu(\partial_\nu \phi) \right],$$

$$= a^\nu \partial_\mu \left[\frac{\partial L}{\partial(\partial_\mu \phi)} \partial_\nu \phi \right]. \quad [as\ a^\mu = a^\nu \delta^\mu_\nu]$$

Therefore,

$$\partial_\mu \left[L\delta^\mu_\nu - \frac{\partial L}{\partial(\partial_\mu \phi)} \partial_\nu \phi \right] a^\nu = 0.$$

Defining energy-momentum tensor T^μ_ν as

$$T^\mu_\nu = \frac{\partial L}{\partial(\partial_\mu \phi)} - L\delta^\mu_\nu,$$

One can see it is conserved, i.e., $\partial_\mu T^\mu_\nu = 0$.

Exercise 3.7

If the Lagrangian density for a scalar field is

$$L = \frac{1}{2} \phi_{,i} \phi_{,k} g^{ik},$$

derive energy-momentum tensor T_{ik}.

Hint: Using $T_{\mu\nu} = 2\frac{\partial L_\phi}{\partial g^{\mu\nu}} - L_\phi g_{\mu\nu}$, one finds

$$T_{ik} = \phi_{,i}\phi_{,k} - \frac{1}{2}\phi_{,j}\phi_{,j}g_{ik}.$$

Exercise 3.8

If the Lagrangian density for a scalar field is

$$L = \frac{1}{2}\left[\phi_{,\mu}\phi_{,\nu}g^{\mu\nu} - m^2\phi^2\right].$$

Find E–L equation and derive energy-momentum tensor $T_{\mu\nu}$.

Hints:

Here for flat spacetime,

$$\partial^\mu = \left(\frac{\partial}{\partial t}, -\frac{\partial}{\partial x^i}\right) = g^{\mu\nu}\partial_\nu, \quad g^{\mu\nu} = diag(1, -1, -1, -1).$$

Now,

$$\frac{\partial L}{\partial\left(\frac{\partial\phi}{\partial x^\mu}\right)} = \partial^\mu\phi$$

and

$$\frac{\partial L}{\partial\phi} = -m^2\phi.$$

Therefore, using E–L equation, we get

$$\partial_\mu\partial^\mu\phi + m^2\phi = 0.$$

This is Klein–Gordon equation in flat space.

However, for curved space, the above equation reads

$$\Box^2\phi + m^2\phi = 0, \text{ i.e., } \frac{1}{\sqrt{-g}}\frac{\partial}{\partial x^k}\left(\sqrt{-g}g^{ik}\frac{\partial\phi}{\partial x^i}\right) + m^2\phi = 0.$$

Exercise 3.9

If the Lagrangian density for a scalar field is

$$L = -\frac{1}{2}\phi_{,\mu}\phi_{,\nu}g^{\mu\nu} - V(\phi).$$

Derive energy-momentum tensor $T_{\mu\nu}$.

Hints: Here, the energy-stress tensor is

$$T_{\mu\nu} = \frac{\partial\phi}{\partial x^\mu}\frac{\partial\phi}{\partial x^\nu} - g_{\mu\nu}\left(\frac{1}{2}\frac{\partial\phi}{\partial x^\mu}\frac{\partial\phi}{\partial x^\nu}g^{\mu\nu} + V(\phi)\right).$$

If the scalar field depends on time only (as for homogeneous and isotropic model of the universe), the energy-stress tensor takes the form

$$T_{\mu\nu} = diag\left(\frac{1}{2}\dot{\phi}^2 + V(\phi), \frac{1}{2}\dot{\phi}^2 - V(\phi), \frac{1}{2}\dot{\phi}^2 - V(\phi), \frac{1}{2}\dot{\phi}^2 - V(\phi)\right) = diag(\rho, p, p, p).$$

This implies

$$\rho = \frac{1}{2}\dot{\phi}^2 + V(\phi), \quad p = \frac{1}{2}\dot{\phi}^2 - V(\phi).$$

Thus the equation of state is

$$\frac{p}{\rho} = \frac{\frac{1}{2}\dot{\phi}^2 - V(\phi)}{\frac{1}{2}\dot{\phi}^2 + V(\phi)}.$$

This type of matter distribution is known as **Quintessence**.

3.7　Some Modified Theories of Gravity

In this chapter, we just provide an outline of some modified theories of gravity.

1. *f(R)* **theory of gravity:** In the action of Einstein general theory of relativity, we use Lagrangian for geometry as $L_G = R$, where R is a Ricci scalar. Now, it can be generalized by taking $L_G = f(R)$. This newly developed theory of gravity is known as *f(R)* theory of gravity. For a purely phenomenological thought, *f(R)* could be expanded in a power series with positive as well as negative powers of the curvature scalar as

$$f(R) = \ldots\ldots + \frac{\alpha_2}{R^2} + \frac{\alpha_1}{R} - 2\Lambda + R + \frac{R^2}{\beta_2} + \frac{R^3}{\beta_3} + \ldots\ldots\ldots$$

where the coefficients α_i and β_i have the appropriate dimensions.

Following the same point of view, we can write the action for *f(R)* gravity as

$$I = \int \sqrt{-g}[f(R) + 2kL_m]d^4x.$$

Here, L_m is the matter Lagrangian. Varying with respect to $g_{\mu\nu}$, we get modified Einstein field equation as

$$\Xi_{\mu\nu} \equiv F(R)R_{\mu\nu} - \frac{1}{2}g_{\mu\nu}f(R) - [\nabla_\mu\nabla_\nu - g_{\mu\nu}\Box]F(R) = -kT_{\mu\nu}.$$

Here, $F(R) = \frac{df(R)}{dR}$, ∇_μ is the covariant derivative with $\Box F = \nabla^\mu \nabla_\mu F = \frac{1}{\sqrt{-g}} \partial_\mu (\sqrt{-g} g^{\mu\nu} \partial_\nu F)$. As usual $T_{\mu\nu}$ is the energy–momentum tensor of the matter fields. Conservation equation $\nabla^\mu T_{\mu\nu} = 0$ implies $\nabla^\mu \Xi_{\mu\nu} = 0$. Also trace of the above field equation in $f(R)$ gravity is

$$3\Box F(R) + F(R)R - 2f(R) = -kT,$$

where $T = g^{\mu\nu} T_{\mu\nu}$.

We can recover Einstein gravity without the cosmological constant by taking $f(R) = R$. This implies $F(R) = 1$ with $\Box F = 0$ and we get $R = kT$. In general, the trace of field equation in $f(R)$ gravity indicates that the field equations of $f(R)$ theory will provide various different solutions than Einstein's theory. For a simple example, one can notice here that the Birkhoff's theorem does not hold good, which states that the Schwarzschild solution is the unique spherically symmetric vacuum solution. Note that for $T = 0$ no longer implies that $R = 0$, or is even constant.

2. **Gauss–Bonnet Gravity:** Gauss–Bonnet (GB) gravity is a modification of the Einstein–Hilbert action, which includes the GB term

$$G = R^2 - 4R^{\mu\nu} R_{\mu\nu} + R^{\mu\nu\rho\sigma} R_{\mu\nu\rho\sigma},$$

proposed by Carl Friedrich Gauss and Pierre Ossian Bonnet. In this theory, the action is

$$I = \int \sqrt{-g} [R - 2\Lambda + \alpha_{GB} L_{GB} + 2k L_m] d^4 x,$$

where $L_{GB} = R^2 - 4R^{\mu\nu} R_{\mu\nu} + R^{\mu\nu\rho\sigma} R_{\mu\nu\rho\sigma}$ and α_{GB} is a coupling constant. Sometimes it is known as GB parameter. We note that as $\alpha_{GB} \to 0$, we will come back to Einstein gravity. In the context of Riemannian geometry, GB term is the next higher order correction to R. It is actually a topological invariant term in four dimensions and for the addition of this term to the Ricci scalar term in Einstein–Hilbert action, it has no contribution to the equations of motion.

Varying with respect to $g_{\mu\nu}$, we get field equation in GB gravity as

$$G_{\mu\nu} + H_{\mu\nu} = -kT_{\mu\nu}.$$

Here,

$$G_{\mu\nu} = R_{\mu\nu} - \frac{1}{2} g_{\mu\nu} R + \Lambda g_{\mu\nu},$$

$$H_{\mu\nu} = 2(R_{\mu\lambda\rho\sigma} R_\nu^{\lambda\rho\sigma} - R_{\mu\rho\nu\sigma} R^{\rho\sigma} - 2R_{\mu\sigma} R_\nu^\sigma + RR_{\mu\nu}) - \frac{1}{2} g_{\mu\nu} L_{GB}.$$

As usual $T_{\mu\nu}$ is the energy-momentum tensor of the matter fields.

3. $f(G)$ **theory of gravity or modified GB gravity:** As we know the GB term G is a topological invariant term in four dimensions and it has no contribution to the equations of motion, i.e., it has no dynamics to couple it with the matter field, therefore, it is obvious that some function of G, i.e., $f(G)$ is also a topological invariant term in four dimensions and it has no contribution to the equations of

motion. In the modified GB gravity theory, the action is

$$I = \int \sqrt{-g}[R + f(G) + 2kL_m]d^4x.$$

Varying with respect to $g_{\mu\nu}$, we get the field equation in modified GB gravity as

$$G_{\mu\nu} - H_{\mu\nu} = -kT_{\mu\nu}.$$

Here,

$$G_{\mu\nu} = R_{\mu\nu} - \frac{1}{2}g_{\mu\nu}R,$$

$$H_{\mu\nu} = \frac{1}{2}fg_{\mu\nu} - 2FRR_{\mu\nu} + 4FR^\sigma_\mu R_{\sigma\nu} - 2FR_{\mu\sigma\lambda\theta}R_\nu^{\sigma\lambda\theta} - 4FR_{\mu\sigma\lambda\nu}R^{\sigma\lambda}$$
$$-4R^\sigma_\mu\nabla_\nu\nabla_\sigma F - 4R^\sigma_\nu\nabla_\mu\nabla_\sigma F + 4R_{\mu\nu}\nabla^2 F + 4g_{\mu\nu}R^{\sigma\lambda}\nabla_\sigma\nabla_\lambda F - 4R^{\mu\sigma\nu\lambda}\nabla^\sigma\nabla^\lambda F,$$

and $F = \frac{df}{dG}$, which is not dimensionless unlike in $f(R)$ gravity. As usual $T_{\mu\nu}$ is the energy-momentum tensor of the matter fields. Also trace of the above field equation in $f(G)$ gravity is

$$-R = -kT + 2f - 2FG - 2R\square F + 4R_{\mu\nu}\nabla^\mu\nabla^\nu,$$

where $T = g^{\mu\nu}T_{\mu\nu}$ and $\square = g^{\mu\nu}\nabla_\mu\nabla_\nu$.

Thus, curvature-related quantities (R, G, F) are determined by the energy-momentum tensor of the matter fields.

4. $f(T)$ **theory of gravity:** Teleparallel gravity is a theory that adopts Weitzenbock connection without considering the Levi-Civita connection and this produces a null curvature with a nonvanishing torsion. Similar to $f(R)$ gravity, f(T) gravity assumes a generalization of the action of teleparallel gravity. In this gravitational theory, tetrads are the key fields instead of the metric as in general relativity. Teleparallel gravity is defined in the Weitzenbock's spacetime where the line element is designated by

$$dS^2 = g_{\mu\nu}dx^\mu dx^\nu, \tag{3.11}$$

where the symmetric components of the metric, $g_{\mu\nu}$ have 10 degrees of freedom. In this case, the line element (3.11) can be rewritten as

$$dS^2 = g_{\mu\nu}dx^\mu dx^\nu = \eta_{ij}\theta^i\theta^j, \tag{3.12}$$

$$dx^\mu = e_i{}^\mu\theta^i, \; \theta^i = e^i{}_\mu dx^\mu, \tag{3.13}$$

where $\eta_{ij} = diag[1, -1, -1, -1]$ and $e_i{}^\mu e^i{}_\nu = \delta^\mu_\nu$ or $e_i{}^\mu e^j{}_\mu = \delta^j_i$. Here, the metric determinant is given by $\sqrt{-g} = \det\left[e^i{}_\mu\right] = e$ and the matrix $e^i{}_\mu$ are called **tetrads** and express the dynamic fields of the theory.

The **Weitzenbock's connection** is defined as

$$\Gamma^{\alpha}_{\mu\nu} = e_i{}^{\alpha}\partial_{\nu}e^i{}_{\mu} = -e^i{}_{\mu}\partial_{\nu}e_i{}^{\alpha}. \tag{3.14}$$

All the geometrical quantities of the spacetime are made from this connection. The constituents of the **torsion tensor** are described through the antisymmetric part of this connection

$$T^{\alpha}_{\mu\nu} = \Gamma^{\alpha}_{\nu\mu} - \Gamma^{\alpha}_{\mu\nu} = e_i{}^{\alpha}\left(\partial_{\mu}e^i{}_{\nu} - \partial_{\nu}e^i{}_{\mu}\right) \tag{3.15}$$

The constituents of the **contorsion** are defined as

$$K^{\mu\nu}{}_{\alpha} = -\frac{1}{2}\left(T^{\mu\nu}{}_{\alpha} - T^{\nu\mu}{}_{\alpha} - T_{\alpha}{}^{\mu\nu}\right). \tag{3.16}$$

To construct a new scalar, which is comparable to the curvature scalar of general relativity, we express a different tensor $S_{\alpha}{}^{\mu\nu}$, created from the constituents of the torsion and contorsion tensors as

$$S_{\alpha}{}^{\mu\nu} = \frac{1}{2}\left(K^{\mu\nu}{}_{\alpha} + \delta^{\mu}_{\alpha}T^{\beta\nu}{}_{\beta} - \delta^{\nu}_{\alpha}T^{\beta\mu}{}_{\beta}\right). \tag{3.17}$$

Now, we will describe the **torsion scalar** using the following contraction

$$T = T^{\alpha}_{\mu\nu}S_{\alpha}{}^{\mu\nu}. \tag{3.18}$$

By generalizing the teleparallel theory, we can define the action of the $f(T)$ gravity theory as

$$S = \int e\left[f(T) + 2k\mathcal{L}_{Matter}\right]d^4x. \tag{3.19}$$

Here $f(T)$ is an arbitrary function of T (torsion scalar). Varying the action (3.19) in regard to the tetrads, one can obtain the resulting field equations

$$S_{\mu}{}^{\nu\rho}\partial_{\rho}Tf_{TT} + \left[e^{-1}e^i{}_{\mu}\partial_{\rho}\left(ee_i{}^{\alpha}S_{\alpha}{}^{\nu\rho}\right) + T^{\alpha}_{\lambda\mu}S_{\alpha}{}^{\nu\lambda}\right]f_T + \frac{1}{4}\delta^{\nu}_{\mu}f = -\frac{k}{2}\mathcal{T}^{\nu}_{\mu}, \tag{3.20}$$

where $\mathcal{T}^{\nu}_{\mu}$ is the energy-momentum tensor, $f_T = df(T)/dT$ and $f_{TT} = d^2f(T)/dT^2$. For $f(T) = a_1T + a_0$, the field equations (3.20) get back to the **teleparallel theory** in the presence of a cosmological constant, which is dynamically identical to general relativity. The above equations depend on the tetrads under consideration.

5. $f(R, T)$ **theory of gravity:** $f(R, T)$ is a newly proposed modified theory of gravity. Here, the gravitational Lagrangian can be considered as an arbitrary function of the Ricci scalar R and of the trace of the stress-energy tensor T. This new proposal has been thought because the gravitational field equations are influenced by the nature of the matter source. In $f(R, T)$ theory, the action is taken as (with geometrical units $G = c = 1$)

$$S = \frac{1}{16\pi}\int d^4xf(R, T)\sqrt{-g} + \int d^4x\mathcal{L}_m\sqrt{-g}, \tag{3.21}$$

where $f(R, T)$ is an arbitrary function of the Ricci scalar R and the trace of the energy-momentum tensor T and $\mathcal{L}_m$ is the Lagrangian for matter fields. As usual, g is the determinant of the metric tensor $g_{\mu\nu}$.

Now, if we vary the above action (3.21), we get

$$\delta S = \frac{1}{16\pi} \int \left[f_R(R, T)\delta R + f_T(R, T)\frac{\delta T}{\delta g^{\mu\nu}}\delta g^{\mu\nu} - \frac{1}{2}g_{\mu\nu}f(R, T)\delta g^{\mu\nu} + 16\pi\delta(\sqrt{-g}L_m) \right] d^4x,$$

where

$$\delta R = R_{\mu\nu}\delta g^{\mu\nu} + g_{\mu\nu}\Box\delta g^{\mu\nu} - \nabla_\mu\nabla_\nu\delta g^{\mu\nu}.$$

The variation with respect to the metric $g_{\mu\nu}$ yields the following field equations of $f(R, T)$ gravity:

$$f_R(R, T)R_{\mu\nu} - \frac{1}{2}f(R, T)g_{\mu\nu} + (g_{\mu\nu}\Box - \nabla_\mu\nabla_\nu)f_R(R, T)$$
$$= 8\pi T_{\mu\nu} - f_{T(R,T)}T_{\mu\nu} - f_{T(R,T)}\Theta_{\mu\nu}, \tag{3.22}$$

where

$$f_R(R, T) = \frac{\partial f(R, T)}{\partial R}, \quad f_T(R, T) = \frac{\partial f(R, T)}{\partial T},$$

$$\Box \equiv \frac{1}{\sqrt{-g}}\partial_\mu(\sqrt{-g}g^{\mu\nu}\partial_\nu),$$

$R_{\mu\nu}$ is the Ricci tensor, ∇_μ the covariant derivative with respect to the symmetric connection associated to $g_{\mu\nu}$,

$$\Theta_{\mu\nu} = g^{\alpha\beta}\frac{\delta T_{\alpha\beta}}{\delta g^{\mu\nu}},$$

and the stress-energy tensor can be defined as

$$T_{\mu\nu} = g_{\mu\nu}\mathcal{L}_m - 2\frac{\partial\mathcal{L}_m}{\partial g^{\mu\nu}}.$$

One can notice that when $f(R, T) = f(R)$, Eq. (3.22) yields the field equations of $f(R)$ gravity.

Contraction of (3.22) provides a relationship between the Ricci scalar R and the trace T of the stress-energy tensor as

$$f_R(R, T) + 3\Box f_R(R, T) - 2f(R, T) = 8\pi T - f_T(R, T)T - f_T(R, T)\Theta. \tag{3.23}$$

Substituting the term $\Box f_R(R, T)$ from the Eq. (3.23), the gravitational field equation (3.22) takes the form

$$f_R(R, T)(R_{\mu\nu} - \frac{1}{3}Rg_{\mu\nu}) + \frac{1}{6}f(R, T)g_{\mu\nu} = 8\pi(T_{\mu\nu} - \frac{1}{3}Tg_{\mu\nu})$$

$$-f_T(R, T)(T_{\mu\nu} - \frac{1}{3}Tg_{\mu\nu}) - f_T(R, T)(\Theta_{\mu\nu} - \frac{1}{3}\Theta g_{\mu\nu}) + \nabla_\mu\nabla_\nu f_R(R, T). \tag{3.24}$$

Taking the mathematical identity

$$\nabla^{\mu}[f_R(R,T)R_{\mu\nu} - \frac{1}{2}f(R,T)g_{\mu\nu} + (g_{\mu\nu}\Box - \nabla_{\mu}\nabla_{\nu})f_R(R,T)] = 0.$$

The covariant divergence of the above stress-energy tensor yields

$$\nabla^{\mu}T_{\mu\nu} = \frac{f_T(R,T)}{8\pi - f_T(R,T)}[(T_{\mu\nu} + \Theta_{\mu\nu})\nabla^{\mu}\ln f_T(R,T)$$

$$+ \nabla^{\mu}\Theta_{\mu\nu} - (1/2)g_{\mu\nu}\nabla^{\mu}T]. \tag{3.25}$$

6. **Brans–Dicke theory of gravity:** The Brans–Dicke theory of gravitation was established in 1961 by Robert H. Dicke and Carl H. Brans. It is a scalar–tensor theory where the gravitational interaction is reconciled by a scalar field along with a tensor field of general relativity. The gravitational constant G is not constant, rather replaced by $\frac{1}{\phi}$ where ϕ is a scalar field, which can vary from place to place and with time. ϕ has the dimension $ML^{-3}T^2$ and plays the role analogous to G^{-1}. Brans–Dicke theory of gravity is a generalization of general relativity, however, it is not a complete geometrical theory of gravitation as gravitational effects are described by a scalar field in a Riemannian manifold. Thus, gravitational effects are in a part of geometrical, and in part due to scalar interaction. In this new theory of gravitation, the action is written as

$$I = \frac{1}{16\pi}\int\left[\phi R - \frac{\omega\phi_i\phi^i}{\phi}\right]\sqrt{-g}d^4x - 16\pi\int L_F\sqrt{-g}d^4x, \tag{3.26}$$

where, ω is known as the Brans–Dicke coupling constant, which is dimensionless and $\phi_i \equiv \frac{\partial\phi}{\partial x^i}$.

Variation with respect to g_{ik}, we get the field equations in Brans–Dicke theory of gravitation,

$$R_{ik} - \frac{1}{2}g_{ik}R = \frac{8\pi}{\phi}T_{ik} + \frac{\omega}{\phi^2}\left(\phi_i\phi_k - \frac{1}{2}g_{ik}\phi_l\phi^l\right) + \frac{1}{\phi}(\nabla_i\nabla_k\phi - g_{ik}\Box\phi). \tag{3.27}$$

As usual, T_{ik} is the energy-momentum tensor of the matter fields. Also, trace of the above field equation is given by

$$-R = \frac{8\pi}{\phi}T - \frac{\omega}{\phi^2}\phi_i\phi^i - 3\phi^{-1}\Box\phi.$$

Here $T = g^{ik}T_{ik}$ is the trace of the stress-energy tensor and $\Box\phi = g^{ik}\nabla_i\nabla_k\phi$.

Variation with respect to ϕ, one gets,

$$2\omega\phi^{-1}\Box\phi - \frac{\omega}{\phi^2}\phi_i\phi^i + R = 0.$$

Combining these two equations, we can write the wave equation for ϕ in the following form

$$\Box\phi = \frac{8\pi}{3+2\omega}T. \tag{3.28}$$

Usually it is believed that general relativity can be obtained from the Brans–Dicke theory in the limit $\omega \to \infty$. But recently it is claimed that this does not happen when the trace of the stress-energy momentum vanishes, i.e., $T_i^i = 0$.

7. **Weyl gravity:** In Weyl gravity theory, the Lagrangian is considered as

$$L_{weyl} = C_{\mu\nu\rho\sigma} C^{\mu\nu\rho\sigma},$$

where $C_{\mu\nu\rho\sigma}$ is the conformally invariant Weyl tensor.

Thus for the fourth order Weyl theory of gravity, the Einstein–Hilbert action is written as

$$I = \alpha \int C_{\mu\nu\rho\sigma} C^{\mu\nu\rho\sigma} \sqrt{-g} d^4 x. \tag{3.29}$$

The Weyl tensor $C_{\mu\nu\rho\sigma}$ is described as the anti symmetric part of the Riemann tensor and α is a dimensionless parameter. This action is actually invariant under the conformal transformation

$$g_{\mu\nu} \to g'_{\mu\nu} = e^{\Omega(x)} g_{\mu\nu}, \tag{3.30}$$

as the Weyl tensor $C_{\mu\nu\rho\sigma}$ itself is invariant under this conformal transformation.

Now, we can express $C_{\mu\nu\rho\sigma} C^{\mu\nu\rho\sigma}$ in terms of Riemann and Ricci tensors, and the scalar curvature as

$$C_{\mu\nu\rho\sigma} C^{\mu\nu\rho\sigma} = R_{\mu\nu\rho\sigma} R^{\mu\nu\rho\sigma} - 2 R_{\mu\nu} R^{\mu\nu} + \frac{1}{3} R^2.$$

In four dimensions, the GB term is given by

$$G = R_{\mu\nu\rho\sigma} R^{\mu\nu\rho\sigma} - 4 R_{\mu\nu} R^{\mu\nu} + R^2,$$

which is topological and as a result it does not take part in the field equations. Hence, only the part of $C_{\mu\nu\rho\sigma} C^{\mu\nu\rho\sigma}$ modulo GB term,

$$C_{\mu\nu\rho\sigma} C^{\mu\nu\rho\sigma} = 2 \left(R_{\mu\nu} R^{\mu\nu} - \frac{1}{3} R^2 \right) (Mod\ G),$$

will play the role to the field equations.

Thus, the action of Weyl gravity can be rewritten as

$$I = 2\alpha \int \left(R_{\mu\nu} R^{\mu\nu} - \frac{1}{3} R^2 \right) \sqrt{-g} d^4 x. \tag{3.31}$$

Now, variation with respect to metric tensor yield the vacuum field equations of the Weyl gravity as

$$B_{\mu\nu} = K_{\mu\nu} - \frac{1}{3} H_{\mu\nu} = 0, \tag{3.32}$$

where

$$K_{\mu\nu} = \Box \left(R_{\mu\nu} + \frac{1}{2} g_{\mu\nu} \right) - \nabla_\lambda \nabla_\mu R_\nu^\lambda - \nabla_\lambda \nabla_\nu R_\mu^\lambda + 2 R_{\mu\lambda} R_\nu^\lambda - \frac{1}{2} g_{\mu\nu} R_{\alpha\beta} R^{\alpha\beta}$$

$$H_{\mu\nu} = 2R \left(R_{\mu\nu} - \frac{1}{4} g_{\mu\nu} \right) + 2(g_{\mu\nu} \Box - \nabla_\mu \nabla_\nu) R.$$

Here, $B_{\mu\nu}$ is known as **Bach tensor**. The trace of the Bach tensor is zero as it is conformally invariant. If one includes the matter distribution, then, field equations in Weyl gravity is taken the following form

$$B_{\mu\nu} = 8\pi G T_{\mu\nu}. \tag{3.33}$$

Linearized Gravity

4.1 Newtonian Gravity

Newton's theory of gravitation can be treated as a three-dimensional field theory. The gravitational field is characterized by a scalar field $\phi(x, y, z)$. This satisfies

$$\nabla^2 \phi = 4\pi G \rho(x, y, z), \tag{4.1}$$

where $G = 6.67 \times 10^{-8} \text{cm}^3 \text{ gm}^{-1} \text{ sec}^{-2}$ is the gravitational constant and ρ is the mass density of matter in space that produces the gravitational field. The above equation is known as the **Poisson equation**.

Proof of Poisson Equation

Let us consider a mass M occupying a volume V, which is enclosed by a surface S. The gravitational flux passing through the elementary surface dS is given by $\mathbf{g}.\mathbf{n}dS$, where $\mathbf{g}$ is gravitational vector field (also known as gravitational acceleration) and $\mathbf{n}$ is the unit outward normal vector to S. Now, the total gravitational flux through S is

$$\int_S \mathbf{g}.\mathbf{n}dS = -GM \int_S \frac{\mathbf{e}_r.\mathbf{n}}{r^2} dS.$$

We know $\frac{\mathbf{e}_r.\mathbf{n}}{r^2} dS = \frac{\cos\theta dS}{r^2}$ is the elementary solid angle $d\Omega$ subtended at M by the elementary surface dS, where $\mathbf{e}_r$ is the radial unit vector. Thus,

$$\int_S \mathbf{g}.\mathbf{n}dS = -GM \int_S d\Omega = -4\pi GM = -4\pi G \int_V \rho(r)dV.$$

Applying Gauss divergence theorem to the left-hand side, we get

$$\int_V \nabla.\mathbf{g}dV = -4\pi G \int_V \rho(r)dV.$$

Thus, we get

$$\nabla.\mathbf{g} = -4\pi G\rho.$$

Using $\mathbf{g} = -\nabla\phi$ (gravity is a conservative force, therefore, it can be written as the gradient of a scalar potential ϕ, known as the gravitational potential), we finally obtain

$$\nabla^2\phi = 4\pi G\rho.$$

The gravitational field (F) is proportional to the negative of $\nabla\phi$, i.e.,

$$F = -m\nabla\phi.$$

This is the force acting on a particle of mass m.

For a single mass M that produces the potential ϕ, then the solution of Poisson equation is given by

$$\phi = \frac{-GM}{r}.$$

The force acting on another particle with mass m will be

$$F = GMm\nabla\left(\frac{1}{r}\right) = \frac{-GMm}{r^2}.$$

The ratio of gravitational force and electrical force between two electrons is given by

$$\frac{F_{grav}}{F_{elec}} = \frac{Gm_e^2}{k_e e^2} = 0.24 \times 10^{-42}.$$

Here, k_e is Coulomb's constant with m_e and e are mass and charge of the electron, respectively. This indicates that the gravitational force is very weak.

Exercise 4.1

Find the gravitational potential inside and outside of a sphere of uniform mass density having a radius R and a total mass M. Normalize the potential so that it vanishes at infinity.

Hints:

The mass density in the sphere is (see Fig. 8)

$$\rho = \frac{M}{\frac{4\pi R^3}{3}} = \frac{3M}{4\pi R^3} = constant. \tag{4.2}$$

For spherically symmetric distribution of matter, the gravitational potential ϕ is a function of radius r only. Therefore, from Poisson equation $\nabla^2\phi = 4\pi G\rho$, we get

$$\frac{1}{r^2}\frac{d}{dr}\left(r^2\frac{d\phi}{dr}\right) = 4\pi\rho G. \tag{4.3}$$

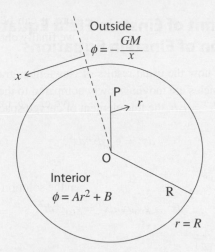

Outside
$$\phi = -\frac{GM}{x}$$

x

P

r

O

Interior
$$\phi = Ar^2 + B$$

R

$r = R$

Figure 8 Sphere of uniform mass.

The general solution ($r < R$) of the above equation

$$\phi(r) = Ar^2 + B, \tag{4.4}$$

where A and B are integration constants.

Equations (4.1) and (4.2) yield A as

$$A = \frac{2\pi}{3} G\rho = \frac{1}{2}\left(\frac{GM}{R}\right)\frac{1}{R^2}, \tag{4.5}$$

B can be found from the matching condition to the exterior solution

$$\phi(r) = -\frac{GM}{r},$$

at $r = R$, as

$$B = -\frac{3GM}{2R}. \tag{4.6}$$

Thus,

$$\phi(r) = \frac{GM}{2R}\left[\left(\frac{r}{R}\right)^2 - 3\right], \quad r < R$$

$$= -\frac{GM}{r}, \quad r > R.$$

4.2 Newtonian Limit of Einstein Field Equations or Weak Field Approximation of Einstein Equations

We now try to take a look at how the usual results of Newtonian gravity fit Einstein's theory. At first, we consider that the particles are moving slowly compared to the velocity of light.

For the consideration of $dx^0 = cdt$, the line element of curved spacetime

$$ds^2 = g_{\mu\nu}dx^\mu dx^\nu, \tag{4.7}$$

can be written approximately as

$$ds^2 \cong g_{00}dx^0 dx^0 = g_{00}c^2 dt^2. \tag{4.8}$$

[Since $cdt >> dx^k$, i.e., $c >> \frac{dx^k}{dt}$]

Thus,

$$g_{00}c^2 dt^2 >> g_{\mu\nu}dx^\mu dx^\nu \quad (\mu, \nu \neq 0).$$

We know the other form of Einstein field equation is

$$R_{\mu\nu} = k\left(T_{\mu\nu} - \frac{1}{2}g_{\mu\nu}T\right). \tag{4.9}$$

From the above, we can conclude that it will be enough to use only the 00 component of this equation. Here,

$$T = T_{\mu\nu}g^{\mu\nu} \cong T_{\mu\nu}\eta^{\mu\nu} \cong T_{00}\eta^{00} = T_{00}. \tag{4.10}$$

Thus,

$$R_{00} = k\left(T_{00} - \frac{1}{2}g_{00}T\right) \cong k\left(T_{00} - \frac{1}{2}\eta_{00}T\right) = \frac{1}{2}kT_{00} = \frac{1}{2}k\rho c^2. \tag{4.11}$$

Here, ρ denotes the mass density of the matter distribution that generates the gravitational field. In fact, the gravitational field is said to be **weak** when metric of curved spacetime, g_{ij}, departs slightly from Minkowski metric $\eta_{ij} \equiv (+1, -1, -1, -1)$, i.e., from flat spacetime. Or, equivalently, the metric tensor $g_{\mu\nu} \approx \eta_{\mu\nu} + h_{\mu\nu}$, $|h_{\mu\nu}| << 1$.

Also,

$$R_{00} = \frac{\partial \Gamma^\rho_{00}}{\partial x^\rho} - \frac{\partial \Gamma^\rho_{0\rho}}{\partial x^0} + \Gamma^\sigma_{00}\Gamma^\rho_{\rho\sigma} - \Gamma^\sigma_{0\rho}\Gamma^\rho_{0\sigma} \cong \frac{\partial \Gamma^\rho_{00}}{\partial x^\rho}. \tag{4.12}$$

[Neglecting nonlinear terms and terms that are time derivatives and using $\Gamma^\lambda_{\mu\nu} \approx \frac{1}{2}\eta^{\lambda\rho}(\partial_\mu h_{\rho\nu} + \partial_\nu h_{\mu\rho} - \partial_\rho h_{\mu\nu})$]

Now,

$$\Gamma^k_{00} = \frac{1}{2} g^{k\lambda} \left(2 \frac{\partial g_{\lambda 0}}{\partial x^\lambda} - \frac{\partial g_{00}}{\partial x^\lambda} \right),$$

$$\cong \frac{-1}{2} \eta^{k\lambda} \frac{\partial h_{00}}{\partial x^\lambda} = \frac{1}{2} \delta^{kl} \frac{\partial h_{00}}{\partial x^l},$$

i.e.

$$\Gamma^k_{00} \cong \frac{1}{2} \frac{\partial h_{00}}{\partial x^k}. \quad [k = 1, 2, 3 \ but \ not \ 0] \tag{4.13}$$

Therefore,

$$R_{00} \equiv \frac{1}{2} \frac{\partial h_{00}}{\partial x^m \partial x^m} = \frac{1}{2} \nabla^2 h_{00}, \tag{4.14}$$

where ∇^2 = Laplacian.

Also, we know the motion of test particles in the gravitational field of any curved spacetime geometry is characterized by the geodesic equation,

$$\frac{d^2 x^\mu}{d\tau^2} + \Gamma^\mu_{\rho\sigma} \left(\frac{dx^\rho}{d\tau} \right) \left(\frac{dx^\sigma}{d\tau} \right) = 0. \tag{4.15}$$

Here, $x^\mu(\tau)$ is the world line of the particle in the spacetime. If the particle's speed $\frac{dx^a}{d\tau}$ is much lower than the speed of light, then we may approximate $\frac{dx^a}{d\tau}$ as $(c, 0, 0, 0)$ in the second term. Here, proper time τ is equivalent to the time coordinate t, i.e.,

$$d\tau \equiv dt \quad when \quad v << c. \tag{4.16}$$

Thus we find

$$\frac{d^2 x^a}{dt^2} = -\frac{1}{2} \frac{c^2 \partial h_{00}}{\partial x^a} \quad [a = 1, 2, 3], \tag{4.17}$$

We know the Newtonian equation of motion

$$\frac{d^2 x^a}{dt^2} = -\frac{\partial \phi}{\partial x^a}, \tag{4.18}$$

where ϕ is the Newtonian gravitational potential. At a large distance, $\phi \longrightarrow 0$ and $g_{00} \longrightarrow 1$, i.e., $h_{00} \longrightarrow 0$.

Now comparing the above two equations, we get

$$h_{00} \cong \frac{2\phi}{c^2}, \quad i.e., \quad g_{00} \cong 1 + \frac{2\phi}{c^2}. \tag{4.19}$$

This is called a **weak field limit**.

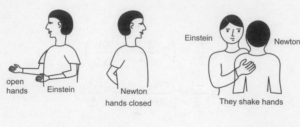

open
hands   Einstein   Newton   They shake hands
                              hands closed

Strong Gravitational field   Weak Gravitational field
such as Relativistic Astrophysics,   such as planet's motion, etc.
Black holes physics, etc.

Figure 9   Phenomenological comparison of Einstein and Newtonian theories.

Using the result in Eq. (4.19), one obtains from Eq. (4.14) as

$$R_{00} = \frac{1}{c^2} \nabla^2(\phi), \tag{4.20}$$

and then Einstein's field equation becomes,

$$\nabla^2 \phi(x) = \frac{1}{2} kc^4 \rho(x). \tag{4.21}$$

We know Newton's gravitational theory as

$$\nabla^2 \phi = 4\pi G\rho. \tag{4.22}$$

Identifying the above two equations, we find

$$k = \frac{8\pi G}{c^4}, \tag{4.23}$$

which is **Einstein's gravitational constant**.

Thus, in the weak gravitational fields such as planetary motion, etc., Newtonian and Einstein gravitational theories provide almost the same results. However, for a strong gravitational field, such as compact stars, black holes, etc., two theories are significantly different. Rather, one can say that for a strong gravitational field. Newtonian theory does not work good, whereas Einstein's theory works in excellent manner. The Schematic diagram is shown in Fig. 9.

4.3   Poisson Equation as an Approximation of Einstein Field Equations

For the consideration of $dx^0 = cdt$, the line element of curved spacetime

$$ds^2 = g_{\mu\nu} dx^\mu dx^\nu$$

can be written approximately, i.e., in the weak field approximation as

$$ds^2 \cong g_{00}dx^0 dx^0 = g_{00}c^2 dt^2,$$

$$where, \quad g_{00} = 1 + \frac{2\phi}{c^2}.$$

We know the other form of Einstein field equation is

$$R_{\mu\nu} = k\left(T_{\mu\nu} - \frac{1}{2}g_{\mu\nu}T\right).$$

From the above, we can conclude that it will be enough to use only the 00 component of this equation.
    Here,

$$T = T_{\mu\nu}g^{\mu\nu} \cong T_{\mu\nu}\eta^{\mu\nu} \cong T_{00}\eta^{00} = T_{00}.$$

Thus,

$$R_{00} = k\left(T_{00} - \frac{1}{2}g_{00}T\right) \cong k\left(T_{00} - \frac{1}{2}\eta_{00}T\right) = \frac{1}{2}kT_{00} = \frac{1}{2}k\rho c^2.$$

Here, ρ denotes the mass density of the matter distribution that generates the gravitational field.
    Also,

$$R_{00} = \frac{\partial \Gamma^{\rho}_{00}}{\partial x^{\rho}} - \frac{\partial \Gamma^{\rho}_{0\rho}}{\partial x^0} + \Gamma^{\sigma}_{00}\Gamma^{\rho}_{\rho\sigma} - \Gamma^{\sigma}_{0\rho}\Gamma^{\rho}_{0\sigma},$$

$$\cong \frac{\partial \Gamma^{\rho}_{00}}{\partial x^{\rho}}.$$

[Neglecting nonlinear terms and terms that are time derivatives]
    Now,

$$\Gamma^{k}_{00} = \frac{1}{2}g^{k\lambda}\left(2\frac{\partial g_{\lambda 0}}{\partial x^{\lambda}} - \frac{\partial g_{00}}{\partial x^{\lambda}}\right)$$

$$\cong \frac{-1}{2}\eta^{k\lambda}\frac{\partial g_{00}}{\partial x^{\lambda}} = \frac{1}{2}\delta^{kl}\frac{\partial g_{00}}{\partial x^l},$$

$$= \frac{1}{2}\frac{\partial g_{00}}{\partial x^k}. \quad [k = 1, 2, 3 \; but \; not \; 0]$$

Therefore,

$$R_{00} \equiv \frac{1}{2}\frac{\partial g_{00}}{\partial x^m \partial x^m} = \frac{1}{2}\nabla^2 g_{00}.$$

Equating R_{00} in the above two expressions, we get

$$\frac{1}{2}\nabla^2 g_{00} = \frac{1}{2}k\rho c^2.$$

Putting the value of g_{00}, we get

$$\nabla^2 \phi = 4\pi G\rho.$$

4.4 Gravitational Wave

We consider that the gravitational field is very weak, i.e., spacetime is nearly flat. Therefore, the fundamental tensor can be expressed as

$$g_{\mu\nu} = \eta_{\mu\nu} + h_{\mu\nu}, \tag{4.24}$$

where $\eta_{\mu\nu} = diag(1,-1,-1,-1)$ is Minkowski metric and $h_{\mu\nu}$ are small perturbations, i.e., $|h_{\mu\nu}| \ll 1$.

We can rewrite the Einstein field equations as,

$$R_{\mu\nu} = -8\pi\left(T_{\mu\nu} - \frac{1}{2}g_{\mu\nu}T\right) = -8\pi T^*_{\mu\nu}, \tag{4.25}$$

where, $T^*_{\mu\nu} = T_{\mu\nu} - \frac{1}{2}g_{\mu\nu}T$.

Now we are trying to express the field equations in terms of $h_{\mu\nu}$ and keep only the linear terms. The Ricci tensor takes the form

$$R_{\mu\nu} \approx \partial_\lambda \Gamma^\lambda_{\nu\mu} - \partial_\nu \Gamma^\lambda_{\lambda\mu}.$$

Now,

$$\Gamma^\alpha_{\mu\nu} \approx \frac{1}{2}\eta^{\alpha\beta}[h_{\mu\beta,\nu} + h_{\beta\nu,\mu} - h_{\mu\nu,\beta}] = \frac{1}{2}[h^\alpha_{\mu,\nu} + h^\alpha_{\nu,\mu} - h^\alpha_{\mu\nu}].$$

Therefore, the Ricci tensor takes the following form:

$$R_{\mu\nu} \approx \frac{1}{2}\eta^{\lambda\sigma}\left(\frac{\partial^2 h_{\lambda\sigma}}{\partial x^\nu \partial x^\mu} + \frac{\partial^2 h_{\mu\nu}}{\partial x^\sigma \partial x^\lambda} - \frac{\partial^2 h_{\mu\sigma}}{\partial x^\nu \partial x^\lambda} - \frac{\partial^2 h_{\lambda\nu}}{\partial x^\sigma \partial x^\mu}\right). \tag{4.26}$$

We follow the convention for raising and lowering the tensor by $\eta^{\mu\nu}$ and using $h^\nu_\mu = \eta^{\nu\alpha}h_{\mu\alpha}$, $h = h^\mu_\mu = \eta^{\mu\nu}h_{\mu\nu}$ in Eq. (4.26) to yield

$$R_{\mu\nu} \approx \frac{1}{2}\left(\frac{\partial^2 h_{\mu\nu}}{\partial x^\sigma \partial x^\sigma} + \frac{\partial^2 h}{\partial x^\nu \partial x^\mu} - \frac{\partial^2 h^\lambda_\mu}{\partial x^\nu \partial x^\lambda} - \frac{\partial^2 h^\sigma_\nu}{\partial x^\sigma \partial x^\mu}\right),$$

$$= \frac{1}{2}\left[\frac{\partial^2 h_{\mu\nu}}{\partial x^\sigma \partial x^\sigma} + \frac{1}{2}\left(\frac{\partial^2}{\partial x^\nu \partial x^\mu} + \frac{\partial^2}{\partial x^\nu \partial x^\mu}\right)h - \frac{\partial^2 h^\lambda_\mu}{\partial x^\nu \partial x^\lambda} - \frac{\partial^2 h^\sigma_\nu}{\partial x^\sigma \partial x^\mu}\right],$$

$$= \frac{1}{2} \left(\frac{\partial^2 h_{\mu\nu}}{\partial x^\sigma \partial x^\sigma} + \frac{1}{2} \frac{\partial^2}{\partial x^\nu \partial x^\lambda} (\eta_\mu^\lambda h) - \frac{\partial^2 h_\mu^\lambda}{\partial x^\nu \partial x^\lambda} + \frac{1}{2} \frac{\partial^2}{\partial x^\mu \partial x^\sigma} (\eta_\nu^\sigma h) - \frac{\partial^2 h_\nu^\sigma}{\partial x^\sigma \partial x^\mu} \right),$$

$$= \frac{1}{2} \left[\frac{\partial^2 h_{\mu\nu}}{\partial x^\sigma \partial x^\sigma} - \frac{\partial^2}{\partial x^\nu \partial x^\lambda} \left(h_\mu^\lambda - \frac{1}{2} \eta_\mu^\lambda h \right) - \frac{\partial^2}{\partial x^\sigma \partial x^\mu} \left(h_\nu^\sigma - \frac{1}{2} \eta_\nu^\sigma h \right) \right]. \tag{4.27}$$

Let us define

$$\gamma_\mu^\nu = h_\mu^\nu - \frac{1}{2} \delta_\mu^\nu h. \tag{4.28}$$

Now, Ricci tensor can be written in terms of $h_{\mu\nu}$ and newly defined $\gamma_{\mu\nu}$ as

$$R_{\mu\nu} = \frac{1}{2} \Box^2 h_{\mu\nu} - \frac{1}{2} \frac{\partial}{\partial x^\nu} \left(\frac{\partial}{\partial x^\lambda} \gamma_\mu^\lambda \right) - \frac{1}{2} \frac{\partial}{\partial x^\mu} \left(\frac{\partial}{\partial x^\sigma} \gamma_\nu^\sigma \right), \tag{4.29}$$

where, $\Box^2 = \frac{\partial^2}{\partial t^2} - \nabla^2$ is the wave operator. Hence, Eq. (4.25) takes the following form

$$\Box^2 h_{\mu\nu} - \frac{\partial}{\partial x^\nu} \left(\frac{\partial}{\partial x^\lambda} \gamma_\mu^\lambda \right) - \frac{\partial}{\partial x^\mu} \left(\frac{\partial}{\partial x^\sigma} \gamma_\nu^\sigma \right) = -16\pi T_{\mu\nu}^*. \tag{4.30}$$

For further simplification, one can impose the Lorentz-Gauge condition

$$\frac{\partial}{\partial x^\nu} \gamma_\mu^\nu = 0 = \partial_\nu \left(h_\mu^\nu - \frac{1}{2} \delta_\mu^\nu h \right). \tag{4.31}$$

Therefore, Eq. (4.30) becomes

$$\Box^2 h_{\mu\nu} = -16\pi T_{\mu\nu}^*. \tag{4.32}$$

Note that if $\gamma_{\mu\nu}$ does not follow the Lorentz-Gauge condition, i.e.,

$$\frac{\partial}{\partial x^\nu} \gamma_\mu^\nu = q_\mu,$$

then one can propose a new coordinate transformation as

$$\gamma_{\mu\nu}' = \gamma_{\mu\nu} - \xi_{\mu,\nu} - \xi_{\nu,\mu} + \eta_{\mu\nu} (\partial_\rho \xi^\rho).$$

Now imposing the condition

$$\Box^2 \xi_\nu = q_\nu,$$

we get

$$\partial_\mu \gamma'^{\mu\nu} = 0.$$

[Here, ξ^μ is an infinitesimal vector field]

The retarded solution with the subsidiary Lorentz-Gauge condition is given by

$$h_{\mu\nu}(t,x) = -4 \int \frac{T^*_{\mu\nu}(t - |x - x'|, x')}{|x - x'|} d^3x'.$$

(4.33)

In the weak field approximation, Eqs. (4.31) and (4.32) are the linearized equation of general relativity.

These linearized equations indicate that the gravitational field propagates with unit velocity (in the gravitational units). In other words, gravitational field propagates at the speed of light. When the energy-momentum tensor $T_{\mu\nu}$ is zero, i.e., in linearized theory in a vacuum, Eqs. (4.31) and (4.32) assume the following form

$$\frac{\partial}{\partial x^\nu} \gamma^\nu_\mu = 0,$$

$$\Box^2 h_{\mu\nu} = 0.$$

It is obvious that $h_{\mu\nu}$ follows the wave equation and the gravitational waves move with the velocity of light.

Lie Derivatives and Killing's Equation

5.1 Introduction

The Lie derivative is a significant concept of differential geometry, named after the discovery by Sophus Lie in the late nineteenth century. It estimates the modification of a tensor field (containing scalar function, vector field), along the flow defined by an additional vector field. Lie derivative can be defined on any differentiable manifold as this change is coordinate invariant.

A vector field X is a linear mapping from C^∞ function to C^∞ function on a manifold, satisfying

$$X(fg) = (Xf)g + fX(g), \quad \forall f, g \in C^\infty(M)$$

$$X(f) = \sum \frac{dx^\mu}{dv} \frac{\partial f}{\partial x^\mu}$$

such that

$$X = \sum a^\mu \frac{\partial}{\partial x^\mu}.$$

Suppose $X^\mu(x)$ is a vector field defined over a manifold M. Trajectory of X^μ is obtained by solving

$$\frac{dx^\mu}{dv} = X^\mu(x(v)). \tag{5.1}$$

Let us consider a coordinate transformation

$$\bar{x}^\mu = \bar{x}^\mu(\epsilon, x^\gamma), \tag{5.2}$$

where ϵ is a parameter. This is known as one parameter set of transformation. This transformation designates a mapping of the spacetime onto itself. If the transformation takes the form

$$\bar{x}^\mu = x^\mu + \epsilon\, \xi^\mu(x), \tag{5.3}$$

then it is called **infinitesimal one parameter transformation or infinitesimal mapping**.

Here, $\xi^\mu(x)$ is a contravariant vector field defined by

$$\xi^\mu(x) = \left. \frac{\partial \bar{x}^\mu}{\partial \epsilon} \right|_{\epsilon=0}. \tag{5.4}$$

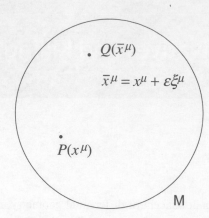

Figure 10 Two neighboring points under the infinitesimal one parameter transformation.

[Infinitesimal mapping means that for each point $P(x^\mu)$ of the spacetime there corresponds another neighboring point $Q(x^\mu + \epsilon \xi^\mu)$ in the same coordinate system (see Fig. 10).]

Suppose a tensor $T(x)$, defined in our spacetime. At the point Q, one can assess the tensor $T(x)$ in two different ways:

1. One can have the value of T at Q, i.e., $T(\bar{x}^\mu)$.

2. The $\bar{T}(\bar{x}^\mu)$ can be obtained as transmuted tensor $\bar{T}$ using normal coordinate.

Thus, transformation for tensor at the point Q has two techniques. The difference among these two values of the tensor calculated at the point $Q(\bar{x}^\mu)$ hints to the concept of Lie derivative of the tensor T. One can differentiate the functions, tensor fields with respect to a vector field.

5.2 Lie Derivative of a Scalar

Let at point Q, the value of scalar $\phi(x)$ be $\phi(\bar{x})$, i.e.,

$$\phi(\bar{x}) = \phi(x + \epsilon \xi) = \phi(x) + \epsilon \frac{\partial \phi}{\partial x^\alpha} \xi^\alpha.$$

[neglecting higher power of ϵ]

Also, transformation of scalar function $\phi(x)$ remains unaffected. Thus,

$$\bar{\phi}(\bar{x}) = \phi(x).$$

One can define the Lie derivative of scalar function $\phi(x)$ (denoted as $L_\xi \phi$) as

$$L_\xi \phi(x) = \lim_{\epsilon \to 0} \frac{\phi(\bar{x}) - \bar{\phi}(\bar{x})}{\epsilon}$$

$$= \xi^\alpha(x) \frac{\partial \phi(x)}{\partial x^\alpha}.$$

As ϕ is a scalar, therefore, then partial derivative can be replaced by a covariant derivative to yield,

$$L_\xi \phi(x) = \xi^\alpha(x) \nabla_\alpha \phi(x).$$

5.3 Lie Derivative of Contravariant Vector

The transformation of V^α is

$$\bar{V}^\alpha(\bar{x}) = \frac{\partial \bar{x}^\alpha}{\partial x^\beta} V^\beta(x). \tag{5.5}$$

Also,

$$\bar{x}^\alpha = x^\alpha + \epsilon \xi^\alpha \Rightarrow \frac{\partial \bar{x}^\alpha}{\partial x^\beta} = \delta^\alpha_\beta + \epsilon \frac{\partial \xi^\alpha}{\partial x^\beta}. \tag{5.6}$$

Now,

$$V^\alpha(\bar{x}) = V^\alpha(x + \epsilon \xi) = V^\alpha(x) + \epsilon \xi^\beta(x) \frac{\partial V^\alpha(x)}{\partial x^\beta}. \tag{5.7}$$

Putting (5.6) in (5.5), one can get,

$$\bar{V}^\alpha(\bar{x}) = \left(\delta^\alpha_\beta + \epsilon \frac{\partial \xi^\alpha}{\partial x^\beta} \right) V^\beta(x),$$

i.e.,

$$\bar{V}^\alpha(\bar{x}) = V^\alpha(x) + \epsilon \frac{\partial \xi^\alpha}{\partial x^\beta} V^\beta(x).$$

Now,

$$L_\xi V^\alpha = \lim_{\epsilon \to 0} \frac{V^\alpha(\bar{x}) - \bar{V}^\alpha(\bar{x})}{\epsilon}$$

$$= \xi^\beta \frac{\partial V^\alpha}{\partial x^\beta} - \frac{\partial \xi^\alpha}{\partial x^\beta} V^\beta. \tag{5.8}$$

5.4 Lie Derivative of Covariant Vector

The transformation of covariant vector V_α is

$$\bar{V}_\alpha(\bar{x}) = \frac{\partial x^\mu}{\partial \bar{x}^\alpha} V_\mu(x).$$

From,

$$\bar{x}^\mu = x^\mu + \epsilon \xi^\mu,$$

we have,

$$\delta_\alpha^\mu = \frac{\partial x^\mu}{\partial \bar{x}^\alpha} + \epsilon \frac{\partial \xi^\mu}{\partial \bar{x}^\alpha},$$

Therefore,

$$\frac{\partial x^\mu}{\partial \bar{x}^\alpha} = \delta_\alpha^\mu - \epsilon \frac{\partial \xi^\mu}{\partial \bar{x}^\alpha}.$$

Also,

$$V_\alpha(\bar{x}) = V_\alpha(x + \epsilon \xi) = V_\alpha(x) + \epsilon \xi^\mu \frac{\partial V_\alpha(x)}{\partial x^\mu}.$$

Now,

$$L_\xi V_\alpha = \lim_{\epsilon \to 0} \frac{V_\alpha(\bar{x}) - \bar{V}_\alpha(\bar{x})}{\epsilon}$$

$$= \xi^\mu \frac{\partial V_\alpha}{\partial x^\mu} + V_\mu \frac{\partial \xi^\mu}{\partial x^\alpha}.$$

$[\epsilon \to 0, \bar{x}_\alpha \to x_\alpha]$

Similar to the scalar function case, one can replace the partial derivatives by covariant derivatives.

Thus,

$$L_\xi V^\alpha = \xi^\beta \, \nabla_\beta \, V^\alpha - \nabla_\beta \xi^\alpha V^\beta, \tag{5.9}$$

$$L_\xi V_\alpha = \xi^\mu \, \nabla_\mu \, V_\alpha + V_\mu \nabla_\alpha \xi^\mu. \tag{5.10}$$

5.5 Lie Derivative of Covariant and Contravariant Tensors of Order Two

The transformation of $T_{\alpha\beta}$ is

$$\bar{T}_{\alpha\beta}(\bar{x}) = \frac{\partial x^\mu}{\partial \bar{x}^\alpha} \frac{\partial x^\nu}{\partial \bar{x}^\beta} T_{\mu\nu}(x),$$

$$\bar{x}^\mu = x^\mu + \epsilon \xi^\mu$$

$$\Rightarrow$$

$$\delta_\alpha^\mu = \frac{\partial x^\mu}{\partial \bar{x}^\alpha} + \epsilon \frac{\partial \xi^\mu}{\partial \bar{x}^\alpha}.$$

Also, we can write

$$\delta_\beta^\nu = \frac{\partial x^\nu}{\partial \bar{x}^\beta} + \epsilon \frac{\partial \xi^\nu}{\partial \bar{x}^\beta}.$$

Hence

$$\bar{T}_{\alpha\beta} = \left(\delta_\alpha^\mu - \epsilon \frac{\partial \xi^\mu}{\partial \bar{x}^\alpha} \right) \left(\delta_\beta^\nu - \epsilon \frac{\partial \xi^\nu}{\partial \bar{x}^\beta} \right) T_{\mu\nu}(x)$$

$$= \delta_\alpha^\mu \delta_\beta^\nu T_{\mu\nu}(x) - \epsilon \delta_\alpha^\mu \frac{\partial \xi^\nu}{\partial \bar{x}^\beta} T_{\mu\nu}(x) - \epsilon \frac{\partial \xi^\mu}{\partial \bar{x}^\alpha} \delta_\beta^\nu T_{\mu\nu}(x) + \epsilon^2 \frac{\partial \xi^\mu}{\partial \bar{x}^\alpha} \frac{\partial \xi^\nu}{\partial \bar{x}^\beta} T_{\mu\nu}(x),$$

[put first term $\mu = \alpha$ and $\nu = \beta$, and neglecting the last term]

$$= T_{\alpha\beta} - \epsilon \left(T_{\mu\beta} \frac{\partial \xi^\mu}{\partial \bar{x}^\alpha} + T_{\alpha\nu} \frac{\partial \xi^\nu}{\partial \bar{x}^\beta} \right).$$

Also

$$T_{\alpha\beta}(\bar{x}) = T_{\alpha\beta}(x + \epsilon \xi) = T_{\alpha\beta}(x) + \epsilon \xi^\rho \left(\frac{\partial T_{\alpha\beta}}{\partial x^\rho} \right).$$

Hence

$$L_\xi T_{\alpha\beta} = \lim_{\epsilon \to 0} \frac{T_{\alpha\beta}(\bar{x}) - \bar{T}_{\alpha\beta}(\bar{x})}{\epsilon},$$

as $\epsilon \to 0$, $\bar{x}_\alpha \to x_\alpha$

$$= \xi^\rho \frac{\partial T_{\alpha\beta}}{\partial x^\rho} + T_{\alpha\nu} \frac{\partial \xi^\nu}{\partial x^\beta} + T_{\mu\beta} \frac{\partial \xi^\mu}{\partial x^\alpha}.$$

Similarly,

$$L_\xi T^{\alpha\beta} = \xi^\rho \frac{\partial T^{\alpha\beta}}{\partial x^\rho} - T^{\alpha\nu} \frac{\partial \xi^\beta}{\partial x^\nu} - T^{\mu\beta} \frac{\partial \xi^\alpha}{\partial x^\mu}.$$

Once again, one can replace partial derivatives by covariant derivatives. Thus, we get,

$$L_\xi T_{\alpha\beta} = \xi^\rho \nabla_\rho T_{\alpha\beta} + T_{\alpha\nu} \nabla_\beta \xi^\nu + T_{\mu\beta} \nabla_\alpha \xi^\mu L_\xi T^{\alpha\beta}, \tag{5.11}$$

$$L_\xi T^{\alpha\beta} = \xi^\rho \nabla_\rho T^{\alpha\beta} - T^{\alpha\nu} \nabla_\nu \xi^\beta - T^{\mu\beta} \nabla_\mu \xi^\alpha. \tag{5.12}$$

Also

$$L_\xi (V^\alpha T_{\beta\nu}) = V^\alpha L_\xi (T_{\beta\nu}) + T_{\beta\nu} L_\xi (V^\alpha). \tag{5.13}$$

Note 5.1

Let us consider a curve passing through a point P such that only x^1 will vary and other components $x^2, \ldots, x^n$ remain constant along the curve. Therefore, we get

$$X^a = \delta_1^a = (1, 0, 0, \ldots, 0).$$

Then the vector field

$$X = X^a \partial_a = \partial_1,$$

Hence,

$$L_X T^{ab} = \partial_1 T^{ab}.$$

Thus, ordinary differentiation is a special case of Lie differentiation in a particular coordinate system. Suppose $X^i(x)$ is a vector field defined over a manifold M, then

$$X^i(A^{\alpha\beta\cdots}_{\mu\nu\cdots})_{;\,i} = X^i \nabla_i(A^{\alpha\beta\cdots}_{\mu\nu\cdots}),$$

is called the **directional (tensor) derivative** of the tensor $A^{\alpha\beta\cdots}_{\mu\nu\cdots}$ in the direction of the vector $X^i(x)$.

Note 5.2

Lie derivative of a mixed tensor

$$L_X T^{a\cdots}_{b\cdots} = X^c \partial_c T^{a\cdots}_{b\cdots} - T^{c\cdots}_{b\cdots} \partial_c X^a - \cdots + T^{a\cdots}_{c\cdots} \partial_b X^c + \cdots$$

[negative sign for contravariant vector and positive sign for covariant vector]

Note 5.3

The Lie derivative can be defined in various equivalent ways. Let us define two vector fields X and Y on a smooth manifold M. Lie derivative of Y along X is defined by the Lie bracket $[X, Y]$. Thus, Lie derivative of Y along the flow generated by X is defined by the Lie bracket of X and Y at p in terms of local coordinates by the formula $\mathcal{L}_X Y(p) = [X, Y](p) = \partial_X Y(p) - \partial_Y X(p)$, where ∂_X and ∂_Y are the directional derivatives with respect to X and Y, respectively. This actually stipulates to the Lie derivative of any tensor field along the flow generated by X.

Exercise 5.1

Let the expansion scalar θ of velocity field **u** be given by

$$\theta = \nabla.\mathbf{u}.$$

Show that in an arbitrary coordinate system

$$\theta = \frac{1}{2} g^{ij} L_{\mathbf{u}} g_{ij},$$

where the Lie derivative $L_{\mathbf{u}}$ along the flow generated by the velocity field **u**.

Hints: Lie derivative for the fundamental tensor g_{ij} along the flow generated by the velocity field **u** is given by

$$L_{\mathbf{u}}g_{ij} = u^k \frac{\partial}{\partial x^k}g_{ij} + g_{ik}\frac{\partial}{\partial x^j}u^k + g_{kj}\frac{\partial}{\partial x^i}u^k.$$

$$Thus, \; g^{ij}L_{\mathbf{u}}g_{ij} = u^k g^{ij}\frac{\partial}{\partial x^k}g_{ij} + g^{ij}g_{ik}\frac{\partial}{\partial x^j}u^k + g^{ij}g_{kj}\frac{\partial}{\partial x^i}u^k$$

$$= \frac{1}{2}u^k\frac{\partial}{\partial x^k}\left(g^{ij}g_{ij}\right) + \delta^j_k\frac{\partial}{\partial x^j}u^k + \delta^i_k\frac{\partial}{\partial x^i}u^k$$

$$= \frac{1}{2}u^k\frac{\partial}{\partial x^k}(4) + \frac{\partial}{\partial x^k}u^k + \frac{\partial}{\partial x^k}u^k = 2\nabla.u$$

Therefore

$$\theta = \frac{1}{2}g^{ij}L_{\mathbf{u}}g_{ij}.$$

5.6 Killing Equation

The structure of the metric tensor implies the structure of the spacetime.

Question: Does the metric tensor $g_{\mu\nu}$ change its value under the infinitesimal coordinate transformation

$$\bar{x}^\mu = x^\mu + \epsilon\xi^\mu(x)?$$

To search the answer to this question, one has to check whether Lie derivative of $g_{\mu\nu}$ vanish or not.
A mapping of the spacetime onto itself of the form

$$\bar{x}^\mu = x^\mu + \epsilon\xi^\mu,$$

[i.e., infinitesimal transformation] is known as **isometric mapping** if the Lie derivative of the metric tensor vanishes, i.e.,

$$L_\xi g_{\mu\nu} = 0,$$

$$\Rightarrow$$

$$\xi^\rho\nabla_\rho g_{\alpha\beta} + g_{\alpha\nu}\nabla_\beta\xi^\nu + g_{\mu\beta}\nabla_\alpha\xi^\mu = 0,$$

$$\Rightarrow$$

$$\nabla_\beta\xi_\alpha + \nabla_\alpha\xi_\beta = 0 \equiv A_{\alpha\beta}.$$

The equation

$$L_\xi g_{\mu\nu} = \nabla_\beta\xi_\alpha + \nabla_\alpha\xi_\beta = 0.$$

is known as **Killing equation**. The solutions $\xi^\mu(x)$ of the Killing equation are termed as **Killing vectors (KVs)**.

KV exist $\Rightarrow \exists$ solution of Killing equations $\Rightarrow$ presence of a definite intrinsic symmetry in that spacetime.

No solution of the Killing equation $\Rightarrow$ does not exist KV $\Rightarrow$ the spacetime has no symmetry whatsoever.

Exercise 5.1

Find the KV for the metric

$$ds^2 = dx^2 + x^2 dy^2.$$

Hints:

For KV we have,

$$L_\xi g_{ij} = 0. \tag{i}$$

Here, we denote $x^1 = x$ and $x^2 = y$.

Now (i) implies

$$\frac{\partial}{\partial x^k} g_{ij}\xi^k + g_{kj}\frac{\partial}{\partial x^i}\xi^k + g_{ik}\frac{\partial}{\partial x^j}\xi^k = 0, \tag{ii}$$

where ξ^k is the KV, $k = 1, 2$.

From the metric, we have,

$$ds^2 = dx^2 + x^2 dy^2$$

$$\Rightarrow g_{11} = 1,\ g_{12} = g_{21} = 0,\ g_{22} = x^2.$$

Now for $i = j = 1$

$$\text{(ii)} \Rightarrow \frac{\partial}{\partial x^k} g_{11}\xi^k + g_{k1}\frac{\partial}{\partial x^1}\xi^k + g_{1k}\frac{\partial}{\partial x^1}\xi^k = 0,$$

$$\left(\frac{\partial}{\partial x^1} g_{11}\right)\xi^1 + \left(\frac{\partial}{\partial x^2} g_{11}\right)\xi^2 + g_{11}\frac{\partial}{\partial x^1}\xi^1 + g_{21}\frac{\partial}{\partial x^1}\xi^2 + g_{11}\frac{\partial}{\partial x^1}\xi^1 + g_{12}\frac{\partial}{\partial x^1}\xi^2 = 0.$$

Substituting the values of x^i and g_{ij}, we have

$$\frac{\partial 1}{\partial x}\xi^1 + \frac{\partial 1}{\partial y}\xi^2 + 1\frac{\partial \xi^1}{\partial x} + 0\frac{\partial \xi^2}{\partial x} + 1\frac{\partial \xi^1}{\partial x} + 0\frac{\partial \xi^2}{\partial x} = 0,$$

$$or,\quad \frac{\partial \xi^1}{\partial x} = 0,$$

$$\Rightarrow \quad \xi^1 = f(y), \tag{iii}$$

when $i = j = 2$ then (*ii*) reduces to

$$\frac{\partial}{\partial x^k} g_{22} \xi^k + g_{k2} \frac{\partial}{\partial x^2} \xi^k + g_{2k} \frac{\partial}{\partial x^2} \xi^k = 0,$$

$$\left(\frac{\partial}{\partial x^1} g_{22}\right) \xi^1 + \left(\frac{\partial}{\partial x^2} g_{22}\right) \xi^2 + g_{12} \frac{\partial}{\partial x^2} \xi^1 + g_{22} \frac{\partial}{\partial x^2} \xi^2 + g_{21} \frac{\partial}{\partial x^2} \xi^1 + g_{22} \frac{\partial}{\partial x^2} \xi^2 = 0.$$

Substituting the values of x^i and g_{ij} and ξ^1 we have

$$\left(\frac{\partial x^2}{\partial x}\right) f(y) + \left(\frac{\partial x^2}{\partial y}\right) \xi^2 + 0\frac{\partial \xi^1}{\partial y} + x^2 \frac{\partial \xi^2}{\partial y} + 0\frac{\partial \xi^1}{\partial y} + x^2 \frac{\partial \xi^2}{\partial y} = 0,$$

$$2xf(y) + 2x^2 \frac{\partial \xi^2}{\partial y} = 0,$$

$$xf(y) + x^2 \frac{\partial \xi^2}{\partial y} = 0, \tag{iv}$$

when $i = 1, j = 2$ then (ii) reduces to

$$\frac{\partial}{\partial x^k} g_{12} \xi^k + g_{k2} \frac{\partial}{\partial x^1} \xi^k + g_{1k} \frac{\partial}{\partial x^2} \xi^k = 0,$$

$$\left(\frac{\partial}{\partial x^1} g_{12}\right) \xi^1 + \left(\frac{\partial}{\partial x^2} g_{12}\right) \xi^2 + g_{12} \frac{\partial}{\partial x^1} \xi^1 + g_{22} \frac{\partial}{\partial x^1} \xi^2 + g_{11} \frac{\partial}{\partial x^2} \xi^1 + g_{12} \frac{\partial}{\partial x^2} \xi^2 = 0.$$

Substituting the values of x^i and g_{ij} and ξ^1 we have

$$x^2 \frac{\partial \xi^2}{\partial x} + f'(y) = 0, \tag{v}$$

$$\Rightarrow \frac{\partial \xi^2}{\partial x} = -\frac{f'(y)}{x^2},$$

$$\Rightarrow \xi^2 = \frac{f'(y)}{x} + g(y). \tag{vi}$$

Hence from (iii) we have

$$f(y) + x \frac{\partial \xi^2}{\partial y} = 0$$

$$\Rightarrow f(y) + x \left[\frac{f''(y)}{x} + g'(y)\right] = 0,$$

$$\Rightarrow [f(y) + f''(y)] + xg'(y) = 0.$$

$$\Rightarrow g'(y) = 0$$

and

$$f(y) + f''(y) = 0,$$

$$\Rightarrow g(y) = c,$$

and

$$f(y) = A \sin y + B \cos y \Rightarrow f'(y) = A \cos y - B \sin y.$$

Therefore,

$$\xi^1 = A \sin y + B \cos y, \quad and, \quad \xi^2 = \frac{A \cos y - B \sin y}{x} + c.$$

Exercise 5.2

Find the KV for the metric

$$ds^2 = d\theta^2 + \sin^2 \theta d\phi^2.$$

Hints:
Consider the KV for the metric

$$ds^2 = d\theta^2 + \sin^2 \theta d\phi^2,$$

as

$$\xi^i = (\xi^\theta, \xi^\phi)$$

which satisfy the Killing equations as

$$\frac{\partial \xi^\theta}{\partial \theta} = 0,$$

$$\frac{\partial \xi^\theta}{\partial \phi} + \sin^2 \theta \frac{\partial \xi^\phi}{\partial \theta} = 0,$$

$$\frac{\partial \xi^\phi}{\partial \phi} + \cot \theta \xi^\theta = 0.$$

$$1st \Rightarrow \xi^\theta = f(\phi)$$

$$2nd \Rightarrow \sin^2 \theta \frac{\partial \xi^\phi}{\partial \theta} = -f'(\phi),$$

$$\Rightarrow$$

$$\xi^\phi = f'(\phi) \cot \theta + g(\phi).$$

Using the above two results, $3rd \Rightarrow$

$$g'(\phi) + [f''(\phi) \cot \theta + f(\phi) \cot \theta] = 0.$$

Since this equation must hold $\forall \theta$ and ϕ, we must have, $g'(\phi) = 0$;

$$f''(\phi) + f(\phi) = 0,$$

$$\Rightarrow$$

$$\xi^\theta = A \sin \phi + B \cos \phi; \quad \xi^\phi = (A \cos \phi - B \sin \phi) \cot \theta + c.$$

Exercise 5.3

Let X_μ be a KV and $T^{\mu\nu}$, the energy-stress tensor, then show that the current $J^\mu = T^{\mu\nu} X_\nu$ is a conserved current, i.e., $J^\mu_{;\mu} = 0$.

Hints:

The covariant derivative of J^μ is

$$J^\mu_{;\mu} = (T^{\mu\nu} X_\nu)_{;\mu} = (T^{\mu\nu}_{;\mu}) X_\nu + T^{\mu\nu} (X_{\nu;\mu}).$$

We know stress-energy tensor is conserved, i.e., $T^{\mu\nu}_{;\mu} = 0$, therefore,

$$J^\mu_{;\mu} = T^{\mu\nu} (X_{\nu;\mu}).$$

Also stress-energy tensor is symmetric, above equation can be rewritten as

$$J^\mu_{;\mu} = \frac{1}{2}(T^{\mu\nu}(X_{\nu;\mu}) + T^{\nu\mu}(X_{\mu;\nu})) = \frac{1}{2}T^{\mu\nu}(X_{\nu;\mu} + X_{\mu;\nu}) = 0.$$

Exercise 5.4

Let $u^i = \phi \xi^i$ be a contravariant vector, where ξ^i is a time-like KV. Also ϕ is so chosen that u^i is a unit vector. Show that

$$u^i u_{k;i} = \frac{\partial}{\partial x^k}(\ln \phi).$$

Hints:

We write the given equation as

$$\xi_i = u_i \phi^{-1}.$$

Taking covariant derivative, we obtain,

$$\xi_{i;k} = u_{i;k}\phi^{-1} - u_i \phi_{,k} \phi^{-2}.$$

The Killing equation, $\xi_{i;k} + \xi_{k;i} = 0$ implies

$$(u_{i;k} + u_{k;i})\phi - (u_i\phi_{,k} + u_k\phi_{,i}) = 0. \tag{i}$$

Since, $u^i u_i = 1$, therefore, we get $u^i u_{i;k} = 0$. Now, multiplying Eq. (i) with u^i and using the above result, we get

$$\phi u^i u_{k;i} = \phi_{,k} + u^i u_k \phi_{,i}. \tag{ii}$$

Again, multiplying Eq. (ii) by u^k, we get

$$u^k \phi u^i u_{k;i} = u^k \phi_{,k} + u^k u^i u_k \phi_{,i}.$$

Using the result, $u^k u_{k;i} = 0$, we get

$$0 = u^k \phi_{,k} + u^i \phi_{,i},$$

or

$$u^i \phi_{,i} = 0.$$

Therefore, the Eq. (ii) yields

$$\phi u^i u_{k;i} = \phi_{,k}.$$

Hence the result.

Exercise 5.5

If u^i is a tangent vector to a geodesic C and ξ_i is a KV then show that $\xi_i u^i = constant\ along$ with C.
Hints: We know if **u** is a tangent vector, then rate of change of **u** along **u** is zero, i.e.,

$$\nabla_{\mathbf{u}}\mathbf{u} = 0\ or\ u^i u^k_{;i} = 0.\quad \text{(See Section 2.5)}$$

In other words, the above equation is the geodesic equation. Now the covariant directional derivative of $\xi_k u^k$ along the geodesic curve is

$$\nabla_{\mathbf{u}}(\xi_k u^k) = u^i u^k_{;i}\xi_k + u^i u^k \xi_{k;i} = u^i u^k \xi_{k;i} = 0.$$

[as $u^i u^k$ symmetric and $\xi_{k;i}$ is antisymmetric]

Exercise 5.6

Show that

$$ds^2 = dt^2 - e^{2Ht}[dr^2 + r^2 d\theta^2 + r^2 \sin^2 \theta]$$

has a time-like KV.

Hints:

We know the Killing equation

$$\xi^\rho g_{\alpha\beta,\rho} + g_{\alpha\nu}\xi^\nu_{,\beta} + g_{\mu\beta}\xi^\mu_{,\alpha} = 0.$$

Now, we assume the KV with vanishing third and fourth components without any loss of generality as $\xi^\mu = (\xi^0, \xi^1, 0, 0)$. Then using the above equation, we get,

$$\frac{\partial \xi^0}{\partial t} = 0,$$

$$\frac{\partial \xi^0}{\partial r} = \frac{\partial \xi^1}{\partial t} e^{2Ht},$$

$$\frac{\partial \xi^1}{\partial r} + H\xi^0 = 0.$$

Solutions of these equation

$$\xi^0 = 1, \quad \xi^1 = -Hr$$

Any vector ξ^μ is time-like if

$$\xi^\mu \xi_\mu > 0,$$

$$or, \ g_{\mu\nu}\xi^\mu \xi^\nu > 0,$$

$$or, \ g_{00}\xi^{0^2} + g_{11}\xi^{1^2} > 0 \ or, \ 1 - H^2 r^2 e^{2Ht} > 0.$$

Therefore, the vector ξ^μ is timelike for $H^2 r^2 e^{2Ht} < 1$.

Exercise 5.7

If ξ is a KV, prove that

$$\xi_{\mu;\alpha\beta} = R_{\gamma\beta\alpha\mu}\xi^\gamma.$$

Hints:

The definition of curvature tensor $R_{\alpha\beta\gamma\delta}$ implies that

$$\xi_{\sigma;\rho\mu} - \xi_{\sigma;\mu\rho} = R^{\lambda}_{\sigma\mu\rho}\xi_{\lambda}, \tag{i}$$

where ξ be any arbitrary vector. Also the property of curvature tensor

$$R^{\lambda}_{\sigma\rho\mu} + R^{\lambda}_{\mu\sigma\rho} + R^{\lambda}_{\rho\mu\sigma} = 0,$$

yields

$$0 = \xi_{\sigma;\rho\mu} - \xi_{\sigma;\mu\rho} + \xi_{\mu;\sigma\rho} - \xi_{\mu;\rho\sigma} + \xi_{\rho;\mu\sigma} - \xi_{\rho;\sigma\mu}. \tag{ii}$$

We know

$$\xi_{\sigma;\rho} = -\xi_{\rho;\sigma} \implies \xi_{\sigma;\rho\mu} = -\xi_{\rho;\sigma\mu},$$

therefore, (ii) implies

$$0 = \xi_{\sigma;\rho\mu} - \xi_{\sigma;\mu\rho} - \xi_{\mu;\rho\sigma}. \tag{iii}$$

Equation (*i*) and (*iii*) yield

$$\xi_{\mu;\rho\sigma} = R^{\lambda}_{\sigma\rho\mu}\xi_{\lambda}.$$

5.7 Stationary and Static Spacetimes

A spacetime is said to be **stationary** if it asserts a time-like KV field $\xi^{\mu}(x)$. Thus, the Killing equation

$$\nabla_{\mu}\xi_{\nu} + \nabla_{\nu}\xi_{\mu}(x) = 0$$

possesses a solution ξ_{μ} such that

$$\xi^2 = \xi_{\mu}\xi^{\mu} > 0.$$

It is conceivable to build world lines (trajectories) of the vector field $\xi^{\mu}(x)$ in such a way that only time coordinate x^0 changes along these trajectories whereas the spatial coordinates x^1, x^2, x^3 are not altered. This is feasible as the vector $\xi^{\mu}(x)$ is time-like. Thus, directions of these trajectories of ξ^{μ} coincide with x^0 axis (see Fig. 11).

Hence, in this new coordinate system, the spatial components of ξ^{α} are zero, i.e., $\xi^k = 0$, $k = 1, 2, 3$. Thus, $\xi^{\mu}(x) = (1, 0, 0, 0)$ is a nonzero KV, which is time-like. Now from Killing equation,

$$L_{\xi}g_{\mu\nu} \equiv \xi^{\rho}\frac{\partial g_{\mu\nu}}{\partial x^{\rho}} + g_{\mu\rho}\frac{\partial \xi^{\rho}}{\partial x^{\nu}} + g_{\rho\nu}\frac{\partial \xi^{\rho}}{\partial x^{\mu}} = 0,$$

we get

$$\frac{\partial g_{\mu\nu}}{\partial x^0} = 0.$$

This is the required condition for a spacetime to be stationary. However, in relation to black holes, stationary only requires a time-like KV in an asymptotically flat region.

A typical situation of a stationary spacetime is called **static** if the trajectories of the KV ξ^μ are orthogonal to a family of hypersurfaces.

The conditions for static spacetime are

$$\frac{\partial g_{\mu\nu}}{\partial x^0} = 0, \quad g_{0k} = 0.$$

In other words: A spacetime is said to be static if it admits a hypersurface, which has an orthogonal time-like KV field.

5.8 Spherically Symmetric Spacetime

A spacetime is said to be spherically symmetric if the Killing equation possesses three linearly independent space-like KV fields X^μ whose orbits are closed (i.e., topological circles) and obeying the following conditions

$$[X^1, X^2] = X^3, \quad [X^2, X^3] = X^1, \quad [X^3, X^1] = X^2.$$

Thus, in a spherically symmetric spacetime, a coordinate x^a exists such that the KV fields X^a take the following forms

$$X^0 = 0, \quad X^\alpha = \omega^\alpha_\beta x^\beta, \quad \omega_{\alpha\beta} = -\omega_{\beta\alpha}.$$

The quantity $\omega_{\alpha\beta}$ is characterized by three parameters, which specify three space-like rotations.

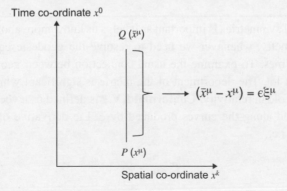

Figure 11 Direction of Killing vector along the time axis.

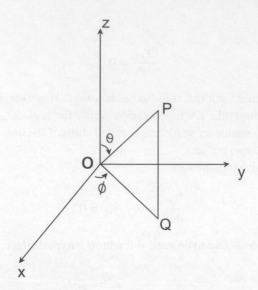

Figure 12 Spherical symmetry.

Intuitively, if a system is invariant under spatial rotations about a fixed point, called the origin O, then this is known as spherical symmetry (see Fig. 12).

5.9 Cylindrically Symmetric Spacetime (Axially Symmetry)

If a system is such that a rotation through any angle about an axis is invariant, then it is called cylindrically symmetric spacetime. The conditions for cylindrically symmetric spacetime are as follows:

(i) ∃ two KV in which one is time-like and other space-like

(ii) Orbits of space-like KV are closed.

Note 5.4

The study of conformal symmetries is important as it helps us know more about the internal structure of the spacetime geometry, whenever we need to resolve the geodesic equations of motion for the concerning spacetimes. To examine the usual connection between geometry and matter, this symmetry also helps a lot. The deportment of the metric is significant while progressed together with curves on a manifold in relativity. **Conformal KV** ξ is defined as a vector field on a manifold when a metric is pulled along the curves produced by ξ. Lie derivative of the metric is straight proportionate to itself, i.e.,

$$\mathcal{L}_\xi g_{ik} = \psi g_{ik},$$

where the scalar field ψ is termed as a conformal factor and $\mathcal{L}$ is the Lie derivative operator. The physical significance of this requirement is that when the metric is pulled along precise congruence of curves it perseveres itself modulo certain scale factor, ψ, that might fluctuate from location to location within the manifold. It is to be noticed that ψ is not haphazardly chosen rather it is dependent on the conformal KV ξ as $\psi(x^k) = \frac{1}{4}\xi^i_{;i}$ for four-dimensional Riemannian space. The vector ξ illustrates the conformal symmetry, however, the metric tensor g_{ik} is conformally haggard against itself along ξ.

Here, the conformal KV ξ are called homothetic motions or **homothetic vector** (HV) fields if ψ is constant, and for $\psi = 0$ one will get KV fields.

Exercise 5.8

Find the conformal KV for the spacetime

$$ds^2 = A(r)dt^2 - B(r)dr^2 - r^2 d\theta^2.$$

Hints: We know the conformal KV equation is

$$L_\xi g_{ij} = \psi g_{ij} \ or \ \frac{\partial}{\partial x^k}g_{ij}\xi^k + \frac{\partial}{\partial x^i}g_{kj}\xi^k + \frac{\partial}{\partial x^j}g_{ik}\xi^k = \psi g_{ij}.$$

Let us consider the conformal KV ξ as

$$\xi^\mu = (\xi^0, \xi^1, 0) = \xi^0\delta^\mu_0 + \xi^1\delta^\mu_1 \equiv \alpha(r)\delta^\mu_0 + \beta(r)\delta^\mu_1.$$

In the following, we try to find nontrivial conformal KV equations. For $i = j = 1$, we have

$$\frac{\partial}{\partial x^k}g_{ij}\xi^k + \frac{\partial}{\partial x^i}g_{kj}\xi^k + \frac{\partial}{\partial x^j}g_{ik}\xi^k = \psi g_{ij},$$

$$or, \ \frac{\partial}{\partial r}g_{11}\xi^1 + 2g_{11}\frac{\partial}{\partial x^1}\xi^1 = \psi g_{11},$$

$$or, \ B'(r)\beta + 2B(r)\beta' = \psi B(r).$$

Again for $i = j = 0$, we have

$$\frac{\partial}{\partial r}g_{00}\xi^1 = -\psi A(r), \ or, \ A'(r)\beta = \psi A(r).$$

Using $i = j = 2$, we get

$$\frac{\partial}{\partial r}g_{22}\xi^1 = -\psi g_{22}, \ or, \ \beta = \frac{\psi r}{2}.$$

$i = 0, j = 1$ yields

$$g_{00} \frac{\partial \xi^0}{\partial r} = 0,$$

$$or, \ \xi^0 = \alpha = \ constant \ = C_1 \ (say).$$

Eliminating ψ, we get

$$B'(r)\beta + 2B(r)\beta' = \frac{2B(r)}{r}\beta.$$

Solving this equation, we get

$$\beta = \frac{rk}{\sqrt{B(r)}},$$

k is an integration constant.

Using $\beta = \frac{\psi r}{2}$, we get the expression of ψ as

$$\psi = \frac{2k}{\sqrt{B(r)}}.$$

Also we get the metric coefficient after integrating the above equation,

$$A(r) = r^2 C^2, \ C = \ integration \ constant.$$

Exercise 5.9

Find the conformal KV for the spacetime

$$ds^2 = A(r)dt^2 - B(r)dr^2 - r^2(d\theta^2 + \sin^2\theta d\phi^2).$$

Hint:

$$Here, \ g_{00} = A(r), \ g_{11} = -B(r), \ g_{22} = -r^2, \ g_{33} = -r^2 \sin^2\theta.$$

Let the conformal KV be

$$\xi^\mu = (\xi^0, \xi^1, 0, 0) = \alpha(r)\delta_0^\mu + \beta(r)\delta_1^\mu, \ where, \xi^0 = \alpha(r), \xi^1 = \beta(r).$$

In the following, we search for nontrivial conformal KV equations as follows:
For $i = j = 0$, the conformal KV equations yield

$$\xi^1 \left(\frac{\partial g_{00}}{\partial x^1} \right) = \psi g_{00}, \ or \ \beta A' = \psi A.$$

For $i = j = 1$, we have

$$\xi^1 \left(\frac{\partial g_{11}}{\partial x^1} \right) + 2g_{11} \frac{\partial \xi^1}{\partial x^1} = \psi g_{11}, \ \ or \ \beta B' + 2B\beta' = \psi B.$$

For $i = j = 2$, we get

$$\xi^1 \left(\frac{\partial g_{22}}{\partial x^1} \right) = \psi g_{22}, \ \ or \ 2r\beta = \psi r^2, \ \ or \ \beta = \frac{\psi r}{2}.$$

$i = j = 3$ gives the same result as $i = j = 2$

$$\xi^1 \left(\frac{\partial g_{33}}{\partial x^1} \right) = \psi g_{33}, \ \ or \ 2r \sin^2 \theta \beta = \psi r^2 \sin^2 \theta, \ or \ \beta = \frac{\psi r}{2}.$$

From the above equations, we get

$$\beta = \frac{rk}{\sqrt{B}}, \ \psi = \frac{2k}{\sqrt{B}}, \ A = r^2 C^2.$$

[k, C are integration constants]

The case $i = 1, j = 0$ yields

$$g_{00} \frac{\partial \xi^0}{\partial x^1} = \psi g_{01} = 0 \ or \ \xi^0 = \alpha = constant.$$

Exercise 5.10

Find the nonstatic conformal KV for the spacetime

$$ds^2 = e^{\nu(r)} dt^2 - e^{\lambda(r)} dr^2 - r^2 (d\theta^2 + \sin^2 \theta d\phi^2).$$

Hint: Here, one takes four vectors ξ as nonstatic but conformal factor, ψ is static as

$$\xi = \alpha(t, r) \partial_t + \beta(t, r) \partial_r,$$

$$\psi = \psi(r)$$

The above equations give the following set of expressions from $\mathcal{L}_\xi g_{ik} = \psi g_{ik}$

$$\alpha = A + \frac{1}{2}kt,$$

$$\beta = \frac{1}{2}Bre^{-\frac{\lambda}{2}},$$

$$\psi = Be^{-\frac{\lambda}{2}},$$

$$e^\nu = C^2 r^2 exp\left[-2kB^{-1}\int \frac{e^{\frac{\lambda}{2}}}{r}dr\right],$$

where C, k, A, B are constants.

Spacetimes of Spherically Symmetric Distribution of Matter and Black Holes

6.1 Spherically Symmetric Line Element

Spherically symmetric means an invariance under any arbitrary rotation of axes at a particular point, called the center of symmetry. Using θ and ϕ (polar coordinates) and choosing the center of symmetry at origin, we have the general form of the line element with spherical symmetry.

$$ds^2 = A(r,t)dt^2 + 2H(r,t)drdt - B(r,t)dr^2 - F(r,t)(d\theta^2 + \sin^2\theta d\phi^2). \qquad (6.1)$$

For the surfaces $r = $ constant and $t = $ constant, the line elements reduces to form two spheres on which a typical point is labeled by coordinate θ and ϕ and line element takes the form

$$ds^2 = d\theta^2 + \sin^2\theta d\phi^2. \qquad (6.2)$$

This spherical symmetric line element is invariant when θ and ϕ are varied. The center of symmetry is the point O, which is given by $r = 0$.

Now we introduce new coordinates by the transformations:

$$r = r', \quad t = K(r', t'), \qquad (6.3)$$

where the function K will be chosen later.

From the above transformation equations, we have

$$dr = dr'; \; dt = \frac{\partial K}{\partial r'}dr' + \frac{\partial K}{\partial t'}dt'. \qquad (6.4)$$

Then the line element becomes

$$ds^2 = A\left(\frac{\partial K}{\partial r'}dr' + \frac{\partial K}{\partial t'}dt'\right)^2 + 2Hdr'\left(\frac{\partial K}{\partial r'}dr' + \frac{\partial K}{\partial t'}dt'\right) - Bd(r')^2 - F(d\theta^2 + \sin^2\theta d\phi^2).$$

Now we choose K such that coefficient of $dr'dt'$ is zero.

Thus, we have

$$A\frac{\partial K}{\partial r'} + H = 0. \qquad (6.5)$$

Hence we get general line element on

$$ds^2 = e^\sigma dt'^2 - e^\omega dr'^2 - e^\mu(d\theta^2 + \sin^2\theta d\phi^2),$$

where σ, ω, and μ are functions of r' and t'.

Now we take another transformation,

$$R = e^{\frac{\mu}{2}} \text{ and } t' = q(R, T), \tag{6.6}$$

where q is so chosen that the coefficient of $dR\, dT$ in ds^2 is zero.

Thus, finally, we get the line element

$$ds^2 = e^\nu dT^2 - e^\lambda dR^2 - R^2(d\theta^2 + \sin^2\theta d\phi^2), \tag{6.7}$$

where ν and λ are functions of R and T.

Note 6.1

Here R coordinate has specific significance:

The area of the surface of the sphere, $R = $ constant is given by $A = 4\pi R^2$.

$$A = \int_{sphere} (Rd\theta)(R\sin\theta d\phi) = \int_0^\pi d\theta \int_0^{2\pi} d\phi R^2 \sin\theta = 4\pi R^2.$$

Three volume of the sphere with radius R

$$V = \int_{sphere} (e^{\frac{\lambda}{2}} dr)(r d\theta)(r \sin\theta d\phi) = \int_0^R dr \int_0^\pi d\theta \int_0^{2\pi} d\phi r^2 e^{\frac{\lambda}{2}} \sin\theta.$$

e.g., for $e^\lambda = (1 - ar^2)^2$, one can get

$$V = \frac{4\pi R^3}{3}\left(1 - \frac{3aR^2}{5}\right).$$

For the four-dimensional tube that is bounded by the sphere with radius R and two planes, $t = $ constant, separated by a time T, the four volume is

$$V_4 = \int_{tube} (e^{\frac{\nu}{2}} dt)(e^{\frac{\lambda}{2}} dr)(rd\theta)(r\sin\theta d\phi) = \int_0^T dt \int_0^R dr \int_0^\pi d\theta \int_0^{2\pi} d\phi r^2 e^{\frac{\nu}{2}} e^{\frac{\lambda}{2}} \sin\theta.$$

e.g., for $e^\lambda = e^\nu = (1 - ar^2)^2$, one can get

$$V_4 = \frac{4\pi R^3 T}{3}\left(1 - \frac{6aR^2}{5} + \frac{3a^2R^4}{7}\right).$$

6.2 Schwarzschild Solution or Exterior Solution

The exact solution of the Einstein field equation in empty space was obtained by Schwarzschild in 1916, which describes the geometry of spacetime outside a spherically symmetric distribution of matter.

Consider the metric of the empty spacetime outside of a spherically symmetric distribution of matter of mass M as

$$ds^2 = e^\nu dt^2 - e^\lambda dr^2 - r^2 d\theta^2 - r^2 \sin^2\theta d\phi^2, \tag{6.8}$$

where λ and ν are functions of r alone.

A metric of this type is known as spherically symmetric.

Let us write

$$g_{11} = -e^\lambda; \quad g_{22} = -r^2; \quad g_{33} - r^2\sin^2\theta; \quad g_{44} = e^\nu,$$

$$then, \quad g^{11} = -e^{-\lambda}; \quad g^{22} = -\frac{1}{r^2}; \quad g^{33} = -\frac{1}{r^2\sin^2\theta}; \quad g^{44} = e^{-\nu}.$$

The nonvanishing Christoffel symbols of second kind are

$$\Gamma^1_{11} = \frac{1}{2}\lambda'; \Gamma^2_{12} = \frac{1}{r}; \Gamma^3_{31} = \frac{1}{r}, \Gamma^1_{22} = -re^{-\lambda}, \Gamma^2_{33} = -\sin\theta\cos\theta,$$

$$\Gamma^3_{23} = \cot\theta; \Gamma^1_{33} = -r\sin^2\theta e^{-\lambda}; \Gamma^1_{44} = \frac{1}{2}\nu' e^{-\lambda+\nu}; \Gamma^4_{14} = \frac{1}{2}\nu'$$

$$\left[\Gamma^a_{aa} = \left[\ln\sqrt{|g_{aa}|}\right]_{,a} ; \Gamma^a_{ba} = \left[\ln\sqrt{|g_{aa}|}\right]_{,b} ; \Gamma^b_{aa} = -\frac{1}{2g_{bb}}g_{aa,b}\right],$$

where $A_{ab,c} \equiv \frac{\partial}{\partial x^c}A_{ab}$.

The complements of the Ricci tensor are

$$R_{11} = -\frac{\lambda'}{r} - \frac{1}{4}\lambda'\nu' + \frac{1}{2}\nu'' + \frac{1}{4}\nu'^2, \tag{6.9}$$

$$R_{22} = \csc^2\theta R_{33} = -1 + e^{-\lambda}\left[1 - \frac{1}{2}r\lambda' + \frac{1}{2}r\nu'\right], \tag{6.10}$$

$$R_{44} = e^{-\lambda+\nu}\left[\frac{1}{4}\lambda'\nu' - \frac{1}{2}\nu'' - \frac{1}{r}\nu' - \frac{1}{4}\nu'^2\right], \tag{6.11}$$

$$R_{\alpha\beta} = 0, \ \alpha \neq \beta. \tag{6.12}$$

The Ricci scalar takes the form

$$R = g^{ik}R_{ik} = \frac{2}{r^2} + e^{-\lambda}\left[-\frac{2}{r^2} + \frac{2}{r}\lambda' + \frac{1}{2}\lambda'\nu' - \nu'' - \frac{2}{r}\nu' - \frac{1}{2}\nu'^2\right]. \tag{6.13}$$

The gravitational field equations for the spherically symmetric metric in vacuum are

$$-\frac{1}{r^2} + e^{-\lambda}\left(\frac{1}{r^2} + \frac{v'}{r}\right) = 0, \tag{6.14}$$

$$e^{-\lambda}\left[-\frac{1}{2r}\lambda' - \frac{1}{4}\lambda'v' + \frac{1}{2}v'' + \frac{1}{2r}v' + \frac{1}{4}v'^2\right] = 0, \tag{6.15}$$

$$-\frac{1}{r^2} + e^{-\lambda}\left[\frac{1}{r^2} - \frac{1}{r}\lambda'\right] = 0. \tag{6.16}$$

From Eq. (6.16), we get

$$e^{-\lambda} = 1 - \frac{m}{r}. \tag{6.17}$$

Here, the constant of integration m is related to the rest mass of the gravitating body. Subtracting (6.16) from (6.14), one can find

$$\frac{e^{-\lambda}(v' + \lambda')}{r} = 0,$$

i.e.,

$$v' + \lambda' = 0.$$

From which we have

$$\lambda + v = k, \tag{6.18}$$

where k is a constant of integration.

Thus,

$$e^v = e^k\left(1 - \frac{m}{r}\right). \tag{6.19}$$

One can find the constant k using the boundary condition that at a large distance from the source $r \longrightarrow \infty$, the spacetime becomes Minkowski space, i.e., $e^v \longrightarrow 1$ *and* $e^\lambda \longrightarrow 1$. Therefore, k should be equal to zero.

Thus, we obtain the Schwarzschild solutions as

$$e^v = e^{-\lambda} = \left(1 - \frac{m}{r}\right). \tag{6.20}$$

Here, one can take $m = \frac{2GM}{c^2}$, where M denotes the mass of the body under consideration.

Thus, the required Schwarzschild line element

$$ds^2 = \left(1 - \frac{2GM}{c^2 r}\right)dt^2 - \left(1 - \frac{2GM}{c^2 r}\right)^{-1} dr^2 - r^2(d\theta^2 + \sin^2\theta d\phi^2). \tag{6.21}$$

Note 6.2

For $r >> GM$, Schwarzschild metric represents Newtonian gravitation in the weak field approximation. Here, we may recognize the Newtonian potential ϕ as $\phi = \frac{1}{2}(e^v - 1)$. Since

$$g_{00} = 1 - \frac{2\phi}{c^2}, \quad c \equiv 1,$$

we have

$$\phi \sim \frac{GM}{r}. \tag{6.22}$$

Hence, one may identify the constant M with the gravitational mass of the system located within the radius r.

Note 6.3

One can note that the metric becomes unusual for $\frac{2GM}{c^2 r} = 1$. The limiting radius $r = R_S = \frac{2GM}{c^2}$ is known as **Schwarzschild radius** of the given mass M of a body. For sun, the Schwarzschild radius is 3 km, whereas for earth it is 1 cm.

$$[M_{sun} = 1.99 \times 10^{30} kg, \ M_{earth} = 5.972 \times 10^{24} kg, \ G = 6.66 \times 10^{-11} \ m^3 \ kg \ sec^2.]$$

Notice that the actual radius of the sun is $\cong 7 \times 10^5$ km and for earth $R_E = 6000$ km. In fact, the radius of a physical object is much higher than its Schwarzschild radius.

The hypothetical objects whose radius equals their Schwarzschild radius are known as **black holes**.

Note 6.4

The stellar evolution theory states that a star with equivalent sun's mass can achieve its final equilibrium state as a white dwarf or a neutron star. However, it is not possible to achieve this equilibrium for bigger mass stars. In this situation, the internal pressure and stresses are not able to cope with the inward pressure due to gravity, and as a result, stars gradually contract further. According to general relativity, a spherically symmetric star will inevitably contract till the entire matter confined within the star reaches a singularity at the center of symmetry. Actually, if the spherically symmetric nonrotating star starts collapsing, it will continue until the surface of the star approaches its Schwarzschild radius.

A Classical Argument

The concept of a black hole, in a lucid sense, is that of a star whose gravitational field is very powerful, so that even light cannot seep to different areas. In the Newtonian regime, let us assume a particle of mass m traveling radially from a spherically symmetric distribution of matter having total mass M, radius R, and uniform density ρ. Let us also assume that at a distance r from the center, the particle possesses a velocity v, then the conservation of energy E yields

$$E = K.E. + P.E. = \frac{1}{2}mv^2 - \frac{GMm}{r}.$$ (i)

The escape velocity v_0 is the minimum velocity desired to take a body from the surface of the distribution of matter to infinity. This demands $v \to 0$ as $r \to \infty$. Hence from (i) we have $E = 0$. Solving for v, we obtain, $v^2 = \frac{2GM}{r}$; therefore, the escape velocity is

$$v_0^2 = \frac{2GM}{R}.$$ (ii)

This indicates that when the particle's radial velocity at the surface is less than v_0, it will ultimately come back to the surface of the distribution of matter due to the gravitational attraction. For the case of a light ray (with velocity c), which would just escape to infinity, one gets a relation between c with mass and radius of the distribution as

$$c^2 = \frac{2GM}{R}.$$ (iii)

Thus, we have two possibilities for which light could no longer escape: either when the mass M is increased with a fixed radius or the radius R is decreased with fixed mass. Note that the limiting condition (iii) yields the radius R as

$$R = \frac{2GM}{c^2},$$ (iv)

which is known as **Schwarzschild radius**.

Construction of Black Hole from a Human Being — A Toy Model

Now we convince a man named Kajal to transform him to a black hole. Let us consider the mass of Kajal is 60 kg. He would become a black hole if one can compress him to a body of radius

$$R_S = \frac{2 \times 6.67 \times 10^{-11} \times 60}{(3 \times 10^8)^2} \, m = 88.9 \times 10^{-33} \, m.$$

Now, Kajal is a **living black hole** and his density will become

$$d = \frac{M}{V} = 2 \times 10^{94} \, kg/m^3.$$

If one wants to change the earth to a black hole, one has to compress the earth of a body of radius just 0.9 cm.

It is not possible to observe a black hole directly unless one has a great fortune to see a star disappear. However, observers search for double stars with one unseen companion. It is believed that a black hole extracts matter from its companion and as a result, an accretion disc surrounding the black hole is formed. The hot inner portions will generate powerful explosions of X-rays made by synchrotron radiation soon before the spiraling matter falls back into the hole.

In 1971, telescopes in the Uhuru satellite first observed the quick variants of the X-ray source Cygnus X-1 and this is the first indication of the supportable presence of black holes. A specific study of the visible component of the supergiant star and X-ray source confirms that the unseen distribution of matter is a compact object whose mass is greater than $9M_\odot$. It is believed that white dwarfs and neutron stars can have at most $1.4M_\odot$ and $9M_\odot$ masses, respectively. Hence, one could decide that the object is a black hole. After 1971, a large quantity of new black hole contenders have been observed in X-ray binaries.

Note 6.5

One can see that the line element develops a singularity at $r = \frac{2GM}{c^2}$. In fact, this singularity is not really a singularity of the gravitational field. It occurs entirely due to the specific choice of our coordinate system. Actually, if one writes the Schwarzschild line element in a different coordinate system (U, r, θ, ϕ) with

$$dt = dU + \frac{r^2 dr}{r^2 - \frac{2GMr}{c^2}}, \tag{6.23}$$

then

$$ds^2 = \left(1 - \frac{2GM}{rc^2}\right) dU^2 + 2dU dr - r^2 d\theta^2 - r^2 \sin^2 \theta d\phi^2. \tag{6.24}$$

Note that this redefined Schwarzschild line element does not show any singularity at $r = \frac{2GM}{c^2}$.

Note 6.6

For the following coordinate transformation

$$T = t + f(r), \quad where \quad \frac{df}{dr} = \left(1 - \frac{2m}{r}\right)^{-1} \sqrt{\frac{2m}{r}},$$

the Schwarzschild metric assumes the following form in (T, r, θ, ϕ) coordinates as

$$ds^2 = dT^2 - \left[dr + \sqrt{\frac{2m}{r}} dT\right]^2 - r^2(d\theta^2 + \sin^2 \theta d\phi^2).$$

This is the famous **Painlevé–Gullstard form** of Schwarzschild metric.
Here, the three-dimensional hypersurface, $r = 2m$, yields

$$ds^2_{r=2m} = (2m)^2(d\theta^2 + \sin^2 \theta d\phi^2).$$

The event horizon, $r = 2m$ is a null hypersurface. The Painlevé–Gullstard metric is regular for $r > 0$.

Note 6.7

The **event horizon** satisfies the horizon equation $g^{ab}\partial_a F\partial_b F = 0$, where g_{ab} is the spacetime metric and $F(x^a)$ is a level surface (hypersurface) function with normal $n_a = \partial_a F$. Actually, the event horizon is defined as a **null surface** (a surface in spacetime whose normal vector is everywhere null). For spherically symmetric metric, $g^{ab}\partial_a F\partial_b F = g^{ab}n_a n_b = g^{rr} = 0$ gives event horizon (the surface defined by $F = r - r_0 = 0$ has a normal $n_\mu = \delta_r^\mu$, i.e., $n_\mu = (0,1,0,0)$). In 1965, Roger Penrose defined a new idea known as trapped surface. A **trapped surface** is described by the inner region of an event horizon of a black hole. It is one where light is not moving away from the black hole. We define **apparent horizon** as the boundary of the union of all trapped surfaces around a black hole. The term trapped surfaces are used only when the null vector fields give rise to null surfaces. However, by **marginally trapped surfaces** we mean that it may be space-like, time-like, or null.

6.3 Vacuum Solution or Exterior Solution with Cosmological Constant

The gravitational field equations with cosmological constant ($G_{\mu\nu} = \Lambda g_{\mu\nu}$) for the spherically symmetric metric

$$ds^2 = e^{\nu(r)}dt^2 - e^{\lambda(r)}dr^2 - r^2 d\theta^2 - r^2 \sin^2\theta d\phi^2, \tag{6.25}$$

in vacuum are,

$$-\frac{1}{r^2} + e^{-\lambda}\left(\frac{1}{r^2} + \frac{\nu'}{r}\right) = -\Lambda, \tag{6.26}$$

$$e^{-\lambda}\left[-\frac{1}{2}\lambda' - \frac{1}{4}\lambda'\nu' + \frac{1}{2}\nu'' + \frac{1}{2r}\nu' + \frac{1}{4}\nu'^2\right] = -\Lambda, \tag{6.27}$$

$$-\frac{1}{r^2} + e^{-\lambda}\left[\frac{1}{r^2} - \frac{1}{r}\lambda'\right] = -\Lambda. \tag{6.28}$$

From the last equation, we get

$$e^{-\lambda} = 1 - \frac{m}{r} - \frac{1}{3}\Lambda r^2. \tag{6.29}$$

Here, the constant of integration m is related to the rest mass of gravitating body and can be taken as $m = \frac{2MG}{c^2}$. Subtracting last equation from first, we obtain

$$\frac{e^{-\lambda}(\nu' + \lambda')}{r} = 0, \quad i.e., \quad \nu' + \lambda' = 0.$$

From which we have

$$\lambda + \nu = k, \tag{6.30}$$

where k is a constant of integration.

$$\textit{Thus,} \quad e^\nu = e^k \left(1 - \frac{2GM}{rc^2} - \frac{1}{3}\Lambda r^2 \right). \tag{6.31}$$

Exercise 6.1

Prove that $m = \frac{2GM}{c^2}$, where m is the constant appearing in the Schwarzschild solution for a gravitating body of mass M.

Hint: For large r, the Schwarzschild metric represents Newtonian gravitation in the weak field approximation. Here, we may recognize the Newtonian potential ϕ as $\phi = \frac{1}{2}(e^\nu - 1)$.

$$\textit{Since,} \quad g_{00} = 1 - \frac{m}{r} = 1 + \frac{2\phi}{c^2},$$

$$\textit{we have,} \quad \phi = -\frac{mc^2}{2r}.$$

We know gravitational field intensity $g = -\frac{GM}{r^2}$ is related to gravitational potential ϕ as

$$g = -\nabla \phi,$$

$$\textit{or,} \quad -\frac{GM}{r^2} = -\frac{\partial \phi}{\partial r} = -\frac{mc^2}{2r^2}.$$

This yields

$$m = \frac{2GM}{c^2}.$$

6.4 Birkhoff's Theorem

In 1923, G. D. Birkhoff proposed and proved an important proposal in general relativity: any spherically symmetric solution of the vacuum field equations must be static and asymptotically flat. This proposal is known as Birkhoff's theorem. More precisely the statement of this theory is as follows:

The spherically symmetric vacuum spacetime has a Schwarzschild solution even if the metric is not explicitly assumed to be static (i.e., outside the spherical objects, the metric is static).

Proof: Consider the metric of the empty spacetime outside of a spherically symmetric distribution of matter of mass M as

$$ds^2 = e^\nu dt^2 - e^\lambda dr^2 - r^2 d\theta^2 - r^2 \sin^2\theta d\phi^2, \tag{6.32}$$

where λ and v are functions of r and t. Now, we write down the field equation for the above metric as

$$G_0^0 = e^{-\lambda}\left(\frac{1}{r^2} - \frac{\lambda'}{r}\right) - \frac{1}{r^2} = 0, \tag{6.33}$$

$$G_1^1 = e^{-\lambda}\left(\frac{v'}{r} + \frac{1}{r^2}\right) - \frac{1}{r^2} = 0, \tag{6.34}$$

$$G_2^2 = G_3^3 = -\frac{1}{2}e^{-\lambda}\left(v'' + \frac{v'^2}{2} + \frac{(v' - \lambda')}{r} - \frac{\lambda'v'}{2}\right)$$

$$+ \frac{1}{2}e^{-v}\left(\ddot{\lambda} + \frac{1}{2}\dot{\lambda}^2 - \frac{\dot{\lambda}\dot{v}}{2}\right) = 0, \tag{6.35}$$

$$G_4^1 = -\frac{1}{2}e^{-\lambda}\frac{\dot{\lambda}}{r} = 0. \tag{6.36}$$

The field Eq. (6.36) shows that λ is a function of radial coordinate r only.
From (6.33) and (6.34), we get

$$v' = -\lambda' \Rightarrow v = -\lambda + f(t).$$

Here, $f(t)$ is an arbitrary function of t. Hence, we get

$$e^v = e^{f(t)}e^{-\lambda}. \tag{6.37}$$

Now, consider the transformation of the time coordinate

$$t' = \int e^{\frac{f(t)}{2}}\, dt. \tag{6.38}$$

This transformation does not affect the spatial coordinate. Therefore, we again get Schwarzschild static solution

$$g_{00} = (g_{11})^{-1} = \left(1 - \frac{m}{r}\right) = \left(1 - \frac{2GM}{c^2 r}\right). \tag{6.39}$$

Therefore, we have an obvious conclusion that the spherically symmetric gravitational field in a vacuum is necessarily static and of Schwarzschild form.

Note 6.8

According to Birkhoff's theorem, the spacetime of the inside of a self-gravitating hollow sphere is Schwarzschild spacetime. This implies

$$g_{00} = (g_{11})^{-1} = \left(1 - \frac{m}{r}\right).$$

Two items should be noticed here: point $r = 0$ is not a singularity and the mass m is actually an integration constant of the solution of Einstein equations. So, in this case, we should take the integration constant, m to be zero in the metric to avoid the singularity. Hence, the metric in the hollow sphere is a flat space metric, i.e., Minkowski metric. Hence, **a particle experiences no force inside a hollow sphere**.

Show that the following metric

$$ds^2 = dt^2 - \frac{4}{9} \left[\frac{9m}{2(r-t)} \right]^{\frac{2}{3}} dr^2 - \left[\frac{9m}{2}(r-t)^2 \right]^{\frac{2}{3}} (d\theta^2 + \sin^2\theta d\phi^2),$$

is actually static Schwarzschild metric though apparently, it looks nonstatic. This is known as **Lemaitre Coordinates**.

Hint: Let us take the following coordinate transformation

$$R = \left[\frac{9m}{2}(r-t)^2 \right]^{\frac{1}{3}}.$$

$$This\ implies, \quad dt = dr - \sqrt{\frac{R}{2m}}dR.$$

Now, put the value of dt in the given metric to yield

$$ds^2 = \left(1 - \frac{2m}{R}\right) dr^2 - 2\sqrt{\frac{R}{2m}}drdR + \frac{R}{2m}dR^2 - R^2(d\theta^2 + \sin^2\theta d\phi^2).$$

Now, we assume another transformation as

$$r = T + F(R),$$

where the function $F(R)$ is to be determined later. Substituting this in the above equation, we get

$$ds^2 = \left(1 - \frac{2m}{R}\right)(dT^2 - 2F'dTdR + F'^2dR^2) - 2\sqrt{\frac{R}{2m}}dR(dT + F'dR)$$

$$+ \frac{R}{2m}dR^2 - R^2(d\theta^2 + \sin^2\theta d\phi^2).$$

Now, we choose $F(R)$ such that the coefficient of $dR\,dT$ is zero, i.e.,

$$\left(1 - \frac{2m}{R}\right)F' = \sqrt{\frac{R}{2m}}.$$

Plugging the value of F' in the above metric, one can obtain

$$ds^2 = \left(1 - \frac{2m}{R}\right) dT^2 - \left(1 - \frac{2m}{R}\right)^{-1} dR^2 - R^2(d\theta^2 + \sin^2\theta d\phi^2).$$

This is actually static Schwarzschild metric.

6.5 Schwarzschild Interior Solution

We have seen, above, the exterior solution of a spherically symmetric body. Now we try to obtain the gravitational field in the interior region of a spherically symmetric body. For mathematical simplicity, we assume that the spherically symmetric body is in a static state and contains incompressible perfect fluid (ideal fluid). For perfect fluid, the energy-stress tensor can be expressed as

$$T_{\mu\nu} = -pg_{\mu\nu} + (p+\rho)\frac{dx^\mu}{ds}\frac{dx^\nu}{ds} = -pg_{\mu\nu} + (p+\rho)v^\mu v^\nu.$$

We can write the mixed tensor form by raising the index as

$$T^\nu_\mu = (\rho+p)g_{\alpha\mu}\frac{dx^\alpha}{ds}\frac{dx^\nu}{ds} - g^\nu_\mu p. \tag{6.40}$$

As the body is in static state, therefore, all velocity components of fluid matter contained in it must be zero, i.e.,

$$\frac{dr}{ds} = \frac{d\theta}{ds} = \frac{d\phi}{ds} = 0. \tag{6.41}$$

Now from the static spherically symmetric metric

$$ds^2 = e^\nu dt^2 - e^\lambda dr^2 - r^2(d\theta^2 + \sin^2\theta d\phi^2), \tag{6.42}$$

one gets,

$$\frac{dt}{ds} = e^{-\nu/2}. \tag{6.43}$$

The explicit form of energy-stress tensor T^ν_μ for ideal fluid is

$$T^\nu_\mu = \begin{pmatrix} \rho & 0 & 0 & 0 \\ 0 & -p & 0 & 0 \\ 0 & 0 & -p & 0 \\ 0 & 0 & 0 & -p \end{pmatrix}, \; i.e., \quad T^1_1 = T^2_2 = T^3_3 = -p \; and \; T^0_0 = \rho. \tag{6.44}$$

The Einstein field equations are

$$e^{-\lambda}\left(\frac{1}{r^2} - \frac{\lambda'}{r}\right) - \frac{1}{r^2} = -k\rho, \tag{6.45}$$

$$e^{-\lambda}\left(\frac{1}{r^2} + \frac{\nu'}{r}\right) - \frac{1}{r^2} = kp, \tag{6.46}$$

$$\frac{e^{-\lambda}}{2}\left(\nu'' + \frac{1}{2}\nu'^2 + \frac{(\nu' - \lambda')}{r} - \frac{1}{2}\lambda'\nu'\right) = kp. \tag{6.47}$$

Conservation equation $T^\nu_{\mu;\nu} = 0$ gives

$$p' = -\frac{1}{2}\nu'(p + \rho). \tag{6.48}$$

6.6 The Tolman–Oppenheimer–Volkoff Equation

The Tolman–Oppenheimer–Volkoff (TOV) equation is a constraint equation for constructing a spherically symmetric body of isotropic material, which is in static gravitational equilibrium.

After integrating, Eq. (6.45) yields

$$e^{-\lambda} = 1 - \frac{2m(r)}{r}, \tag{6.49}$$

where

$$m(r) = 4\pi G \int_0^r \rho r^2 dr, \tag{6.50}$$

i.e.,

$$\frac{dm}{dr} = 4\pi G r^2 \rho. \tag{6.51}$$

Eqs. (6.45) and (6.46)$\Longrightarrow$

$$8\pi G(p + \rho) = \frac{e^{-\lambda}}{r}(\lambda' + \nu'). \tag{6.52}$$

Eliminating λ from Eq. (6.52) and using Eq. (6.49), we get

$$8\pi G(p + \rho) = \left(1 - \frac{2m}{r}\right)\frac{\nu'}{r} + \frac{1}{r}\left(8\pi G\rho r - \frac{2m}{r^2}\right).$$

Again, putting the value of ν' from Eq. (6.48), we finally get **TOV equation** as

$$\frac{dp}{dr} = -\frac{(p + \rho)(4\pi G p r^3 + m)}{r(r - 2m)}. \tag{6.53}$$

We have to integrate Eq. (6.53) to find the interior solution of the spherically symmetric object. To find the exact solution, we will have to consider an additional equation connecting p and ρ. This equation helps us construct the stellar model.

Note 6.9

For general anisotropic matter distribution, the energy-momentum tensor compatible with spherical symmetry is given by

$$T_{\mu\nu} = (\rho + p_t)U_\mu U_\nu - p_t g_{\mu\nu} + (p_r - p_t)\chi_\mu \chi_\nu,$$

where $U^\mu = \sqrt{\frac{1}{g_{tt}}}\delta^\mu_t$ is the four velocity normalized in such a way that $g_{\mu\nu}U^\mu U^\nu = 1$ and $\chi^\mu = \sqrt{\frac{1}{g_{rr}}}\delta^\mu_r$ is the unit space-like vector in the radial direction, i.e., $g_{\mu\nu}\chi^\mu\chi^\nu = -1$. $\rho(r)$ is the energy density and p_r is the radial pressure measured in the direction of the space-like vector. p_t is the transverse pressure in the orthogonal direction to p_r. Or, equivalently, one can obtain as

$$T^\nu_\mu = diag(\rho, -p_r, -p_t, -p_t).$$

Then the **generalized TOV** equation for general anisotropic matter distribution reads as

$$\frac{dp_r}{dr} = -(\rho + p_r)\left[\frac{m(r) + 4\pi r^3 p_r}{r\{r - 2m(r)\}}\right] + \frac{2}{r}\left(p_t - p_r\right),$$

where

$$m(r) = \int_0^r 4\pi r^2 \rho\, dr.$$

The **generalized TOV equation for general anisotropic charged distribution of matter** can be written as

$$\frac{dp_r}{dr} = -(\rho + p_r)\left[\frac{m(r) + 4\pi r^3 p_r}{r\{r - 2m(r)\}}\right] + \frac{2}{r}\left(p_t - p_r\right) + \sigma\frac{q}{r^2}e^{\frac{\lambda}{2}},$$

where total charge $q(r)$ with proper charge density $\sigma(r)$ of the source is

$$q(r) = 4\pi\int_0^r \sigma(r)e^{\frac{\lambda(r)}{2}}r^2 dr,$$

and the electric field,

$$E(r) = \frac{q(r)}{r^2}.$$

The last TOV equation can be rewritten as

$$-\frac{M_G\left(\rho + p_r\right)}{r^2}e^{\frac{\lambda-\nu}{2}} - \frac{dp_r}{dr} + \sigma\frac{q}{r^2}e^{\frac{\lambda}{2}} + \frac{2}{r}\left(p_t - p_r\right) = 0,$$

$$or, \quad F_g + F_h + F_e + F_a = 0,$$

where the **effective gravitational mass** derived from the Tolman-Whittaker formula and is given by the expression

$$M_G(r) = \frac{1}{2}r^2 e^{\frac{\nu-\lambda}{2}}\nu'.$$

The above equation implies the equilibrium condition for charged fluid elements subject to gravitational (F_g), hydrostatic (F_h), and electric forces (F_e), plus another force due to pressure anisotropy (F_a).

Note 6.10

For different equations of state, one can get different static spherically symmetric stellar models that form a one-parameter system where the parameter is the central density.

When $m(r)$, $p(r)$, and $\rho(r)$ are known, then the surface of star is the distance from the center, where $p(r = R) = 0$, R being the radius of the star.

6.7   The Structure of Newtonian Star

Now, we try to find Newtonian limit of the TOV equation. In Newtonian circumstances, we take $p << \rho$, therefore $4\pi G r^3 p << m$. Moreover, in Newtonian limit the metric is nearly Minkowskian, therefore, in Eq. (6.49), we have $m << r$. These inequalities help to simplify Eq. (6.53) as

$$\frac{dp}{dr} = -\frac{\rho m}{r^2}. \tag{6.54}$$

This equation coincides with the equation of hydrostatic equilibrium for Newtonian stars.

Note 6.11

We can relate the matter energy density ρ, pressure p, temperature T, and entropy S in volume V for a relativistic fluid through the first law of thermodynamics as

$$d(\rho V) = -pdV + TdS.$$

It is known that Baryon number is conserved, so the above equation is expressed in terms of baryon number density n and entropy per baryon s as

$$d\left(\frac{\rho}{n}\right) = -pd\left(\frac{1}{n}\right) + Tds,$$

$$\left[V = \frac{Number\ of\ baryons}{n}\right],$$

or

$$d\rho = (p + \rho)\left(\frac{dn}{n}\right) + nTds.$$

If the fluid flow is **isentropic**, then $\frac{ds}{dt} = 0$, i.e., $s = $ constant. Hence, the first law of thermodynamics yields for a perfect fluid having equation of state $\rho = \rho(n)$ as

$$\frac{d\rho}{\rho} = \frac{p + \rho}{\rho}\frac{dn}{n}.$$

Let us consider the acoustic wave is developed from a perturbation (isentropic) in a uniform static fluid comprising the parameters ρ_0, p_0, and n_0 and perturbations are ρ_1, p_1, and n_1 (i.e., $p = p_0 + p_1$ and $\rho = \rho_0 + \rho_1$). Let us also consider the fluid velocity to be $(1, \vec{v_1})$ in the rest frame of unperturbed fluid. Now, we calculate the first-order perturbation terms in conservation equation, $T^{\mu\nu}_{\;\;;\nu} = 0$ as

$$\nabla.\vec{v_1} = -\frac{\partial\rho}{\partial t}\frac{1}{\rho_0 + p_0}, \quad for \; \mu = 0,$$

$$\frac{\partial\vec{v_1}}{\partial t} = -\nabla p_1\frac{1}{\rho_0 + p_0}, \quad for \; \mu = 1, 2, 3.$$

These two equations yield

$$\frac{\partial^2\rho}{\partial t^2} - \nabla^2 p_1 = 0.$$

Here, p_1 is related with ρ_1 due to isentropic flow as

$$p_1 = \left.\frac{\partial p}{\partial \rho}\right|_s \rho_1.$$

Substituting the value of p_1, we get the wave equation

$$\nabla^2\rho_1 - \frac{1}{v_s^2}\frac{\partial^2\rho}{\partial t^2} = 0,$$

where

$$v_s = \left(\frac{\partial p}{\partial \rho}|_s\right)^{\frac{1}{2}}$$

is characterized by the **velocity of sound** in a relativistic perfect fluid.
One can also write the sound velocity as

$$v_s^2 = \Gamma_1\frac{p}{\rho + p},$$

where

$$\Gamma_1 = \left.\frac{\partial \log p}{\partial \log n}\right|_s$$

is known as **adiabatic index.** The necessary and sufficient condition for the stability of the isotropic and the spherical stellar system against the radial pulsation is that the adiabatic index (Γ_1) of the system should be greater than $\frac{4}{3}$, i.e., $\Gamma_1 > \frac{4}{3}$.

Exercise 6.3

Find interior solution when p and ρ are related as $p = -\rho$
Hint: Eq. (6.48) $\Longrightarrow$

$$p' = -\frac{1}{2}\nu'(p + \rho) = 0$$

$\Longrightarrow$

$$p = constant = -3H_0^2 \ (say).$$

Therefore,

$$p = -\rho = -3H_0^2.$$

(6.45) $\Longrightarrow$

$$-re^{-\lambda}\lambda' + e^{-\lambda} = 1 - k.3H_0^2 r^2,$$

or

$$(re^{-\lambda})' = 1 - 3kH_0^2 r^2$$

or

$$e^{-\lambda} = 1 - kH_0^2 r^2.$$

(6.45) and (6.46) $\Longrightarrow$

$$e^{-\lambda}\left(\frac{\nu' + \lambda'}{r}\right) = k(p + \rho) = 0,$$

or

$$\nu' + \lambda' = 0,$$

$\Longrightarrow$

$$-\lambda = \nu - \nu_0 \; (\nu_0 = constant).$$

Therefore,

$$e^{-\lambda} = e^{\nu - \nu_0} = (1 - kH_0^2 r^2).$$

This gives

$$e^\nu = e^{\nu_0}(1 - kH_0^2 r^2), \;\; etc.$$

This solution is known as **de Sitter line element**.

Exercise 6.4

Find interior solution when pressure is the same in everywhere in the spherically symmetric body. Explain some features of this spacetime.

Hint: For pressure, $p = p_0 = constant$, Eq. (6.48) $\Longrightarrow$

$$p' = -\frac{1}{2}\nu'(p + \rho) = 0,$$

$\Longrightarrow$

$$\nu' = 0 \Rightarrow \nu = constant.$$

Eq. (6.46) implies

$$e^{-\lambda}\left(\frac{1}{r^2}\right) - \frac{1}{r^2} = kp_0,$$

or

$$e^{-\lambda} = 1 + kp_0 r^2.$$

However, Einstein considered the field equations with cosmological constant. Then, Eq. (6.46) can be rewritten as

$$e^{-\lambda}\left(\frac{1}{r^2}\right) - \frac{1}{r^2} = kp_0 - \Lambda,$$

or

$$e^{-\lambda} = 1 - \frac{r^2}{R^2},$$

where $\frac{1}{R^2} = \Lambda - kp_0$.

Thus, the line element after rescaling time coordinate is

$$ds^2 = dt^2 - \frac{dr^2}{1 - \frac{r^2}{R^2}} - r^2(d\theta^2 + \sin^2\theta d\phi^2). \tag{i}$$

This is known as **Einstein's line element in static universe**.

We will discuss the motion of a particle in the gravitational field of this Einstein's the static universe.

The geodesic equation describing the motion of a test particle is

$$\frac{d^2x^i}{ds^2} + \Gamma^i_{jk}\frac{dx^j}{ds}\frac{dx^k}{ds} = 0.$$

Since the universe is static, we assume that the particle is initially at rest. Hence, the velocity components are

$$\frac{dx^0}{ds} = 1, \quad \frac{dx^1}{ds} = 0, \quad \frac{dx^2}{ds} = 0, \quad \frac{dx^3}{ds} = 0.$$

Therefore, the geodesics become

$$\frac{d^2x^i}{ds^2} + \Gamma^i_{00}\frac{dx^0}{ds}\frac{dx^0}{ds} = 0.$$

For the above metric (i),

$$\Gamma^i_{00} = 0, \quad \forall\, i.$$

As a result, we get

$$\frac{d^2r}{ds^2} = \frac{d^2\theta}{ds^2} = \frac{d^2\phi}{ds^2} = 0.$$

This implies that the particle will have no acceleration. In other words, in Einstein static universe, the matter is without motion.

In this spacetime, a particle at rest remains at rest. Let D be the distance of a star from us, then the distance D is constant. Let a light ray leave the star at time t_1 and reach us at time t_2. From geodesic equation of light in the planes $\theta, \phi = constant$, through

$$0 = dt^2 - \frac{dr^2}{1 - \frac{r^2}{R^2}},$$

we have,

$$t_2 - t_1 = \int_{t_1}^{t_2} dt = \int_0^D \frac{dr}{\sqrt{1 - \frac{r^2}{R^2}}} = R\sin^{-1}\frac{D}{R}.$$

This implies

$$dt_1 = dt_2 (as\ D\ is\ a\ constant).$$

In Einstein static universe, the proper time is same as coordinate time. Therefore, the proper time periods of emitted and observed light are same. In other words, the wavelength of emitted and observed light remains the same. Hence, no Doppler shift in the spectral lines is seen. The observed expanding universe does not support the Einstein static universe.

Exercise 6.5

Find interior solution when density is the same everywhere in the spherically symmetric body. To solve the Einstein field equations for the interior region of the spherically symmetric body, we assume the following boundary conditions.

   (i) The pressure is zero at the boundary ($r = r_0$) of the sphere (Fig. 13).

   (ii) The density ρ is uniform throughout the sphere, i.e., ρ is a constant for $r \leq r_0$; $\rho(r) = 0, r > r_0$

   (iii) The exterior and interior solution become identical at the boundary $r = r_0$ of the sphere.

From (6.45), we get

$$e^{-\lambda} = 1 - \frac{k}{3}\rho r^2 + \frac{\alpha}{r},$$

where $\alpha = constant.$

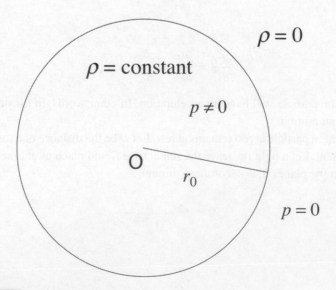

Figure 13 Spherically symmetric body with uniform density.

Since $g_{\mu\nu}$ is regular at $r = 0$, we must take $\alpha = 0$, i.e.,

$$e^{-\lambda} = \left(1 - \frac{r^2}{R_0^2}\right),$$

where

$$R_0^2 = \frac{3}{k\rho}.$$

Other solutions are

$$e^{\frac{\nu}{2}} = A - B\left(1 - \frac{r^2}{R_0^2}\right)^{\frac{1}{2}},$$

$$p = \frac{1}{kR_0^2}\left[\frac{3B(1 - \frac{r^2}{R_0^2})^{\frac{1}{2}} - A}{A - B(1 - \frac{r^2}{R_0^2})^{\frac{1}{2}}}\right],$$

where A and B are constants.

Eq. (6.48) implies

$$p + \rho = A_1 e^{-\frac{\nu}{2}}.$$

Replacing the values of $\rho + p$ and λ, Eq. (6.51) yields

$$e^{\frac{\nu}{2}}\left[\frac{2}{R_0^2} + \frac{1}{r}\left(1 - \frac{r^2}{R_0^2}\right)\nu'\right] = constant.$$

Solving this equation, one can get

$$e^{\frac{\nu}{2}} = A - B\left(1 - \frac{r^2}{R_0^2}\right)^{\frac{1}{2}}.$$

Putting the values of ν' and λ, one gets the value of p as

$$p = \frac{1}{kR_0^2}\left[\frac{3B(1 - \frac{r^2}{R_0^2})^{\frac{1}{2}} - A}{A - B(1 - \frac{r^2}{R_0^2})^{\frac{1}{2}}}\right].$$

Hence, Schwarzschild interior solution is

$$ds^2 = -\left[1 - \frac{r^2}{R_0^2}\right]^{-1} dr^2 - r^2 d\theta^2 - r^2 \sin^2\theta d\phi^2 + \left[A - B\sqrt{1 - \frac{r^2}{R_0^2}}\right]^2 dt^2.$$

Now applying condition (i), i.e., the pressure is zero at the boundary $r = r_0$, we have

$$A = 3B\left[1 - \frac{r_0^2}{R_0^2}\right]^{\frac{1}{2}}.$$

Now using condition (iii), i.e., matching the interior solution at $r = r_0$ with exterior vacuum Schwarzschild solution

$$ds^2 = \left(1 - \frac{2GM}{r}\right) dt^2 - \left(1 - \frac{2GM}{r}\right)^{-1} dr^2 - r^2 d\Omega_2^2,$$

we have

$$B = \frac{1}{2}.$$

$$\left[1 - 2G\frac{4}{3}\frac{\pi r_0^3 \rho}{r_0} = 1 - \frac{8\pi G}{3}r_0^2\rho = 1 - \frac{r_0^2}{R_0^2}, \frac{1}{R_0^2} = \frac{k}{3}\rho\right.$$

$$\left(\text{here, } M = \frac{4}{3}\pi r_0^3\rho\right)$$

$$\left.1 - \frac{2GM}{r_0} = \left[A - B\sqrt{1 - \frac{r_0^2}{R^2}}\right]^2 \Rightarrow B = \frac{1}{2}\right]$$

The interior solution will be real if

$$\frac{r_0^2}{R_0^2} < 1,$$

i.e.,

$$r_0^2 < \frac{3}{8\pi G\rho_0} \text{ or } \frac{2MG}{r_0} < 1.$$

This gives

$$2MG < r_0.$$

This equation indicates an upper bound on the possible size of a sphere of a given density and on the mass of the sphere of a given radius. The solution has an interesting feature that one cannot take all values of r_0 and M. The pressure will become infinite at r_∞, where

$$r_\infty^2 = 9r_0^2 - \frac{4r_0^3}{MG}.$$

For real r_∞, one gets

$$\frac{MG}{r_0} > \frac{4}{9}.$$

This indicates that if the mass of a body exceeds the value $M = \frac{4r_0}{9G}$, relativity admits no solution as a body with fixed radius r_0, then it requires infinite central pressure. This limit is known as **Buchdahl limit**.

Actually, **Buchdahl theorem** states that any spherically symmetric body of mass M and radius R has a realistic interior solution if

$$\frac{M}{R} < \frac{4}{9G}.$$

Exercise 6.6

Find Schwarzschild interior solution when p and ρ are related as $p = w\rho$.

Hint: The basic equations of the interior of a spherically symmetric isotropic fluid sphere are as follows (Einstein field equations, TOV equation, and equation of state):

$$ds^2 = e^{2\phi(r)}dt^2 - \frac{dr^2}{1 - \frac{2m(r)}{r}} - r^2(d\theta^2 + \sin^2\theta d\phi^2),$$

where

$$m(r) = \int_0^r 4\pi r^2 \rho \, dr \text{ or, } m' = 4\pi r^2 \rho,$$

$$\frac{d\phi}{dr} = \frac{(m + 4\pi r^3 p)}{r(r - 2m)},$$

$$\frac{dp}{dr} = -\frac{(p + \rho)(m + 4\pi r^3 p)}{r(r - 2m)}.$$

Here ρ and p are density and pressure of the fluid sphere obeying equation of state $p = w\rho$. Using this equation of state and replacing ρ by m, the last equation yields

$$w(m''r - m')(r - 2m) + (w + 1)m'(m + wrm') = 0.$$

Note that

$$m = ar,$$

is a particular solution of this equation, where

$$a = \frac{w}{4w + 1 + w^2}.$$

The other parameters are found as

$$\rho = \frac{a}{4\pi r^2}, \quad p = \frac{aw}{4\pi r^2}, \quad g_{tt} = e^{2\phi} = \left(\frac{r}{r_0}\right)^{\frac{a(1+w)}{1-2a}}, \quad g_{rr} = \frac{1}{1 - \frac{2m(r)}{r}} = \frac{1}{1 - 2a}.$$

Thus, for $t = constant$, the hypersurface has the metric

$$ds_3^2 = \frac{dr^2}{1 - 2a} + r^2(d\theta^2 + sin^2\theta d\phi^2).$$

Here,

$$a < \frac{1}{2}.$$

One can note that this has a conical singularity at the origin because here the two-sphere with the circumference $2\pi r$ will have a radius of $(\frac{1}{1-2a})^{\frac{1}{2}} r$, which is greater than r.

Exercise 6.7

Suppose a star of mass M has two massless concentric shells. The area of the inner surface is $144\pi M^2$ and the area of the outer surface is $400\pi M^2$. Find the thickness of the shell.

Hint: We know the area of a sphere of radius r in Schwarzschild coordinate is $4\pi r^2$. Therefore, the radii of the inner and outside of the shell are 6 M and 10 M. Hence, the radial distance between two concentric shells is

$$s = \int ds = \int_{6M}^{10M} \left[1 - \frac{2M}{r}\right]^{-\frac{1}{2}} dr = 4.6 \text{ M}.$$

6.8 Isotropic Coordinates

Isotropic coordinate system is a new coordinate system whose spatial distance is proportional to the Euclidean square of the distances. Usually, isotropic means that all three spatial dimensions are treated as identical. Thus, in isotropic coordinate, the line element takes the form

$$ds^2 = A(r)dt^2 - B(r)d\sigma^2.$$

In Cartesian coordinates, the line element of Euclidean three space is

$$d\sigma^2 = dx^2 + dy^2 + dz^2,$$

whereas in spherical polar coordinates

$$x = r \sin \theta \cos \phi, \ y = r \sin \theta \sin \phi, \ z = r \cos \theta,$$

the line element of Euclidean three space is

$$d\sigma^2 = dr^2 + r^2 d\theta^2 + r^2 \sin^2 \theta d\phi^2.$$

Here, the metric with t = constant is conformally related to the metric of Euclidean space. Usually, isotropic coordinates are used when time t = constant hypersurface, i.e., the three-dimensional subspace of spacetime requires to look like Euclidean. Generally, this type of coordinate system is used for modeling the gravitational field of a symmetrical object that does not discriminate between the x, y, or z directions.

Note 6.12

Two metrics, which are conformally related to each other, imply angles between vectors, and ratios of lengths are the same for each metric.

Exercise 6.8

Rewrite the Schwarzschild line element

$$ds^2 = \left(1 - \frac{2m}{r}\right) dt^2 - \frac{dr^2}{1 - \frac{2m}{r}} - r^2 (d\theta^2 + \sin^2 \theta \, d\phi^2),$$

in the isotropic coordinate system.

Hint: We consider a transformation $\rho = \rho(r)$ in such a way that coordinates θ, ϕ, and t do not change. For this transformation, the Schwarzschild line element takes the following form

$$ds^2 = \left(1 - \frac{2m}{r}\right) dt^2 - [\lambda(\rho)]^2 [d\rho^2 + \rho^2 (d\theta^2 + \sin^2 \theta \, d\phi^2)]. \tag{6.55}$$

Now comparing this metric with the given Schwarzschild metric, we get

$$r^2 = \lambda^2 \rho^2, \tag{6.56}$$

$$\left(1 - \frac{2m}{r}\right)^{-1} dr^2 = \lambda^2 d\rho^2. \tag{6.57}$$

Eliminating λ from (6.56) and (6.57), we get

$$(6.57) \div (6.56) \Rightarrow \frac{dr}{(r^2 - 2mr)^{\frac{1}{2}}} = \pm \frac{d\rho}{\rho},$$

$$or, \ \ln[2\sqrt{r^2 - 2mr} + 2r - 2m] = \ln \rho C.$$

(we take positive sign to follow the condition $r \to \infty$, $\rho \to \infty$)

Here, $r = \rho$ for flat space, i.e., when $m = 0$, the integration constant C should take the value as $C = 4$.

$$\Rightarrow r = \rho \left(1 + \frac{1}{2}\frac{m}{\rho}\right)^2. \tag{6.58}$$

$$(6.56) \Rightarrow \quad \lambda^2 = \left(1 + \frac{1}{2}\frac{m}{\rho}\right)^4. \tag{6.59}$$

Hence, finally, we get the Schwarzschild metric in isotropic coordinates that may be treated as a black a hole without hole.

$$ds^2 = \frac{\left(1 - \frac{1}{2}\frac{m}{\rho}\right)^2}{\left(1 + \frac{1}{2}\frac{m}{\rho}\right)^2} dt^2 - \left(1 + \frac{1}{2}\frac{m}{\rho}\right)^4 d\rho^2 [d\rho^2 + \rho^2(d\theta^2 + \sin^2\theta d\phi^2)].$$

Exercise 6.9

Rewrite the Einstein line element for a static universe

$$ds^2 = dt^2 - \frac{dr^2}{1 - \frac{r^2}{R^2}} - r^2(d\theta^2 + \sin^2\theta \, d\phi^2),$$

in isotropic coordinate system.

Hint: We consider a transformation $\rho = \rho(r)$ in such a way that coordinates θ, ϕ, and t do not change. For this transformation, the above line element takes the following form

$$ds^2 = dt^2 - [\lambda(\rho)]^2 [d\rho^2 + \rho^2(d\theta^2 + \sin^2\theta \, d\phi^2)], \tag{6.60}$$

Now comparing this metric with the given metric, we get

$$r^2 = \lambda^2 \rho^2, \tag{6.61}$$

$$\left(1 - \frac{r^2}{R^2}\right)^{-1} dr^2 = \lambda^2 d\rho^2. \tag{6.62}$$

Eliminating λ from (6.61) and (6.62), we get

$$(6.62) \div (6.61) \Rightarrow \frac{R dr}{r\sqrt{R^2 - r^2}} = \pm \frac{d\rho}{\rho}.$$

As before, we take a positive sign and integrate to yield

$$-\ln \left[\frac{R + \sqrt{R^2 - r^2}}{r} \right] = \ln \rho C.$$

Also $r = \rho$ for flat space, i.e., when $R \to \infty$, the integration constant C should take the value as $C = \frac{1}{2R}$.

$$\Rightarrow \quad r = \frac{\rho}{1 + \frac{\rho^2}{4R^2}}. \tag{6.63}$$

Using the value of λ, we get the metric in isotropic coordinates as

$$ds^2 = dt^2 - \frac{1}{\left(1 + \frac{\rho^2}{4R^2}\right)^2} [d\rho^2 + \rho^2 (d\theta^2 + \sin^2 \theta \, d\phi^2)].$$

Exercise 6.10

Rewrite the Reissner–Nordström line element

$$ds^2 = \left(1 - \frac{2m}{r} + \frac{q^2}{r^2} \right) dt^2 - \frac{dr^2}{1 - \frac{2m}{r} + \frac{q^2}{r^2}} - r^2 (d\theta^2 + \sin^2 \theta \, d\phi^2),$$

in the isotropic coordinate system.

Hint: We consider a transformation $\rho = \rho(r)$ in such a way that coordinates θ, ϕ, and t do not change. For this transformation, the above line element takes the following form

$$ds^2 = \left(1 - \frac{2m}{r} + \frac{q^2}{r^2} \right) dt^2 - [\lambda(\rho)]^2 [d\rho^2 + \rho^2 (d\theta^2 + \sin^2 \theta \, d\phi^2)]. \tag{6.64}$$

Now comparing this metric with the given Reissner–Nordström metric, we get

$$r^2 = \lambda^2 \rho^2, \tag{6.65}$$

$$\left(1 - \frac{2m}{r} + \frac{q^2}{r^2} \right)^{-1} dr^2 = \lambda^2 d\rho^2. \tag{6.66}$$

Eliminating λ from (6.65) and (6.66), we get

$$(6.66) \div (6.65) \quad \Rightarrow \quad \frac{dr}{(r^2 - 2mr + q^2)^{\frac{1}{2}}} = \pm \frac{d\rho}{\rho}.$$

As before, we take positive sign and integrate to yield

$$\ln[2\sqrt{r^2 - 2mr + q^2} + 2r - 2m] = \ln \rho C.$$

Here, $r = \rho$ for flat space, i.e., when $m = 0$ and $q = 0$ the integration constant C should take the value as $C = 4$.

$$\Rightarrow \quad r = \rho \left(1 + \frac{1}{2}\frac{m}{\rho}\right)^2 - \frac{q^2}{4\rho}. \tag{6.67}$$

Hence, finally, the Reissner–Nordström metric in isotropic coordinates may be written as

$$ds^2 = \frac{\left[\rho\left(1 + \frac{1}{2}\frac{m}{\rho}\right)^2 - \frac{q^2}{4\rho} - m\right]^2 + q^2 - m^2}{\left[\rho\left(1 + \frac{1}{2}\frac{m}{\rho}\right)^2 - \frac{q^2}{4\rho}\right]^2} dt^2$$

$$- \left[\left(1 + \frac{1}{2}\frac{m}{\rho}\right)^2 - \frac{q^2}{4\rho^2}\right]^2 [d\rho^2 + \rho^2(d\theta^2 + \sin^2\theta d\phi^2)].$$

Exercise 6.11

Rewrite the Extreme Reissner–Nordfström line element

$$ds^2 = \left(1 - \frac{m}{r}\right)^2 dt^2 - \frac{dr^2}{\left(1 - \frac{m}{r}\right)^2} - r^2(d\theta^2 + \sin^2\theta \, d\phi^2),$$

in the isotropic coordinate system.

Hint: We consider a transformation $\rho = \rho(r)$ in such a way that coordinates θ, ϕ, and t do not change. For this transformation, the above line element takes the following form

$$ds^2 = \left(1 - \frac{m}{r}\right)^2 dt^2 - [\lambda(\rho)]^2[d\rho^2 + \rho^2(d\theta^2 + \sin^2\theta \, d\phi^2)].$$

Now comparing this metric with the given Reissner–Nordström metric, we get

$$r^2 = \lambda^2 \rho^2,$$

$$\left(1 - \frac{m}{r}\right)^{-2} dr^2 = \lambda^2 d\rho^2.$$

Here,

$$\lambda(\rho) = 1 + \frac{m}{\rho}, \quad r = m + \rho, \quad etc.$$

Exercise 6.12

Rewrite the following Schwarzschild-Tangherlini black hole line element in the isotropic coordinate system:

$$ds^2 = \left(1 - \frac{\tau}{r^{n-3}}\right) c^2 dt^2 - \frac{dr^2}{\left(1 - \frac{\tau}{r^{n-3}}\right)} - r^2 d\Omega_{n-2}^2,$$

where the parameter τ is related to the mass M of black hole and Ω_{n-2} is the volume of the $(n-2)$ dimensional unit sphere.

Hint: In the isotropic coordinate system, the above line element can be written as

$$ds^2 = \left(1 - \frac{\tau}{r^{n-3}}\right) c^2 dt^2 - [\lambda(\rho)]^2 \left[d\rho^2 + \rho^2 d\Omega_{n-2}^2\right].$$

On comparing this metric with the given metric, we have

$$\rho = \left[\frac{r^{\frac{n-3}{2}} + \sqrt{r^{n-3} - \tau}}{2}\right]^{\frac{2}{n-3}} \quad i.e.\ r = \rho \left[1 + \frac{\tau}{4\rho^{n-3}}\right]^{\frac{2}{n-3}}.$$

Therefore, the isotropic form of the Schwarzschild-Tangherlini metric takes the form:

$$ds^2 = \left[\frac{1 - \frac{\tau}{4\rho^{n-3}}}{1 + \frac{\tau}{4\rho^{n-3}}}\right]^2 c^2 dt^2 - \left[1 + \frac{\tau}{4\rho^{n-3}}\right]^{\frac{4}{n-3}} \left[d\rho^2 + \rho^2 d\Omega_{n-2}^2\right].$$

The quantity $\left[d\rho^2 + \rho^2 d\Omega_{n-2}^2\right]$ has the dimension of the square with the infinitesimal length vector $d\vec{\rho}$.

Note 6.13

The **proper radial distance** $(l(r))$ is linked to the r coordinate as

$$l(r) = \pm \int_{r_0}^{r} \sqrt{g_{rr}(r')} dr'.$$

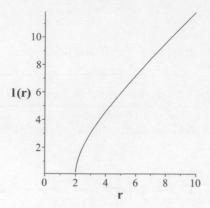

Figure 14 Proper radial distance in Schwarzschild spacetime for $m = 1$ and $r_0 = 2$.

Exercise 6.13

Find proper radial distance for Schwarzschild spacetime.
Hint: We know for Schwarzschild spacetime $g_{rr}(r) = \frac{1}{1-\frac{2m}{r}}$, therefore,

$$l(r) = \pm \int_{r_0}^{r} \frac{dr'}{\sqrt{1 - \frac{2m}{r'}}} = \left[\sqrt{r^2 - 2rm} + 2m^2 \ln(-m + r + \sqrt{r^2 - 2rm}) \right]_{r_0}^{r}.$$

Fig. 14 shows the nature of the proper radial distance in Schwarzschild spacetime.

Note 6.14

The **embedding** of a space-like hypersurface of any spherically symmetric spacetime can be written as follows:
Let us assume the three-dimensional spatial hypersurface given by $t = 0$ of any spherically symmetric spacetime. The metric takes the form for this hypersurface as

$$d\sigma^2 = g_{rr}dr^2 + r^2(d\theta^2 + \sin^2\theta d\phi^2).$$ (i)

One can embed it in a four-dimensional Euclidean space E^4. One can write the flat metric in the embedded space as

$$ds^2 = dz^2 + dr^2 + r^2(d\theta^2 + \sin^2\theta d\phi^2).$$ (ii)

Now, we will have to search the hypersurface in E^4, which is rotationally symmetric around the z axis. The induced metric

$$d\sigma^2 = \left[1 + \left(\frac{dz}{dr}\right)^2\right] dr^2 + r^2(d\theta^2 + \sin^2\theta d\phi^2)$$ (iii)

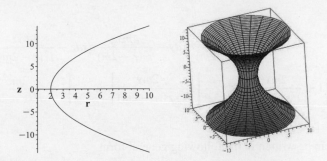

Figure 15 The embedding diagram for Schwarzschild spacetime for $m = 1$ in the left panel and in the right panel we provide the entire imagining of the surface created by the rotation of the embedded curve about the vertical z axis.

should be equal to (i) for the surface given by $z(r)$. Comparing (i) and (iii), we obtain

$$\frac{dz}{dr} = \pm\sqrt{g_{rr} - 1}.$$

Exercise 6.14

Find the embedding of a space-like hypersurface for Schwarzschild spacetime.
Hint: We know for Schwarzschild spacetime $g_{rr}(r) = \frac{1}{1-\frac{2m}{r}}$, therefore,

$$z(r) = \pm\int_{2m}^{r}\sqrt{g_{rr}(r') - 1}\,dr' = \sqrt{8m(r - 2m)}.$$

Here, the three-dimensional hypersurface given in Fig. 15 gives the induced metric. If one writes r in terms of z, then one can find

$$r = \frac{z^2}{8m} + 2m,$$

which is a regular surface everywhere as one can see that $r = 2m$ implies $z = 0$. This expansion is dubbed as **Einstein–Rosen bridge** for Schwarzschild spacetime. Actually, it represents two same copies of Schwarzschild spacetimes with a common horizon. We know no particle (light or any massive) can return from the event horizon, therefore, two exterior Schwarzschild solutions cannot link to each other.

Exercise 6.15

Write Schwarzschild line element in (t, x, y, z) coordinates, where (x, y, z) are defined by

$$x = r\sin\theta\cos\phi,\ y = r\sin\theta\sin\phi,\ z = r\cos\theta.$$

Hint:

$$ds^2 = \left(1 - \frac{2m}{r}\right)dt^2 - \frac{4mxy}{r^2(r-2m)}dxdy - \frac{4mxz}{r^2(r-2m)}dxdz - \frac{4myz}{r^2(r-2m)}dydz$$

$$- \left[1 + \frac{2mx^2}{r^2(r-2m)}\right]dx^2 - \left[1 + \frac{2my^2}{r^2(r-2m)}\right]dy^2 - \left[1 + \frac{2mz^2}{r^2(r-2m)}\right]dz^2,$$

where, $r^2 = x^2 + y^2 + z^2$.

The above metric can be written in the following general form

$$g_{00} = \left(1 - \frac{2m}{r}\right), \quad g_{ij} = -\delta_{ij} - \frac{\frac{2m}{r}}{1 - \frac{2m}{r}} \frac{x^i x^j}{r^2}.$$

6.9 Interaction between Gravitational and Electromagnetic Fields

Electric and magnetic fields are generated when a charge is in motion, and it depends on space and time. This phenomenon is known as **electromagnetism**. The study of time-dependent electromagnetic fields and their behavior is described by a set of equations, known as Maxwell's equations.

Before Maxwell, there were four fundamental equations of electromagnetism prescribed by several researchers. Maxwell improved those equations and composed them in the succeeding compact form known as **Maxwell's equations of electromagnetism**.

(a) $\vec{\nabla} \cdot \vec{E} = \frac{1}{\epsilon_0}\rho, \quad [= 0,$ *without charge in a region*]

(b) $\vec{\nabla} \cdot \vec{B} = 0,$

(c) $\vec{\nabla} \times \vec{E} = -\frac{\partial \vec{B}}{\partial t},$

(d) $\vec{\nabla} \times \vec{B} - \mu_0\epsilon_0\frac{\partial \vec{E}}{\partial t} = \mu_0 \vec{J}.$

(Ampere's law with Maxwell's corrections)

Ampere's law:

$$\vec{\nabla} \times \vec{B} = \mu_0 \vec{J},$$

where, E = electric field strength, B = magnetic field strength, J = current density, ρ = charge density, and μ_0 = magnetic permeability = $4\pi \times 10^{-7}$ weber/amp-meter.

ϵ_0 = electric permittivity of free space = 8.86×10^{-12} coulomb sec^2/ meter and $\epsilon_0\mu_0 = \frac{1}{c^2}$.

One can solve Maxwell's equations for B and E in terms of a scalar function ϕ and a vector function A as

(e) $\vec{B} = curl\ \vec{A}$.

[it comes from (b) as *div curl A* = 0]

(f) $E = -\dfrac{\partial A}{\partial t} - grad\ \phi$.

[It follows from (*c*) as

$$\nabla \times E = -\frac{\partial(\nabla \times A)}{\partial t} = -\nabla \times \frac{\partial A}{\partial t},$$

where A = magnetic potential and ϕ = electric potential.

If charges are moving with velocity $\vec{v}$, then current density

$$\vec{J} = \rho\ \vec{v}\ .$$

From (*a*) and (*d*), we get

$$div\ curl\ B = \mu_0 \epsilon_0\ div\ \left(\frac{\partial E}{\partial t}\right) + \mu_0\ div\ J = 0,$$

$\Rightarrow$

$$\epsilon_0 \frac{\partial(div\ E)}{\partial t} + div\ J = 0,$$

$\Rightarrow$

$$\epsilon_0 \frac{\partial\left(\frac{\rho}{\epsilon_0}\right)}{\partial t} + div\ J = 0,$$

$\Rightarrow$

$$\frac{\partial \rho}{\partial t} + div\ J = 0.$$

This is the **equation of continuity** that states the law of conservation of total charge, i.e., total current flowing out of some volume should be equal to the rate of decrease of charge within the volume.

Let us consider a second rank antisymmetric tensor as

$$F_{\mu\nu} = \frac{\partial A_\nu}{\partial x_\mu} - \frac{\partial A_\mu}{\partial x_\nu}.$$

This is known as **electromagnetic field strength tensor**.

Maxwell's equations given by (a) and (d) can, now, be expressed as

$$\frac{\partial F_{\mu\nu}}{\partial x_\nu} = \mu_0 \, J_\mu.$$

The other set of Maxwell's equations (b) and (c) constitutes another relation as

$$\frac{\partial F_{\mu\nu}}{\partial x_\lambda} + \frac{\partial F_{\nu\lambda}}{\partial x_\mu} + \frac{\partial F_{\lambda\mu}}{\partial x_\nu} = 0.$$

The Lagrangian for the free electromagnetic field can be defined as

$$L_{lm} = -\frac{1}{16\pi} \, F^{\mu\nu} \, F_{\mu\nu},$$

with

$$F_{\mu\nu} = A_{\nu,\mu} - A_{\mu,\nu}.$$

Consequently its energy-momentum tensor via

$$T_{\mu\nu} = 2\frac{\partial L}{\partial g^{\mu\nu}} - Lg_{\mu\nu},$$

is found as

$$(g) \quad (T_{\mu\nu})_{em} = \frac{1}{4\pi}\left[-F_{\mu\alpha}\, F^\alpha_\nu + \frac{1}{4}g_{\mu\nu}F_{\alpha\beta}F^{\alpha\beta} \right].$$

We know the Einstein–Maxwell equations for the outside of a charged body is defined as

$$(h) \quad R_{\mu\nu} - \frac{1}{2}g_{\mu\nu}R = -8\pi(T_{\mu\nu})_{em},$$

where $(T_{\mu\nu})_{em}$ is given in (g).

Since we are considering a static charged body, $J^\mu = \rho U^\mu = 0$ [since, charge is not moving], the field in this coordinate system is purely electrostatic. Thus, only $F_{10} = F_{01} \neq 0$ and all other components of the electromagnetic field vanish. We number the coordinates t, r, θ, ϕ as 0, 1, 2, 3, respectively. Hence, we may take

$$A_0 = \phi \neq 0, \quad A_2 = A_3 = A_1 = 0.$$

Here, the electromagnetic field strength tensor takes the form

$$F_{\mu\nu} = \phi_{,1}\begin{pmatrix} 0 & -1 & 0 & 0 \\ 1 & 0 & 0 & 0 \\ 0 & 0 & 0 & 0 \\ 0 & 0 & 0 & 0 \end{pmatrix}.$$

Now the equation

$$\frac{\partial F_{\mu\nu}}{\partial x_\nu} = \mu_0 J_\mu = 0,$$

$$\Rightarrow \quad (g^{00} g^{11} F_{01} \sqrt{-g})_{,1} = 0,$$

$$or, \quad (g^{00} g^{11} \phi_{,1} \sqrt{-g})_{,1} = 0,$$

$$or, \quad e^{-\frac{(\lambda+\nu)}{2}} \phi_{,1} r^2 = e = constant.$$

Here the energy-stress tensors take the following form

$$T_r^r = T_1^1 = \frac{1}{4\pi} \left[-F^{1\alpha} F_{1\alpha} + \frac{1}{4} g_1^1 F^{\alpha\beta} F_{\alpha\beta} \right]$$

$$= \frac{1}{4\pi} \left[-F^{10} F_{10} + \frac{1}{4}.1(F^{10}F_{10} + F^{01}F_{01}) \right]$$

$$= \frac{1}{4\pi} \left[-\frac{1}{2} F^{10} F_{10} \right] = \frac{1}{8\pi} \left[-g^{11} g^{00} F_{10}^2 \right] = \frac{1}{8\pi} \frac{e^2}{r^4},$$

etc.

Thus, the gravitating body produces the energy-momentum tensor of the electromagnetic field as

$$T_\mu^\nu = \frac{1}{8\pi} \, diag \, \left(\frac{e^2}{r^4} \right) (1, 1, -1, -1).$$

Now the field equation (*h*) for the metric (6.42),

$$e^{-\lambda} \left(\frac{\nu'}{r} + \frac{1}{r^2} \right) - \frac{1}{r^2} = -\frac{e^2}{r^4},$$

$$e^{-\lambda} \left(\frac{\lambda'}{r} - \frac{1}{r^2} \right) + \frac{1}{r^2} = \frac{e^2}{r^4},$$

$$e^{-\lambda} \left(\frac{\nu''}{2} - \frac{\lambda'\nu'}{4} + \frac{\nu'^2}{4} + \frac{\nu'-\lambda'}{2r} \right) = \frac{e^2}{r^4}.$$

The above equations yield

$$\lambda' + \nu' = 0 \ or \ \lambda = -\nu$$

and

$$(re^\nu)' = 1 - \frac{e^2}{r^2}.$$

[we drop the integration constant following the same reason as Schwarzschild case]

Solving these, we get

$$e^{\nu} = e^{-\lambda} = \left(1 - \frac{2m}{r} + \frac{e^2}{r^2} \right).$$

[m is an integration constant]

This metric is called **Reissner–Nordström** metric for a charged particle, i.e., Reissner–Nordström solution is the gravitational field outside of a spherically symmetric charged body. This solution is also known as **Reissner–Nordström black hole**.

For $e = 0$ or for large values of r, the term $\frac{e^2}{r^2}$ falls off quickly and we get Schwarzschild metric, therefore, this m represents the total mass of the gravitating body. Again at a great distance, the geometry tends to Minkowski spacetime, i.e., $\lambda = \nu = 0$, and F_{01} tends to $\frac{e}{r^2}$ showing that e is the charge of the body.

$$F_{01} = F_{10} = \frac{\partial A_0}{\partial r} = \frac{\partial \phi}{\partial r} = \phi_{,1} = -\frac{e}{r^2},$$

i.e.,

$$A_{\mu} = \left(\frac{e}{r}, 0, 0, 0 \right).$$

Here, the Newtonian potential (ψ) is given by

$$\psi = -\frac{m}{r} + \frac{e}{2r^2}.$$

Therefore, the force (F) is obtained as

$$F = \frac{\partial \psi}{\partial r} = \frac{m}{r^2} - \frac{e}{r^3}.$$

Note that the spacetime of Reissner–Nordström black hole solution has two singularities, i.e., it has two event horizons at the points where

$$g_{tt} = (g_{rr})^{-1} = \frac{(r^2 - 2mr + e^2)}{r^2} = 0.$$

Interestingly, we see that for $e^2 > m^2$, there are no real solutions. So it is nonsingular except at $r = 0$, where it is a physical singularity as Schwarzschild black hole. For $e^2 < m^2$, there are two distinct solutions at $r = r_-$ and $r = r_+$, where

$$r_{\pm} = m \pm \sqrt{m^2 - e^2}.$$

For $e^2 = m^2$, two roots coincide, i.e., $r_+ = r_- = m$ and the solution

$$g_{tt} = (g_{rr})^{-1} = \left(1 - \frac{m}{r} \right)^2$$

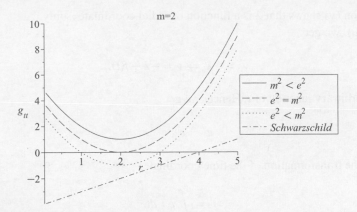

Figure 16 Plots for g_{tt} of Reissner–Nordström black hole and Schwarzschild black hole.

is known as **Extremal Reissner–Nordström black hole solution**. Here, one can imagine that the mass may be balanced by the charge. Fig. 16 indicates that the horizons of Reissner–Nordström black hole are less than the Schwarzschild horizon.

Exercise 6.16

Show that outside of the spherical charged objects, the spacetime metric is Reissner–Nordström solution even if the metric is not explicitly assumed to be static.

Hints: Consider the metric of the empty spacetime outside of a spherically symmetric charged body with charge e and mass M as

$$ds^2 = e^{\nu}dt^2 - e^{\lambda}dr^2 - r^2 d\theta^2 - r^2 \sin^2\theta d\phi^2, \tag{i}$$

where λ and ν are functions of r and t. Now, we write down the field equation for the above metric as

$$e^{-\lambda}\left(\frac{1}{r^2} - \frac{\lambda'}{r}\right) - \frac{1}{r^2} = -\frac{e^2}{r^4}, \tag{ii}$$

$$e^{-\lambda}\left(\frac{\nu'}{r} + \frac{1}{r^2}\right) - \frac{1}{r^2} = -\frac{e^2}{r^4}, \tag{iii}$$

$$-\frac{1}{2}e^{-\lambda}\left(\nu'' + \frac{\nu'^2}{2} + \frac{(\nu' - \lambda')}{r} - \frac{\lambda'\nu'}{2}\right)$$

$$+ \frac{1}{2}e^{-\nu}\left(\ddot{\lambda} + \frac{1}{2}\dot{\lambda}^2 - \frac{\dot{\lambda}\dot{\nu}}{2}\right) = -\frac{e^2}{r^4}, \tag{iv}$$

$$G_4^1 = -\frac{1}{2}e^{-\lambda}\frac{\dot{\lambda}}{r} = 0. \tag{v}$$

The field equation (v) shows that λ is a function of radial coordinate r only.
From (ii) and (iii), we get

$$\nu' = -\lambda' \Rightarrow \nu = -\lambda + f(t).$$

Here, $f(t)$ is an arbitrary function of t. Hence, we get

$$e^{\nu} = e^{f(t)} e^{-\lambda}. \tag{vi}$$

Now, consider the transformation of the time coordinate

$$t' = \int e^{\frac{f(t)}{2}} dt. \tag{vii}$$

This transformation does not affect the spatial coordinate. Therefore, we again get Reissner–Nordström static solution

$$g_{00} = (g_{11})^{-1} = \left(1 - \frac{m}{r} + \frac{e^2}{r^2}\right) = \left(1 - \frac{2GM}{c^2 r} + \frac{e^2}{r^2}\right). \tag{viii}$$

Therefore, we conclude that the spacetime of outside of a spherically symmetric charged object is necessarily static and of Reissner–Nordström form.

Note 6.15

Vaidya Metric

In 1951, Professor P. C. Vaidya discovered a spacetime metric that is nonstatic spherically symmetric describing the radiative field. This is actually a nonstatic generalization of the Schwarzschild metric. It is the solution of Einstein field equations for a spherically symmetric radiating nonrotating body. This type of body is characterized by the energy-stress tensor as

$$T_{ij} = \rho k_i k_j,$$

where k_i is the radially directed outward null vector and ρ is the energy density of the radiation. In Schwarzschild coordinates, the Vaidya-radiating metric can be written as

$$ds^2 = \left[\frac{\dot{m}}{f(m)}\right]^2 \left[1 - \frac{2m}{r}\right] dt^2 - \left[1 - \frac{2m}{r}\right]^{-1} dr^2 - r^2(d\theta^2 + \sin^2\theta d\phi^2),$$

where

$$m(r,t) = \int_0^r \rho(r,t) r^2 dr, \quad f(m) = m'\left[1 - \frac{2m}{r}\right], \quad m' = \frac{\partial m}{\partial r}, \quad \dot{m} = \frac{\partial m}{\partial t}.$$

Alternative form: Let us take the following retarded coordinate u (which is null) as

$$u = t - \int \frac{dr}{\left[1 - \frac{2m}{r}\right]}.$$

Using this transformation, the Vaidya metric takes the following form

$$ds^2 = \left[1 - \frac{2m(u)}{r}\right] du + 2dudr - r^2(d\theta^2 + \sin^2\theta d\phi^2).$$

Note that the Vaidya metric appears very similar to Schwarzschild, but m is a function of retarded time u.

For constant m, i.e., in the Schwarzschild geometry, the null coordinate assumes the following form

$$u = t - r - 2m \ln(r - 2m).$$

Note 6.16

The Tolman Metric

The spherically symmetric dust solution of Einstein's field equations is known as Tolman metric. The pressureless fluid, i.e., dust is characterized by the energy-stress tensor as

$$T^{\mu\nu} = \rho u^\mu u^\nu, \quad u^\mu = \textit{four velocity of the particle, and } \rho = \textit{energy density.}$$

We know the general metric describing the spherical symmetry is (see Eq. (6.7))

$$ds^2 = e^\nu dt^2 - e^\mu dR^2 - R^2(d\theta^2 + \sin^2\theta d\phi^2),$$

where ν, μ, and R are functions of r and t.

Now, conservation law of energy momentum is

$$T^{\mu\nu}_{;\nu} = 0 \Rightarrow (\rho u^\nu)_{;\nu} u^\mu + \rho u^\nu u^\mu_{;\nu} = 0,$$

since

$$u^\mu u_\mu = 1 \Rightarrow (u_\mu u^\mu)_{;\nu} = 0.$$

Hence, we have

$$(\rho u^\nu)_{;\nu} = 0 \Rightarrow (\rho\sqrt{-g}u^\nu)_{,\nu} = 0 \text{ } \textit{by result (8) in Section (3.4).}$$

Thus,

$$u^\mu_{;\nu} u^\nu = 0.$$

This means each particle of the dust moves on a geodesic, i.e., coordinates are **comoving** and hence spatial parts are unchanged along the geodesic. Therefore, we can have four velocity as

$$u^\mu = (u^0, 0, 0, 0), \quad where \quad u^0 = \frac{dt}{ds}.$$

The geodesic equation reduces to

$$\frac{du^\mu}{ds} + \Gamma^\mu_{00}(u^0)^2 = 0 \Rightarrow \Gamma^\mu_{00} = 0 \Rightarrow g_{00,t} = 0 \Rightarrow g_{00} = \alpha(t).$$

Now, after rescaling the above metric, we have

$$ds^2 = dt^2 - e^\mu dR^2 - R^2(d\theta^2 + \sin^2\theta d\phi^2).$$

Hence, the Einstein field equations for the dust are written as

$$e^{-\mu}(2RR'' + R'^2 - RR'\mu') - (R\dot{R}\dot{\mu} + \dot{R}^2 + 1) = -8\pi G\rho R^2, \tag{i}$$

$$e^\mu(2R\ddot{R} + \dot{R}^2 + 1) - R'^2 = 0, \tag{ii}$$

$$2\dot{R}' - R'\dot{\mu} = 0. \tag{iii}$$

(iii) implies

$$(2\ln R' - \mu)^{\cdot} = 0 \implies e^\mu = \frac{R'^2}{1+f}, f \text{ is an arbitrary function of } r. \tag{iv}$$

(ii) implies

$$2R\ddot{R} + \dot{R}^2 - f = 0 \implies (R\dot{R}^2 - fR)^{\cdot} = 0 \implies \dot{R}^2 - f = \frac{F(r)}{R}. \tag{v}$$

Using (iv) and (v), (i) yields

$$\frac{R}{R'}(f' - 2\dot{R}\dot{R}') - (\dot{R}^2 - f) = -8\pi G\rho R^2 \text{ or } \frac{F'}{R'} = 8\pi G\rho R^2. \tag{vi}$$

For $f = 0$, one can get an analytic form of R as

$$R(t, r) = \left[R^{\frac{3}{2}}(r) \pm \frac{3}{2} F^{\frac{1}{2}}(r) t \right]^{\frac{2}{3}}, \text{ where } R(r) = R(0, r). \tag{vii}$$

Also, (vi) yields

$$\frac{\partial}{\partial t}(\rho R^2 R') = 0. \tag{viii}$$

Note 6.17

Energy Conditions

For any arbitrary choice of metric, one can find energy-stress tensor through Einstein's field equation $G_{\mu\nu} = 8\pi G T_{\mu\nu}$. This definitely creates a problem that, in the only existence of solutions of Einstein field equations, one cannot get realistic matter source, i.e., energy-stress tensor. Thus, we have to restrict the arbitrariness of matter distribution by imposing energy conditions. Let us consider general anisotropic matter distribution for which the energy-momentum tensor compatible with spherical symmetry is given by

$$T_{\mu\nu} = (\rho + p_t)U_\mu U_\mu - p_t g_{\mu\nu} + (p_r - p_t)\chi_\mu\chi_\nu,$$

where $U^\mu = \sqrt{g_{tt}}\delta_t^\mu$ is the vector four-velocity and $\chi^\mu = \sqrt{g_{rr}}\delta_r^\mu$ is the space-like vector. $\rho(r)$ is the energy density and p_r is the radial pressure measured in the direction of the space-like vector. p_t is the transverse pressure in the orthogonal direction to p_r. Or, equivalently, one can obtain as (here, the signature of the metric is taken as $(+, -, -, -)$)

$$T_\mu^\nu = diag(\rho, -p_r, -p_t, -p_t).$$

[i] **Null energy condition** (NEC): The NEC states that for any null vector, l^μ,

$$T_{\mu\nu}l^\mu l^\nu \geq 0 \Rightarrow \rho + p_j \geq 0, \ \forall j.$$

[ii] **Weak energy condition** (WEC): The WEC states that for any time-like vector, t^μ,

$$T_{\mu\nu}t^\mu t^\nu \geq 0 \Rightarrow \rho \geq 0 \ and \ \rho + p_j \geq 0, \ \forall j.$$

[iii] **Strong energy condition** (SEC): The SEC states that for any time-like vector, t^μ,

$$\left(T_{\mu\nu} - \frac{T}{2}g_{\mu\nu}\right)t^\mu t^\nu \geq 0 \Rightarrow \rho + \sum_j p_j \geq 0 \ and \ \rho + p_j \geq 0, \ \forall j.$$

[here, $T = T_{\mu\nu}g^{\mu\nu}$ = trace of the stress-energy tensor]

[iv] **Dominant energy condition** (DEC): The DEC states that for any time-like vector, t^μ,

$$T_{\mu\nu}t^\mu t^\nu \geq 0 \ and \ T_{\mu\nu}t^\mu \ is \ not \ space-like.$$

This implies

$$\rho \geq 0 \ and \ p_j \in [-\rho, \rho] \ i.e. \ \rho \geq | p_j | .$$

Dominant energy condition implies WEC but not necessarily SEC.

Note 6.18

Let us consider the line element of spherically symmetric objects (stars, sun, etc.) is

$$ds^2 = e^{2\alpha}dt^2 - e^{2\beta}dr^2 - r^2(d\theta^2 + \sin^2\theta d\phi^2).$$

We can express **self energy** of spherical shell surrounding a solid sphere of radius R as

$$U = -G\int_0^R 4\pi r\rho(r)m(r)dr.$$

The **invariant energy** of the object is

$$E = \int_0^R 4\pi r^2\rho(r)e^{\beta(r)}dr.$$

The **mass** of the object is

$$M = \int_0^R 4\pi r^2\rho(r)dr.$$

Notice that

$$M = E + U.$$

The **gravitational mass** is defined as

$$M_G = \int_0^R 4\pi(T_0^0 - T_1^1 - T_2^2 - T_3^3)r^2 e^{\beta(r)+\alpha(r)}dr.$$

This mass M_G is also known as **Tolman–Whittaker mass**. Alternative form of Tolman–Whittaker mass is given by

$$M_G = r^2 e^{\alpha-\beta}\alpha'.$$

The **surface gravity** is defined as

$$\kappa = -\frac{M_G}{r^2}.$$

Note 6.19

Let us assume the line element of spherically symmetric objects (stars, sun, etc.) is

$$ds^2 = e^{2\phi(r)}dt^2 - \frac{dr^2}{1 - \frac{2m(r)}{r}} - r^2(d\theta^2 + \sin^2\theta d\phi^2).$$

Then the **compactness** of the star $u(r)$ is defined by

$$u(r) = \frac{2m(r)}{r}.$$

It is dimensionless.

The **interior redshift** $Z(r)$ is defined by

$$Z(r) = e^{-\phi(r)} - 1.$$

This interior redshift $Z(r)$, which depends only on $\phi(r)$ should decrease with the increase of r. The **surface redshift** Z_s and compactness are related as

$$Z_s = (1 - u_s)^{-\frac{1}{2}} - 1,$$

where $u_s = \frac{2M(R)}{R}$ with R is the radius of the star.

Note 6.20

In this note, a very brief message on Event Horizon Telescope (EHT), black hole merger, and gravitational wave for the beginner are given as follows:

Black holes are among the most interesting entities in the universe. It is encircling huge amounts of mass in comparatively small regions. These compressed objects have gigantic densities that give rise to some of the strongest gravitational fields in the cosmos. This gravitational field is very strong so that nothing (not even light) can escape. Usually, a black hole is made by a supernova explosion. More precisely, black holes can be formed from the stars with initial masses exceeding ~ 20–$25M_\odot$. It is argued that a cluster of stars comprising N stars produces $\sim 6 \times 10^{-4} N$ black holes with masses m_{bh} between $6M_\odot$ and $18M_\odot$. Einstein's general relativity predicted the black holes in 1915. In 1916, Karl Schwarzschild discovered the solutions for black holes. For several decades, this solution remained as a theoretical interest, however, it got attention among the researchers when X-ray observations confirmed highly energetic emission from the matter in the vicinity of these extreme objects. EHT recently observed the first-ever image of a black hole's dark shadow cast beside the light from the matter in its immediate adjacent vicinity. It is expected that black holes expose a dark shadow, produced by gravitational light bending and a photon is captured at the event horizon. To image and study this phenomenon, scientists have constructed the EHT. The EHT is a very lengthy standard interferometry experiment that directly quantifies the luminosities of the radio illumination distribution on the sky. This permits us to rebuild event-horizon-scale images of the massive black hole in the center of the giant galaxy.

In the next year, after the pioneering introduction of the general relativity, Einstein predicted the existence of gravitational waves from his theory. The acceleration of two colossal stellar objects that spiral in towards each other to merge create gravitational waves. These gravitational waves, that travel at the speed of light, are extremely difficult to detect being infinitesimally small around in the order of 10^{-23} *Hz*, a fraction of the size of an atom. Generally, gravitational radiation can be classified according to their origin as given by: (i) due to merging binaries, (ii) continuous gravitational wave due to a single spinning massive object, like a neutron star, and (iii) stochastic gravitational waves, which were possibly generated during the era of big bang. Obviously, the merger of binaries is a significantly more active and plausible event for the detection of gravitational waves till this date. It is believed that when two black holes spiral towards each other, they will eventually merge. In 2015, the Laser Interferometer Gravitational-Wave Observatory (**LIGO**) and **VIRGO** observatories first observed a **black hole merger** and detected the **gravitational waves**, which are fluctuations in the fabric of spacetime produced by the enormous collision. A system comprising of two black holes in orbit around each other is known as a **binary black hole (BBH)**. The merger of BBHs would be one of the firmest recognized sources of gravitational waves in the universe. The typical waveforms can be calculated using general relativity. After merging of two black holes, the single hole becomes tranquil to a stable form through a phase known as ringdown. The distortion in the shape has degenerated as more gravitational waves emit. After the LIGO and VIRGO experiments in 2015, gravitational-wave astronomy is a growing new field of research. Further, the historic detection of neutron star mergers *GW 170817* in 2017 via gravitational wave and the electromagnetic spectrum, seen by 70 observatories, creates a breakthrough by starting a new branch 'multi-messenger astronomy'.

Particle and Photon Orbits in the Schwarzschild Spacetime

7.1 Motion of Test Particle

Let us consider the motion of a massive test particle or a massless particle, i.e., photon in Schwarzschild spacetime. It is known that all massive particles move along time-like geodesics, whereas photons move along null geodesics. We shall consider geodesics of test particles (either time-like or null) in Schwarzschild spacetime.

Let us take the Lagrangian in the following form as (with p is an affine parameter)

$$L = \left(1 - \frac{2m}{r}\right)\left(\frac{dt}{dp}\right)^2 - \left(1 - \frac{2m}{r}\right)^{-1}\left(\frac{dr}{dp}\right)^2 - r^2\left(\frac{d\theta}{dp}\right)^2 - r^2\sin^2\theta\left(\frac{d\phi}{dp}\right)^2. \tag{7.1}$$

We know

$$\delta \int ds = 0 \Rightarrow \int \delta\left(\frac{ds}{dp}\right)dp = 0 \Rightarrow \delta \int L\,dp = 0,$$

where

$$L = \left(\frac{ds}{dp}\right) = \left(g_{\mu\gamma}\frac{dx^\mu}{dp}\frac{dx^\gamma}{dp}\right)^{\frac{1}{2}}.$$

$\Rightarrow$ Euler–Lagrangian equation

$$\frac{d}{dp}\left(\frac{\partial L}{\partial\left(\frac{dx^\mu}{dp}\right)}\right) - \frac{\partial L}{\partial x^\mu} = 0.$$

Thus, the corresponding Euler–Lagrange's equations are

$$\frac{d}{dp}\left(\frac{\partial L}{\partial r^1}\right) - \frac{\partial L}{\partial r} = 0$$

$$\frac{d}{dp}\left(\frac{\partial L}{\partial\theta^1}\right) - \frac{\partial L}{\partial\theta} = 0, \ etc.$$

["1" implies differentiation with respect to p]

We know the first integral of geodesics equation is

$$g_{\mu\gamma}\frac{dx^\mu}{dp}\frac{dx^\gamma}{dp} = \epsilon, \tag{7.2}$$

where $\epsilon = 1$ for the time-like geodesics and $\epsilon = 0$ for null geodesics.

For the above Lagrangian (7.1), we get the following equations of motion

$$\frac{d}{dp}\left(r^2\frac{d\theta}{dp}\right) = r^2\sin\theta\cos\theta\left(\frac{d\phi}{dp}\right)^2, \tag{7.3}$$

$$\frac{d}{dp}\left(r^2\sin^2\theta\frac{d\phi}{dp}\right) = 0, \tag{7.4}$$

$$\frac{d}{dp}\left[\left(1 - \frac{2m}{r}\right)\frac{dt}{dp}\right] = 0. \tag{7.5}$$

Let us consider the motion in equatorial plane $\theta = \frac{\pi}{2}$, then $\frac{d\theta}{dp} = 0$. Also equation (7.3) implies $r^2\frac{d^2\theta}{dp^2} = 0$. This indicates that if initially θ assumes the constant value $\frac{\pi}{2}$, then throughout the geodesic it assumes the same value. This means planar motion is possible in general relativity, in other words, the geodesic is confined to a single plane as in Newtonian mechanics.

Here, ϕ and t are cyclic coordinates.

Equation (7.4) $\Rightarrow$

$$r^2\frac{d\phi}{dp} = h \quad (constant). \tag{7.6}$$

The above equation indicates the **principle of conservation of angular momentum** with h as the angular momentum.

From (7.5), we obtain,

$$\left(1 - \frac{2m}{r}\right)\frac{dt}{dp} = E \quad (constant). \tag{7.7}$$

Here E is the energy per unit mass. This is the **conservation principle of energy**.

(Actually h and E are the conservation principles corresponding to the cyclic nature of the coordinate ϕ and t.).

Using (7.6) and (7.7), we get from Eq. (7.2) as

$$E^2 - \left(\frac{dr}{dp}\right)^2 - \frac{h^2}{r^2}\left(1 - \frac{2m}{r}\right) = \epsilon\left(1 - \frac{2m}{r}\right),$$

$$or, \quad \left(\frac{dr}{dp}\right)^2 = E^2 - \left(1 - \frac{2m}{r}\right)\left(\epsilon + \frac{h^2}{r^2}\right). \tag{7.8}$$

Equations (7.6) and (7.8) yield

$$\frac{1}{r^4}\left(\frac{dr}{d\phi}\right)^2 = \frac{E^2}{h^2} - \frac{\left[\left(1 - \frac{2m}{r}\right)\left(\epsilon + \frac{h^2}{r^2}\right)\right]}{h^2}. \tag{7.9}$$

Changing the r coordinate by a new one as $U = \frac{1}{r}$, we get,

$$\left(\frac{dU}{d\phi}\right)^2 = \frac{(E^2 - \epsilon)}{h^2} + \left(\frac{2m\epsilon}{h^2}\right)U - (1 - 2mU)U^2. \tag{7.10}$$

Now, we differentiate both sides of the above equation with respect to ϕ to yield

$$\frac{d^2U}{d\phi^2} = \frac{m\epsilon}{h^2} - U + 3mU^2,$$

or

$$\frac{d^2U}{d\phi^2} + U = \frac{m\epsilon}{h^2} + 3mU^2. \tag{7.11}$$

This is the relativistic equation of the motion of particles (massive or massless) in the gravitational field of Schwarzschild spacetime. In the theory of gravitation, the Newtonian equation of motion is

$$\frac{d^2U}{d\phi^2} + U = \frac{m}{h^2}.$$

Thus in relativistic treatment, the last term $3mU^2$ is extra. However, this relativistic correction is very small. The ratio of the last two terms in right-hand side of (7.11) is $3h^2U^2$, which is for earth $\approx 3 \times 10^{-6}$ and for mercury $\approx 10^{-7}$.

7.2 Experimental Test for General Relativity

Using the trajectory in the gravitational field of sun (i.e., in the Schwarzschild spacetime), we try to verify the theory of general relativity.

In the weak field limit, general theory of relativity $\cong$ Newtonian gravity, i.e., any observation, which is consistent with Newtonian theory, gives indirect support to general theory of relativity. But some phenomena, which are not explained by Newtonian theory, can be explained by general theory of relativity.

7.2.1 The precession of the perihelion motion of mercury

The point on the orbit nearest to the sun is called perihelion. Observations indicate that the perihelion of mercury's orbit shows a precession of 5599.74±.41 arc sec per century. Among this 5557.18±.85 arc sec can be explained by Newtonian gravitational theory. It is due to the influences of other

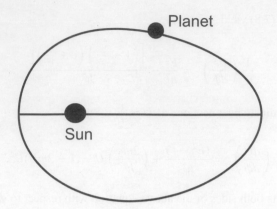

Figure 17   Planets are moving around the sun in an elliptic orbit.

planets. Therefore, an amount $\approx 43.03''$ per century is still unsolved. The unaccounted perihelion motion of $42.56 \pm .94$ arc sec per century can be explained by general theory of relativity.

Consider the motion of a planet. Since mass of a planet is very small compared to the sun, we can regard the planet as a test particle moving in the field of the sun (see Fig. 17). Again, the sun is at least, to a high degree of approximation, a spherical body. So the geometry outside the sun is given by Schwarzschild metric and the planet will move in geodesics (massive particle) in this field. The relativistic equation of planet's orbit is

$$\frac{d^2 U}{d\phi^2} + U = \frac{m}{h^2} + 3mU^2. \tag{7.12}$$

[we have considered planet as a test particle (massive) in the gravitational field of sun and for this $\epsilon = 1$]

Equation (7.12) implies

$$
\begin{aligned}
\frac{d^2 U}{d\phi^2} + U &= \frac{m}{h^2}\left[1 + \frac{3h^2}{r^2}\right] \\
&= \frac{m}{h^2}\left[1 + 3r^2\left(\frac{d\phi}{dp}\right)^2\right] \\
&= \frac{m}{h^2}\left[1 + 3r^2\left(\frac{d\phi}{dt}\right)^2\left(\frac{dt}{dp}\right)^2\right] \\
&= \frac{m}{h^2}\left[1 + 3r^2\frac{v^2}{r^2}\frac{1}{c^2}\right] = \frac{m}{h^2}\left[1 + \frac{3v^2}{c^2}\right],
\end{aligned}
$$

$$\left[\text{since } ds^2 = c^2 dt^2 = dp^2,\ \text{transverse velocity } v = r\frac{d\phi}{dt}\right]$$

i.e.,

$$\frac{d^2U}{d\phi^2} + U \cong \frac{m}{h^2}.$$

[the second term is very small; for mercury $\approx 10^{-7}$]

This is exactly the same of Newtonian equation and has a solution

$$U_0 = \frac{m}{h^2}(1 + e\cos\phi)$$

$$= \frac{1}{l}(1 + e\cos\phi). \quad \left[l = \frac{h^2}{m}\right] \tag{7.13}$$

Using this assumption, one can find approximate solutions of the equation by a perturbation method. We assert the parameter $\epsilon = \frac{3m^2}{h^2}$ and $m = GM$.

Now, Eq. (7.12) yields

$$U^{11} + U = \frac{m}{h^2} + \epsilon\left(\frac{h^2U^2}{m}\right). \quad \left(^1 \equiv \frac{d}{d\phi}\right) \tag{7.14}$$

The solution of this equation is taken in the following form

$$U = U_0 + \epsilon U_1 + O(\epsilon^2) \tag{7.15}$$

Putting this solution in (7.14), we get

$$U_0^{11} + U_0 - \frac{m}{h^2} + \epsilon\left(U_1^{11} + U_1 - \frac{h^2U_0^2}{m}\right) + O(\epsilon^2) = 0.$$

Now, equating the coefficient of different powers of ϵ to zero, we get the zeroth-order solution U_0 as

$$U_0 = \frac{m}{h^2}(1 + e\cos\phi), \quad [\textit{integration constant}, \phi_0 = 0]$$

i.e.,

$$U_0 = \frac{1}{l}(1 + e\cos\phi), \quad \left[l = \frac{h^2}{m}\right]$$

which is usual conic section.

The first-order solution can be obtained from the equation

$$U_1^{11} + U_1 = \frac{h^2}{m}U_0^2 = \frac{m}{h^2}(1 + e\cos\phi)^2$$

$$= \frac{m}{h^2}\left(1 + \frac{1}{2}e^2\right) + \frac{2me}{h^2}\cos\phi + \frac{me^2}{2h^2}\cos 2\phi.$$

We can find the particular solution as

$$U_1 = \frac{1}{D^2 + 1} \left[\frac{m}{h^2} \left(1 + \frac{1}{2}e^2 \right) \right] + \frac{1}{D^2 + 1} \left[\frac{2me}{h^2} \cos\phi \right] + \frac{1}{D^2 + 1} \left[\frac{me^2}{2h^2} \cos 2\phi \right],$$

or

$$U_1 = A + B\phi \sin\phi + C \cos 2\phi,$$

where

$$A = \frac{m}{h^2} \left(1 + \frac{1}{2}e^2 \right); \quad B = \frac{me}{h^2}; \quad C = -\frac{me^2}{6h^2}.$$

Hence, we get the solution in first-order approximation as

$$U \cong U_0 + \epsilon U_1 = U_0 + \epsilon \frac{m}{h^2} \left[1 + e\phi \sin\phi + e^2 \left(\frac{1}{2} - \frac{1}{6} \cos 2\phi \right) \right].$$

The term $e\phi \sin\phi$ provides an important contribution to the correction of U_0. We neglect the third term within the bracket.

[the aperiodic term in the third causes a small change in the latus rectum; the periodic term averaged over a large number of periods vanishes]

Thus,

$$U \cong \frac{m}{h^2} (1 + e\cos\phi + \epsilon e\phi \sin\phi),$$

or

$$U \cong \frac{m}{h^2} [1 + e\cos[\phi(1 - \epsilon)].$$

Hence, we observe the trajectory of the test body is approximately an ellipse, however, orbital period is not 2π, rather it is of period

$$\frac{2\pi}{1 - \epsilon} \cong 2\pi(1 + \epsilon).$$

Thus, we note that planets move in an elliptical way but the axis of the ellipse will rotate by the quantity $2\pi\epsilon$. This is the well-known **precession of the perihelion** (see Fig. 18).

The change in the period of one orbit T is

$$2\pi\epsilon = 6\pi \frac{m^2}{h^2} = \frac{6\pi m^2}{lm} = \frac{6\pi m}{l} = \frac{6\pi m}{a(1 - e^2)}.$$

From Kepler's law, we have

$$T = \frac{2\pi}{\sqrt{m}} a^{\frac{3}{2}}, \quad a = semimajor\ axis\ of\ the\ orbit$$

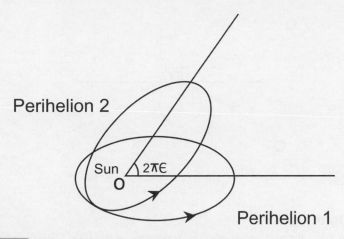

Perihelion 2

Sun $2\pi\epsilon$
O

Perihelion 1

Figure 18 Precession of the perihelion of the planet.

Table I Theoretical prediction and observed values of the advance of perihelion of some planets.

Planet	*General relativity* prediction	Observed
Mercury	43.0	43.1±.5
Venus	8.6	8.4±.8
Earth	3.8	5±1.2
Icarus	10.3	9.8±.8

For nonrelativistic unit, we take into account the speed of light and obtain the advance of perihelion as

$$2\pi\epsilon = \frac{6\pi GM}{c^2 a(1-e^2)} = \frac{24\pi^3 a^2}{c^2 T^2(1-e^2)}. \tag{7.16}$$

For mercury, $a = 0.6 \times 10^8$ km; $e = 0.2$, $T = 88$ days, number of revolutions per century is $= \frac{100 \times 365}{88}$. Therefore, the advance of mercury perihelion per century is given by the amount n where

$$n = \frac{24\pi^3(0.6 \times 10^{13})^2}{(3 \times 10^{10})^2(88 \times 24 \times 3600)^2(1-(0.2)^2)} \times \frac{100 \times 365}{88} \times \frac{180 \times 3600}{\pi} \ sec,$$

or $\approx 43.03''$ per century.

Thus, general theory of relativity resolves the observed irregularity, given before, regarding the advance of perihelion of mercury. Table I indicates the theoretical and observed values of the advance of perihelion of some planets.

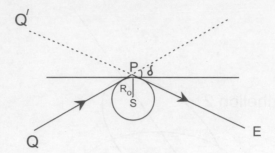

Figure 19 Deviation of the light ray passing near the sun.

7.2.2 Bending of light

The relativistic equation for null geodesics, i.e., trajectory of the light QPE (see Fig. 19) is (putting $\epsilon = 0$ in Eq. (7.11))

$$\frac{d^2U}{d\phi^2} + U = 3mU^2. \tag{7.17}$$

In case of flat spacetime, i.e., when the deflecting source S were absent ($m = 0$), then Eq. (7.17) becomes

$$\frac{d^2U}{d\phi^2} + U = 0.$$

The solution of this equation is

$$U = A\cos\phi + B\sin\phi.$$

Now, we use the following boundary conditions (i) $\phi = 0$, when the value of U is maximum, i.e., when the value of r is minimum, i.e., at closest approach (R_0), $U = \frac{1}{R_0}$ (ii) at $\phi = 0$, one can have turning point, i.e., $\frac{dU}{d\phi} = 0$. Here, R_0 could be solar radius. Hence we get,

$$U = U_0\cos\phi = \frac{1}{R_0}\cos\phi.$$

Substituting this in R.H.S. of (7.17) for U, we get

$$\frac{d^2U}{d\phi^2} + U = 3mU_0^2\cos^2\phi = \frac{3GM}{R_0^2}\cos^2\phi.$$

We can find the particular solution as

$$U = \frac{1}{D^2+1}\left[\frac{3GM}{R_0^2}\cos^2\phi\right] = \frac{1}{D^2+1}\left[\frac{3GM}{2R_0^2}(1+\cos 2\phi)\right],$$

or

$$U = \frac{GM}{2R_0^2}(3 - \cos 2\phi) = \frac{GM}{R_0^2}(2 - \cos^2 \phi).$$

Thus, the general solution is obtained as

$$U = \frac{1}{R_0} \cos \phi + \frac{GM}{R_0^2}(2 - \cos^2 \phi).$$

This curve is described by QPE. The asymptotic directions are given by $U = 0$, i.e.,

$$0 = U_0 \cos \phi + 2mU_0^2 - mU_0^2 \cos^2 \phi.$$

This yields

$$\cos \phi = \frac{1}{2}\left[\frac{1}{mU_0} \pm \frac{1}{mU_0}\sqrt{1 + 8m^2U_0^2}\right],$$

or

$$\cos \phi = \frac{1}{2}\left[\frac{1}{mU_0} \pm \frac{1}{mU_0}(1 + 4m^2U_0^2)\right]. \quad (mU_0 = \frac{GM}{R_0} \text{ is very small})$$

As $\cos \phi$ cannot assume more than unity, we take $-ve$ sign to yield

$$\cos \phi = -2mU_0 = -\sin(2mU_0) = \cos\left(\frac{\pi}{2} + \frac{2m}{R_0}\right).$$

(since mU_0 is small)

$$\textit{Therefore,} \quad \phi = \frac{\pi}{2} + \frac{2m}{R_0}.$$

Hence, net deviation of the light ray is (see Fig. 19)

$$\delta = 2\left(\frac{2m}{R_0}\right) = \frac{4GM}{R_0}.$$

In nonrelativistic unit, net deviation of the light ray is

$$\delta = \frac{4GM}{c^2R_0} \tag{7.18}$$

$$= \frac{4 \times 6.67 \times 10^{-8} \times 2 \times 10^{33}}{(3 \times 10^{10})^2 \times 7 \times 10^{10}} \cong 8.42 \times 10^{-6} \; rad \cong 1.74'' \; \textit{(for sun)}.$$

The results observed by A. Eddington in the year 1919 during solar eclipses gave support to the outcomes predicted by general theory of relativity.

7.2.3 Radar echo delay

In 1964, I. Sharpiro proposed a classical test of general theory of relativity, namely time delay, i.e., a light ray originally takes more time to travel in the curved spacetime than through flat space. Now consider the path of a light ray in the Schwarzschild spacetime. We observe the path in the equatorial plane $\theta = \frac{\pi}{2}$, therefore, the path of the light ray is obtained from (7.7) and (7.8) as (putting $\epsilon = 0$)

$$\left(1 - \frac{2m}{r}\right)\frac{dt}{dp} = E \ (constant),$$

$$\left(\frac{dr}{dp}\right)^2 = E^2 - \left(1 - \frac{2m}{r}\right)\left(\frac{h^2}{r^2}\right),$$

$$or, \quad \left(\frac{dr}{dt}\right) = \pm\left(1 - \frac{2m}{r}\right)\left[1 - \left(1 - \frac{2m}{r}\right)\frac{\beta^2}{r^2}\right]^{\frac{1}{2}}.$$

where $\beta^2 = \frac{h^2}{E^2}$.

We will calculate the time required for light signal passing through the gravitational field of the sun from earth to the planet and back after being reflected from the planet. Let r_c be the closest approach to the sun, therefore, velocity, $\frac{dr}{dt} = 0$ at $r = r_c$. Hence, β can be obtained as

$$\beta^2 = \frac{r_c^2}{1 - \frac{2m}{r_c}}.$$

The total time required for light signal to go from earth to the planet and back after being reflected from the planet is (see Fig. 20)

$$t_{E,P} = 2t(r_E, r_c) + 2t(r_P, r_c),$$

where, r_E and r_P are the distances of earth and planet, respectively, from the sun.

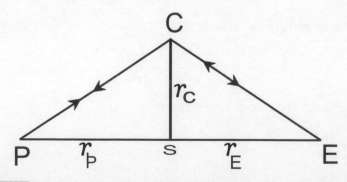

Figure 20 Light signal passing through the gravitational field of the sun, from earth to the planet and back after being reflected from the planet.

Here,

$$t(r_E, r_c) = \int_{r_c}^{r_E} \left(1 - \frac{2m}{r}\right)^{-1} \left[1 - \left(1 - \frac{2m}{r}\right)\frac{\beta^2}{r^2}\right]^{-\frac{1}{2}} dr$$

$$= \int_{r_c}^{r_E} \left(1 - \frac{2m}{r}\right)^{-1} \left[1 - \frac{\left(1 - \frac{2m}{r}\right)r_c^2}{\left(1 - \frac{2m}{r_c}\right)r^2}\right]^{-\frac{1}{2}} dr.$$

Similarly, $t(r_P, r_c)$ is defined. To evaluate the integral, we use some approximations (as $r_E, r_P, r_c \gg 2m = 2GM$) in the integrand ($\alpha$) as

$$\alpha = \left(1 - \frac{2m}{r}\right)^{-1} \left[1 - \frac{\left(1 - \frac{2m}{r}\right)r_c^2}{\left(1 - \frac{2m}{r_c}\right)r^2}\right]^{-\frac{1}{2}}$$

$$\approx \left(1 + \frac{2m}{r}\right)\left[1 - \left(1 - \frac{2m}{r} + \frac{2m}{r_c}\right)\frac{r_c^2}{r^2}\right]^{-\frac{1}{2}}$$

$$= \left(1 + \frac{2m}{r}\right)\left(1 - \frac{r_c^2}{r^2}\right)^{-\frac{1}{2}} \left[1 - \frac{2m\left(1 - \frac{r_c}{r}\right)\left(\frac{r_c}{r^2}\right)}{1 - \frac{r_c^2}{r^2}}\right]^{-\frac{1}{2}}$$

$$= \left(1 + \frac{2m}{r}\right)\left(1 - \frac{r_c^2}{r^2}\right)^{-\frac{1}{2}} \left[1 - \frac{2mr_c}{r(r + r_c)}\right]^{-\frac{1}{2}}.$$

Thus, $\quad \alpha \approx \left(1 - \frac{r_c^2}{r^2}\right)^{-\frac{1}{2}} \left(1 + \frac{2m}{r} + \frac{mr_c}{r(r + r_c)}\right).$

Hence, the time required for light to go from r_c to r_E is

$$t(r_E, r_c) \approx \int_{r_c}^{r_E} \alpha \, dr = \sqrt{r_E^2 - r_c^2} + 2m \ln\left[\frac{r_E + \sqrt{r_E^2 - r_c^2}}{r_c}\right] + m\sqrt{\frac{r_E - r_c}{r_E + r_c}}.$$

Hence, the total time required for light signal to go from earth to the planet and back after being reflected from the planet is

$$t_{E,P} = 2t(r_E, r_c) + 2t(r_P, r_c) = 2\sqrt{r_E^2 - r_c^2} + 4m \ln \left[\frac{r_E + \sqrt{r_E^2 - r_c^2}}{r_c} \right]$$

$$+ 2m\sqrt{\frac{r_E - r_c}{r_E + r_c}} + 2\sqrt{r_P^2 - r_c^2} + 4m \ln \left[\frac{r_P + \sqrt{r_P^2 - r_c^2}}{r_c} \right] + 2m\sqrt{\frac{r_P - r_c}{r_P + r_c}}.$$

In the absence of gravitational field (i.e., $m = 0$), the time is

$$2\sqrt{r_E^2 - r_c^2} + 2\sqrt{r_P^2 - r_c^2}.$$

Therefore, we can estimate the delay in time due to the gravitational field of the sun as

$$\Delta t_{E,P} = 4m \ln \left[\frac{r_E + \sqrt{r_E^2 - r_c^2}}{r_c} \right] \left[\frac{r_P + \sqrt{r_P^2 - r_c^2}}{r_c} \right]$$

$$+ 2m\sqrt{\frac{r_E - r_c}{r_E + r_c}} + 2m\sqrt{\frac{r_P - r_c}{r_P + r_c}}.$$

This excess delay is maximum when r_c is very close to solar radius, i.e., when light signals just scratch the sun, then, $\frac{r_E}{r_c} \gg 1$ and $\frac{r_P}{r_c} \gg 1$. Again, using these approximation, we get

$$\Delta t_{E,P} \approx 4m \left[\ln \left(\frac{4r_E r_P}{r_c^2} \right) + 1 \right].$$

In nonrelativistic unit, this value is

$$\Delta t_{E,P} \approx \frac{4GM}{c^3} \left[\ln \left(\frac{4r_E r_P}{r_c^2} \right) + 1 \right], \tag{7.19}$$

$$\approx 240 \mu \ sec \ (for \ Mercury),$$

$$\approx 200 \mu \ sec \ (for \ Venus).$$

The above prediction of general theory of relativity was confirmed by the experiment performed by Sharpiro et al. in 1971. They obtained time delay from mercury to earth.

7.3 Gravitational Redshift

Another effect predicted by the general theory of relativity is the displacement of the atomic spectral lines in presence of a gravitational field. Let us consider two points denoted by 1 and 2 where two observers are sitting with clocks. Also we assume that their world lines are $x^\alpha = x_1^\alpha$ and $x^\alpha = x_2^\alpha$ (see Fig. 21). The line elements at these points are given by

$$ds^2 = g_{\alpha\beta}dx^{\alpha^2}dx^{\beta^2} = g_{00}c^2 dt^2.$$

(since all spatial infinitesimal displacements vanish)

Hence at two points, we have

$$ds(1) = [g_{00}(1)]^{\frac{1}{2}} c dt; \ ds(2) = [g_{00}(2)]^{\frac{1}{2}} c dt.$$

The proper time is defined by $d\tau = \frac{ds}{c}$, therefore,

$$d\tau(1) = [g_{00}(1)]^{\frac{1}{2}} dt; \ d\tau(2) = [g_{00}(2)]^{\frac{1}{2}} dt.$$

Let the observer 1 transmit radiation to observer 2. Let the time separation between two consecutive wave crests measured by observer 1 be the proper time $d\tau(1)$ and the corresponding interval of reception noted by observer 2 be $d\tau(2)$. Therefore, we have

$$\frac{d\tau(2)}{d\tau(1)} = \left[\frac{g_{00}(2)}{g_{00}(1)}\right]^{\frac{1}{2}}.$$

Let the frequency of the wave sent by observer 1 be ν_1 and when this wave is measured by an observer 2 have the frequency ν_2.

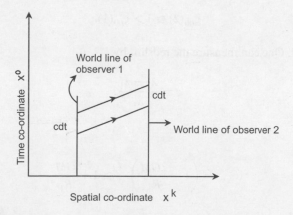

Figure 21 Two observers are sitting with clocks. Observer 1 is sending radiation to observer 2.

Using the relation $v = \frac{1}{T}$, we get,

$$v_2 = v_1 \left[\frac{g_{00}(1)}{g_{00}(2)} \right]^{\frac{1}{2}}.$$

(frequency is expressed in term of proper time)

Also we know $v\lambda = c$, therefore,

$$\frac{\lambda_2}{\lambda_1} = \sqrt{\frac{g_{00}(2)}{g_{00}(1)}}.$$

($\lambda_{1,2}$ = wavelength of the same wave as measured by observers 1 and 2, respectively)

If $g_{00}(1) < g_{00}(2)$, the wave is redshifted ($\lambda_2 > \lambda_1$) while $g_{00}(1) > g_{00}(2)$, it is blue shifted. If "2" is an observer on the earth and "1" is an emitter on the surface of a compact star, 2 will observe that the light from the star to be redshifted. This phenomena can be explained as follows:

Let us consider a spherical star with mass M and radius R_1; then

$$g_{00}(1) = 1 - \frac{2GM}{R_1}$$

Let R_2 be the radius of the earth in the Schwarzschild spacetime centered on the star, then (if we consider the star as sun, then R_2 will be the radius of the earth's orbit around the sun)

$$g_{00}(2) = 1 - \frac{2GM}{R_2}.$$

[there is a very negligible effect on spacetime geometry due to earth's gravity.]

Usually $R_2 >>> 2GM$, therefore,

$$g_{00}(2) \cong 1 > g_{00}(1).$$

Hence, we get redshift. One can measure the redshift by,

$$z = \frac{\lambda_2 - \lambda_1}{\lambda_1}.$$

Hence,

$$1 + z = \left(1 - \frac{2GM}{R_1} \right)^{-\frac{1}{2}} \cong 1 + \frac{GM}{R_1}.$$

For nonrelativistic unit,

$$z = \frac{GM}{c^2 R_1}. \tag{7.20}$$

The gravitational redshift for the light from the sun will be $z \sim 2 \times 10^{-6}$. This effect is larger for the light from white dwarf stars, which are more compact than the sun (a white dwarf of one solar mass has a radius of about 5×10^3 *km*).

For Sirius B, $z \sim 3 \times 10^{-4}$.

7.4 Stable Circular Orbits in the Schwarzschild Spacetime

We know that all the test particles, either massive or massless, follow the geodesics in any gravitational field. For such geodesics, we have from (7.8),

$$\dot{r}^2 = \left(\frac{dr}{ds}\right)^2 = E^2 - \left(1 - \frac{2m}{r}\right)\left(\epsilon + \frac{h^2}{r^2}\right). \tag{7.21}$$

Here, $\epsilon = 1$ for massive particle and $\epsilon = 0$ for massless particle. We can write the above equation as

$$\left(\frac{dr}{ds}\right)^2 = E^2 - V^2, \tag{7.22}$$

where the effective potential V assumes the following form

$$V^2 = \left(1 - \frac{2m}{r}\right)\left(1 + \frac{h^2}{r^2}\right), \quad \textit{for massive or time-like particle} \tag{7.23}$$

$$V^2 = \left(1 - \frac{2m}{r}\right)\left(\frac{h^2}{r^2}\right), \quad \textit{for massless particle or photon.} \tag{7.24}$$

The extrema of V, i.e., maximum at $r = r_{max}$ and minimum at $r = r_{min}$, are given by the equation $\frac{\partial V}{\partial r} = 0$,

$$mr^2 - h^2 r + 3mh^2 = 0, \quad \textit{for massive particle}$$

and

$$1 - \frac{3m}{r} = 0, \textit{ for photon}$$

Thus, for massive particle

$$r_{max} = \frac{[h^2 - (h^4 - 12m^2 h^2)^{1/2}]}{2m}, \tag{7.25}$$

$$r_{min} = \frac{[h^2 + (h^4 - 12m^2 h^2)^{1/2}]}{2m}. \tag{7.26}$$

Note that the maximum and the minimum coincide when $h^2 = 12m^2$, hence

$$r_{max} = r_{min} = 6m \tag{7.27}$$

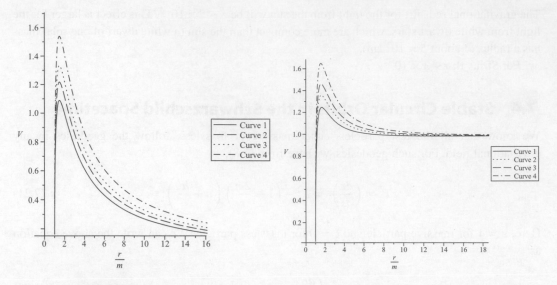

Figure 22 (Left) Effective potential of massless particle against $\frac{r}{m}$ for different values of $\frac{h^2}{m^2}$. Curve 1 for $\frac{h^2}{m^2} = 8$, Curve 2 for $\frac{h^2}{m^2} = 10$, Curve 3 for $\frac{h^2}{m^2} = 12$, Curve 4 for $\frac{h^2}{m^2} = 16$. Note that V has only one extremum point, which is maximum. (Right) Effective potential of massive particle against $\frac{r}{m}$ for different values of $\frac{h^2}{m^2}$. Curve 1 for $\frac{h^2}{m^2} = 8$, Curve 2 for $\frac{h^2}{m^2} = 10$, Curve 3 for $\frac{h^2}{m^2} = 12$, Curve 4 for $\frac{h^2}{m^2} = 16$. Note that V has both maximum and minimum points.

Figure 22 shows the effective potential V of massive and massless particle against $\frac{r}{m}$.

For the particle with constant energy E, one can have two turning points with $\frac{dr}{ds} = 0$. Here the orbit will be elliptic as we know any massive particle moves along an elliptical path in the Schwarzschild spacetime (e.g., planet's path around the sun is elliptic). This orbit will be circular at maximum or minimum values of the effective potential V (i.e. when $\frac{\partial V}{\partial r} = 0$). Obviously, the orbit will be unstable at maximum potential and stable at minimum potential. Rewriting (7.25) and (7.26), we obtain,

$$r_{max} = \frac{6m}{[1 + (1 - \frac{12m^2}{h^2})^{1/2}]}, \quad r_{min} = \frac{6m}{[1 - (1 - \frac{12m^2}{h^2})^{1/2}]}. \quad (7.28)$$

r_{max} will be minimum when $\frac{h^2}{m^2} \to \infty$. Thus $(r_{max})_{min} = 3m$. r_{min} will be minimum when $\frac{h^2}{m^2} = 12$. Thus $(r_{min})_{min} = 6m$.

Hence, one can have closest stable circular orbit with radius

$$r_{csco} = 6m, \quad (7.29)$$

and shortest circular orbit with radius

$$r_{sco} = 3m. \quad (7.30)$$

Thus, we have seen that for massive particle, Schwarzschild solution possesses stable circular orbits for $r > 6m$ and unstable circular orbits for $3m < r < 6m$.

For massless particle or photon, the circular orbit is always unstable as it has only one extremum, which is maximum at $r = 3m$.

Exercise 7.1

Suppose the spacetime metric outside of a star is given by

$$ds^2 = \left(1 - \frac{2M}{r}\right)[dt^2 - dr^2 - r^2(d\theta^2 + \sin^2\theta d\phi^2)].$$

Find the angle of deflection of light by this spherical star.
Hint: Here the Lagrangian is

$$L = \left(1 - \frac{2M}{r}\right)[\dot{t}^2 - \dot{r}^2 - r^2\dot{\theta}^2 - r^2\sin^2\theta\dot{\phi}^2]$$

The first integral of the null geodesics equation is

$$g_{\mu\gamma}\frac{dx^\mu}{dp}\frac{dx^\gamma}{dp} = 0.$$

Let us consider the motion in equatorial plane $\theta = \frac{\pi}{2}$. Using Euler–Lagrangian equation, one can find the equations of motion as in $(7.5 - 7.7)$, which yield

$$\frac{d\phi}{dr} = \frac{1}{r^2}\left(\frac{E^2}{h^2} - \frac{1}{r^2}\right)^{-\frac{1}{2}},$$

where E and h are constants. Using $r = \frac{1}{U}$, the above equation yields

$$\frac{d^2U}{d\phi^2} + U = 0.$$

This equation implies deflection of light is zero as like flat spacetime. We know light rays are moving along null curves for which $ds^2 = 0$. Here the metric is conformally related to the flat space, $ds^2 = (factor) \times ds^2_{flat}$.

Note 7.1

The Parameterized Post-Newtonian Formalism
Parameterized post-Newtonian formalism is actually known as PPN formalism. When one considers the slow motion and weak field limit, then full gravitational theory turns into a simple form. This estimation is recognized as the post-Newtonian limit. It is more precise to the exact phenomena than standard Newtonian gravitational theory. The metric in gravitational theory and the spacetime metric in this limit has the same structure. In PPN formalism, metric can be expressed as the

dimensionless gravitational potentials of varying degrees of smallness, which is an expansion about the flat Minkowskian metric ($h_{ij} = diag(1, -1, -1, -1)$). This formalism is frequently used for calculations of the phenomena where the gravitational field is very weak and the velocities are nonrelativistic, e.g., calculations in the solar system.

In general theory of relativity, consider a spherically symmetric metric

$$ds^2 = A(r)(cdt)^2 - B(r)dr^2 - r^2(d\theta^2 + \sin^2\theta d\phi^2).$$

The spacetime geometry outside of a spherical symmetric body (star) of mass M is Schwarzschild spacetime. In agreement with Newtonian theory, i.e., static weak field metric, the forms of A and B should be

$$A(r) = 1 - \frac{2GM}{c^2 r} + \dots\dots, \quad B(r) = 1 + \dots\dots$$

The first post-Newtonian correction is expressed as

$$A(r) = 1 - \frac{2GM}{c^2 r} + 2(\beta - \gamma)\left(\frac{2GM}{c^2 r}\right)^2 + \dots\dots,$$

$$B(r) = 1 + 2\gamma\left(\frac{GM}{c^2 r}\right) + \dots\dots$$

Here, β and γ are two PPN parameters. Note that these parameters may vary for different gravitational theories. For Schwarzschild metric in general relativity, $\beta = \gamma = 1$.

In PPN framework, the expressions for the precession of perihelion of a planet per orbit, bending of light by the sun, and time delay of the light given in (7.16), (7.18), and (7.19), respectively, are modified and take the following forms

$$2\pi\epsilon = \frac{1}{3}(2 + 2\gamma - \beta)\frac{6\pi GM}{c^2 a(1 - e^2)},$$

$$\delta = \left(\frac{1 + \gamma}{2}\right)\left(\frac{4GM}{c^2 R_0}\right),$$

$$\Delta t_{E,P} \approx \left(\frac{1 + \gamma}{2}\right)\frac{4GM}{c^3}\left[\ln\left(\frac{4r_E r_P}{r_c^2}\right) + 1\right].$$

Note 7.2

A general treatment for the experimental test of general theory of relativity for a general static and spherically symmetric body

I Perihelion Precession

To obtain the desired features, we take the general static and spherically symmetric configuration as

$$ds^2 = A(r)c^2dt^2 - B(r)dr^2 - C(r)(d\theta^2 + \sin^2\theta d\phi^2). \tag{7.31}$$

Now, we begin with the Lagrangian, which can be written as

$$\pounds = Ac^2\dot{t}^2 - B\dot{r}^2 - C\dot{\theta}^2 - C\sin^2\theta\dot{\phi}^2. \tag{7.32}$$

Here over dot indicates the differentiation with respect to the affine parameter τ.

Since the gravitational field is isotropic so angular momentum is conserved. Therefore, geodesic of any arbitrary body (either massive planets or massless photons) are planar. Without any loss of generality, we can select the equatorial plane as $\theta = \frac{\pi}{2}$. Hence, the Lagrangian assumes the following form

$$\pounds = Ac^2\dot{t}^2 - B\dot{r}^2 - C\dot{\phi}^2, \tag{7.33}$$

with massless particle photon, $\pounds = 0$ and for any massive particle, $\pounds = 1$.

Using the generalized coordinates q_i and generalized velocities $\dot{q}_i$, the Euler–Lagrange equations

$$\frac{d}{ds}\left(\frac{\partial L}{\partial \dot{q}_i}\right) - \frac{\partial L}{\partial q_i} = 0. \tag{7.34}$$

give

$$Ac^2\dot{t} = E \text{ and } C\dot{\phi} = L.$$

Here E and L denote the energy and momentum of the particle, respectively, such that

$$\dot{t} = \frac{E}{c^2A} \text{ and } \dot{\phi} = \frac{L}{C}$$

and hence with these notations Eq. (7.33) becomes

$$\pounds = \frac{E^2}{Ac^2} - B\dot{r}^2 - \frac{L^2}{C}, \tag{7.35}$$

which, after some mathematical calculations, is written as

$$\dot{r}^2 = -\frac{\pounds}{B} + \frac{E^2}{ABc^2} - \frac{L^2}{BC}. \tag{7.36}$$

This implies

$$\left(\frac{dr}{d\phi}\right)^2 + \frac{C}{B} = -\frac{\pounds C^2}{BL^2} + \frac{E^2 C^2}{ABL^2 c^2}. \tag{7.37}$$

Now, replacing $r = 1/U$ in Eq. (7.37), we have

$$\left(\frac{dU}{d\phi}\right)^2 + \frac{CU^4}{B} = -\frac{\pounds C^2 U^4}{BL^2} + \frac{E^2 C^2 U^4}{ABL^2 c^2}. \tag{7.38}$$

Letting the subsequent transformations,

$$\frac{1}{B} = 1 - f(U), C = \frac{1}{U^2} + \frac{g(U)}{U^4}. \tag{7.39}$$

the above equation is expressed as

$$\left(\frac{dU}{d\phi}\right)^2 + U^2 = U^2 f(U) + f(U)g(U) - g(U) - \frac{\pounds C^2 U^4}{BL^2} + \frac{E^2 C^2 U^4}{ABL^2 c^2} \equiv G(U), \tag{7.40}$$

Now, differentiation with respect to U yields

$$\frac{d^2 U}{d\phi^2} + U = F(U), \tag{7.41}$$

where,

$$F(U) = \frac{1}{2} \frac{dG(U)}{dU}. \tag{7.42}$$

It is known that the planetary orbits are nearly circular, therefore, the circular orbit $U = U_0$ can be found from the equation

$$F(U_0) = U_0. \tag{7.43}$$

Now, the perihelion precession as a deviation $\delta = U - U_0$ from the circular solution should follow the equation

$$\frac{d^2 \delta}{d\phi^2} + \left[1 - \left(\frac{dF}{dU}\right)_{U=U_0}\right] \delta = 0(\delta^2). \tag{7.44}$$

The trajectory, which is a solution of the above equation as (to the first order in δ)

$$\delta = \delta_0 \cos\left(\sqrt{\left[1 - \left(\frac{dF}{dU}\right)_{U=U_0}\right]} \phi + \beta\right), \tag{7.45}$$

where δ_0 and β are integration constants.

The perihelion precession (from one perihelion to the next), i.e., variation of the orbital angle is

$$\phi = \frac{2\pi}{\left(\sqrt{\left[1 - \left(\frac{dF}{dU}\right)_{U=U_0}\right]}\right)} \equiv \frac{2\pi}{1 - \sigma}. \tag{7.46}$$

From the above equation, the precession σ is given by,

$$\sigma = 1 - \left(\sqrt{\left[1 - \left(\frac{dF}{dU}\right)_{U=U_0}\right]}\right). \tag{7.47}$$

Thus, for $\left(\frac{dF}{dU}\right)_{U=U_0} \ll 1$,

$$\sigma = \frac{1}{2}\left(\frac{dF}{dU}\right)_{U=U_0}. \tag{7.48}$$

Thus, for complete rotation, the advancement of perihelion is

$$\Delta\phi = \phi - 2\pi \approx 2\pi\sigma. \tag{7.49}$$

II Deflection of Light

Now, we would like to observe how the general static and spherically symmetric configuration affects the bending of light. We begin with the Eq. (7.40) (assuming $\pounds = 0$),

$$\left(\frac{dU}{d\phi}\right)^2 + U^2 = U^2 f(U) + f(U)g(U) - g(U) + \frac{E^2 C^2 U^4}{ABL^2 c^2} \equiv P(U). \tag{7.50}$$

Differentiation with respect to U yields

$$\frac{d^2 U}{d\phi^2} + U = Q(U), \tag{7.51}$$

where,

$$Q(U) = \frac{1}{2}\frac{dP(U)}{dU}. \tag{7.52}$$

Now, one can solve the equation using successive approximation. One starts with the straight line (where no gravitating body is present) as zeroth approximation such that $U = \frac{\cos\phi}{R_0}$ where $\phi = 0$ is the point of nearest approach to the sun's surface. Preferably, one can consider R_0 as the solar radius. Replacing this on the right-hand side of Eq. (7.51) for U, one can obtain

$$\frac{d^2 U}{d\phi^2} + U = Q\left(\frac{\cos\phi}{R_0}\right). \tag{7.53}$$

This provides the general solution for $U = U(\phi)$.

Total deflection angle $\Delta\phi = 2\epsilon$ can be obtained from the equation $U(\frac{\pi}{2} + \epsilon) = 0$.

III   Alternative Method of Calculating Deflection of Light

Equation (7.37) implies (with $\pounds = 0$)

$$\left(\frac{dr}{d\phi}\right)^2 = \frac{E^2 C^2}{ABL^2 c^2} - \frac{C}{B}. \tag{7.54}$$

This implies

$$\phi(r) = \phi(\infty) + \int_r^\infty \frac{\sqrt{AB}\,dr}{\sqrt{C}\sqrt{\left[\frac{E^2 C}{L^2 c^2} - A\right]}}, \tag{7.55}$$

At the distance of closest approach $r = r_0$ (which is equivalent to sun's radius), $\frac{dr}{d\phi} = 0$. This gives $\frac{E^2}{c^2 L^2} = \frac{A(r_0)}{C(r_0)}$. Hence we get,

$$\phi(r) = \phi(\infty) + \int_{r_0}^\infty \frac{\sqrt{B}\,dr}{\sqrt{C}\sqrt{\left[\frac{C(r)A(r_0)}{C(r_0)A(r)} - 1\right]}}. \tag{7.56}$$

Taking one transformation $r = r_0 x$, the above equation takes the following as

$$\phi(x) = \phi(\infty) + \int_1^\infty \frac{r_0\sqrt{B(r_0 x)}\,dx}{\sqrt{C((r_0 x))}\sqrt{\left[\frac{C((r_0 x))A(r_0)}{C(r_0)A((r_0 x))} - 1\right]}}. \tag{7.57}$$

The deflection angle is

$$\Delta\phi = 2|\phi(x) - \phi(\infty)| - \pi. \tag{7.58}$$

IV   Gravitational Time Delay

Equation (7.36) implies (with $\pounds = 0$)

$$\dot{r}^2 = \frac{E^2}{ABc^2} - \frac{L^2}{BC}. \tag{7.59}$$

Using $\dot{t} = \frac{E}{Ac^2}$, one will find from the above equation

$$\left(\frac{dr}{dt}\right)^2 = \frac{c^2 A}{B} - \frac{L^2 c^4 A^2}{E^2 BC}. \tag{7.60}$$

At the distance of closest approach $r = r_0$ (which is equivalent to sun's radius), $\frac{dr}{dt} = 0$. This gives $\frac{L^2}{E^2} = \frac{C(r_0)}{c^2 A(r_0)}$.

Hence we get,

$$t(r, r_0) = \int_{r_0}^{r} \frac{dr}{\sqrt{\left[\frac{c^2 A}{B} - \frac{c^2 C(r_0) A^2}{A(r_0) BC}\right]}}. \tag{7.61}$$

When no gravitating body is present, $A = 1$, $B = 1$ and $C = r^2$, which yields

$$t_0 = \int_{r_0}^{r} \frac{dr}{c \sqrt{\left[1 - \frac{r_0^2}{r^2}\right]}}. \tag{7.62}$$

Hence, light ray takes time to travel from mercury to earth is given by

$$t(r, r_0) = \int_{r_0}^{r_1} \frac{dr}{\sqrt{\left[\frac{c^2 A}{B} - \frac{c^2 C(r_0) A^2}{A(r_0) BC}\right]}} + \int_{r_0}^{r_2} \frac{dr}{\sqrt{\left[\frac{c^2 A}{B} - \frac{c^2 C(r_0) A^2}{A(r_0) BC}\right]}}. \tag{7.63}$$

Therefore, time delay for a round trip can be obtained as

$$\Delta t = 2(t - t_0) = \int_{r_0}^{r_1} 2 \left[\frac{1}{\sqrt{\left[\frac{c^2 A}{B} - \frac{c^2 C(r_0) A^2}{A(r_0) BC}\right]}} - \frac{1}{c \sqrt{\left[1 - \frac{r_0^2}{r^2}\right]}} \right] dr$$

$$+ \int_{r_0}^{r_2} 2 \left[\frac{1}{\sqrt{\left[\frac{c^2 A}{B} - \frac{c^2 C(r_0) A^2}{A(r_0) BC}\right]}} - \frac{1}{c \sqrt{\left[1 - \frac{r_0^2}{r^2}\right]}} \right] dr. \tag{7.64}$$

Exercise 7.2

Find the deflection of light as a function of PPN parameters.

Hint: Write the Lagrangian similar to (7.1) as

$$L = \left[1 - \frac{2GM}{c^2 r} + 2(\beta - \gamma)\left(\frac{2GM}{c^2 r}\right)^2\right]\left(\frac{dt}{dp}\right)^2 - \left[1 + 2\gamma\left(\frac{GM}{c^2 r}\right)\right]\left(\frac{dr}{dp}\right)^2$$

$$- r^2 \left(\frac{d\theta}{dp}\right)^2 - r^2 \sin^2\theta \left(\frac{d\phi}{dp}\right)^2,$$

and proceed as before.

Note 7.3

Massless scalar wave equation in the spherically symmetric spacetime

The minimally coupled massless wave equation in a spherically symmetric spacetime background

$$ds^2 = e^{2f(r)} dt^2 - \left(1 - \frac{b(r)}{r}\right)^{-1} dr^2 - r^2(d\theta^2 + \sin^2\theta \, d\phi^2),$$

via **Klein–Gordon equation**

$$\Box\Phi = \frac{1}{\sqrt{-g}} \partial_\mu [\sqrt{-g} g^{\mu\nu} \partial_\nu \Phi] = 0,$$

is given by

$$\frac{\partial^2\Phi}{\partial t^2} \Big/ e^{2f(r)} - \frac{\left(1 - \frac{b(r)}{r}\right)}{e^{2f(r)} r^2} \frac{\partial}{\partial r}\left[e^{2f(r)} r^2 \frac{\partial\Phi}{\partial r}\right] - \frac{1}{r^2 \sin\theta} \frac{\partial}{\partial\theta}\left[\sin\theta \frac{\partial\Phi}{\partial r}\right] - \frac{1}{r^2 \sin\theta} \frac{\partial^2\Phi}{\partial\phi^2} = 0.$$

For the spherically symmetric spacetime, the equation related to scalar field can be solved by using separation of variables as,

$$\Phi_{lm} = Y_{lm}(\theta, \phi) \frac{U_l(r, t)}{r}.$$

Here $Y_{lm}(\theta, \phi)$ are the spherical harmonics and l is the quantum angular momentum.

The possibility of astrophysical observations (such as to detect black holes, wormholes, compact stars, etc.) provides the inspiration for studying the scattering of scalar waves in the spherically symmetric spacetimes. These observations will be interesting to researchers on gravitational radiation, as well as it will be very help full to determine the possible existence of actual astrophysical objects.

Above separable forms yield

$$\left[\frac{1}{\sin\theta} \frac{\partial}{\partial\theta} \sin\theta \frac{\partial}{\partial\theta} + \frac{1}{\sin^2\theta} \frac{\partial^2}{\partial\phi^2}\right] Y_{lm} = l(l+1) Y_{lm}$$

and

$$\frac{\partial^2 U_l}{\partial t^2} + \frac{\partial^2 U_l}{\partial r^{*2}} = V_l U_l,$$

where the potential V_l is given by

$$V_l = e^{2f}\left[\frac{l(l+1)}{r^2} - \frac{b'r - b}{2r^3} + \frac{1}{r}\left(1 - \frac{b}{r}\right)f'\right].$$

Here, the tortoise coordinate transformation r^* is used, i.e.,

$$\frac{\partial}{\partial r^*} = e^f \sqrt{1 - \frac{b}{r}} \frac{\partial}{\partial r}.$$

Here, the dot implies the differentiation with respect to t. Basically, r^* represents the proper distance.
    Considering the time dependence of the wave to be harmonic, one can write

$$U_l(r,t) = \widehat{U}_l(r,\omega) e^{-i\omega t},$$

where ω is the oscillating frequency of the scalar field.
    Using this, one can get the Schrödinger equation

$$\left[\frac{d^2}{dr^{*2}} + \omega^2 - V_l(r) \right] \widehat{U}_l(r,\omega) = 0.$$

Note also that for nonzero spin s, the potential $V_l(r)$ takes the form

$$V_l = e^{2f} \left[\frac{l(l+1)}{r^2} + (1-s) \left\{ \frac{-b'r + (1+2s)b}{2r^3} + \frac{1}{r}\left(1 - \frac{b}{r}\right) f' \right\} \right].$$

For $s = 0$, we come back the above potential with zero spin.
    For $s = 1$, we have vector wave, e.g., electromagnetic field,

$$V_l = e^{2f} \left[\frac{l(l+1)}{r^2} \right].$$

Note 7.4

The motion of test particle using Hamilton–Jacobi (H–J) approach
Let us consider that a test particle with mass m and charge e is moving in a particular gravitational field. The standard H–J equation for the test particle is written as

$$g^{ik} \left(\frac{\partial S}{\partial x^i} + eA_i \right) \left(\frac{\partial S}{\partial x^k} + eA_k \right) + m^2 = 0,$$

where $A_i = \frac{Qdt}{r}$ is the gauge potential (e.g., electromagnetic vector potential) and g_{ik} represents the classical background field and S is the standard Hamilton's characteristic function. We may write the explicit form of H–J equation for the metric

$$ds^2 = f(r)dt^2 - g(r)dr^2 - r^2(d\theta^2 + \sin^2\theta\, d\phi^2),$$

as

$$\frac{1}{f}\left(\frac{\partial S}{\partial t} + \frac{eQ}{r} \right)^2 - \frac{1}{g}\left(\frac{\partial S}{\partial r} \right)^2 - \frac{1}{r^2}\left(\frac{\partial S}{\partial \theta} \right)^2 - \frac{1}{r^2 \sin^2\theta}\left(\frac{\partial S}{\partial \phi} \right)^2 + m^2 = 0.$$

One can solve this partial equation using the method of separation of variables technique. Let us write the *H–J* function *S* in separable form as

$$S(t, r, \theta, \phi) = -E.t + S_1(r) + S_2(\theta) + J.\phi,$$

where *E* and *J* represent the energy and the angular momentum of the particle, respectively.

If we substitute this ansatz for *S* in the above *H–J* equation, then we get the following expression for the unknown function S_1 and S_2:

$$S_1(r) = \epsilon \int \left[\frac{g}{f} \left(E - \frac{eQ}{r} \right)^2 - gm^2 - \frac{gp^2}{r^2} \right]^{\frac{1}{2}} dr,$$

$$S_2(\theta) = 7\epsilon \int \left[p^2 - J^2 cosec^2\theta \right]^{\frac{1}{2}} d\theta,$$

where $\epsilon = \pm 1$, the sign changing whenever *r* passes through a zero of the integral and *p* is a separation constant.

According to *H–J* method to find the trajectory of the particle, we consider $\frac{\partial S}{\partial E} = $ constant; $\frac{\partial S}{\partial p} = $ constant ; $\frac{\partial S}{\partial J} = $ constant.

(without any loss of generality, we have preferred the constants to be zero)

Hence, we get

$$t = \epsilon \int \frac{g}{f} \left(E - \frac{eQ}{r} \right) \left[\frac{g}{f} \left(E - \frac{eQ}{r} \right)^2 - gm^2 - \frac{gp^2}{r^2} \right]^{-\frac{1}{2}} dr,$$

$$S_2(\theta) = \epsilon \int J^2 cosec^2\theta \left[p^2 - J^2 cosec^2\theta \right]^{-\frac{1}{2}} d\theta,$$

$$\cos^{-1}\left(\frac{\cos\theta}{U} \right) = \epsilon \int \frac{g}{r^2} \left[\frac{g}{f} \left(E - \frac{eQ}{r} \right)^2 - gm^2 - \frac{gp^2}{r^2} \right]^{-\frac{1}{2}} dr,$$

with $U^2 = 1 - \frac{J^2}{p^2}$.

From the expression of *t*, we get the radial velocity of the particle as

$$\frac{dr}{dt} = \frac{\left[\frac{g}{f} \left(E - \frac{eQ}{r} \right)^2 - gm^2 - \frac{gp^2}{r^2} \right]^{\frac{1}{2}}}{\frac{g}{f} \left(E - \frac{eQ}{r} \right)}.$$

$\frac{dr}{dt} = 0$ gives the turning points of the trajectory and as a consequence the potential curve are

$$V(r) \equiv \frac{E}{m} = \frac{eQ}{mr} + \sqrt{f(r)} \left(\frac{p^2}{m^2 r^2} + 1 \right)^{\frac{1}{2}}.$$

Note that for uncharged particle, $e = 0$.

Causal Structure of Spacetime

8.1 Introduction

From a physical point of view, an outcome cannot occur before its cause. In special relativity, by causality we mean that an outcome cannot occur from a cause that does not lie in the back (past) light cone of that event. Likewise, a cause cannot affect outside its front (future) light cone. These limitations are expected as causal impacts cannot propagate quicker than the speed of light and/or backward in time. Usually, in relativity, the spacetime is characterized by a Lorentzian manifold. Actually, a Lorentzian manifold is a specific case of a pseudo-Riemannian manifold, where the signature of the metric is $+2$ or -2. Generally, the causal relationships among the points in the manifold are understood by defining which events in spacetime can affect which other events. Causal structures in Minkowski spacetime assume, naturally, a simple form as the spacetime is flat. However, the causal relationships between points of the curved spacetime will be more complicated due to the presence of curvature. The description of the causal structure in curved spacetime is expressed in terms of smooth curves joining pairs of points. The causal relationship is defined by the conditions on the tangent vectors of the curves. In general relativity, the metric of the spacetime up to a conformal factor can be recuperated from the causal structure. In this chapter, we explore different properties of such causal relationships instituting a number of results that will be used for advanced topics in general relativity, particularly to prove the existence of singularities.

8.2 Causality

Causal structure in the special theory of relativity, i.e., in Minkowski spacetime or flat spacetime is characterized by the fact that no massive particle can travel faster than light. In general relativity, also, locally there is no difference between the causality relation with that in Minkowski spacetime. However, globally the causality relation is significantly different due to various spacetime topologies.

Let (M, g_{ab}) be a spacetime that is connected, Hausdorff, paracompact, and admits a Lorentzian metric g_{ab} of signature $(+, -, -, -)$. Let p be an arbitrary event in M, then the tangent space V_p is isomorphic to Minkowski spacetime. As a consequence, one can note that the light cone of p is a subspace of V_p. Similar to Minkowski spacetime, for every event $p \in M$ we can elect half of the light cone as *future* and other half as *past*. But it is not always possible for a non-simply connected spacetime to make a continuous designation of future and past as p varies over M. (M, g_{ab}) is **time orientable** if there is a possibility to create a continuous designation of future and past. A time-like or null vector is **future-directed**, if it lies totally in the *future* half of the light cone. It is

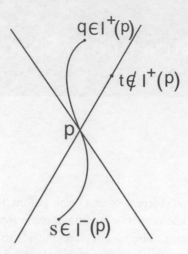

Figure 23   Chronological future and past.

assumed that (M, g_{ab}) is a time-orientable spacetime. A differentiable curve $\lambda(t)$ (t is time parameter) is a **future-directed time-like curve** if for each event p on $\lambda(t)$, the tangent vector at p, t^a is a future-directed time-like vector. Likewise, λ is known as a **future-directed causal curve** if for each event p on $\lambda(t)$, the tangent vector at p, t^a is either future-directed time-like vector or null vector. We can use similar definitions for **past-directed time-like and causal curves**.

An event p **chronologically precedes** another event q, if there exists a smooth future-directed time-like curve from p to q. It is denoted by $p << q$. An event p **causally precedes** another event q, if there exists a nonspace-like curve, i.e., either time-like or null curve from p to q. It is denoted by $p < q$.

The **chronological future** of $p \in M$, denoted by $I^+(p)$, is defined as the set of all events q such that $p << q$. Analogous definition is used for **chronological past** $I^-(p)$ (see Fig. 23).

Thus, we have

$$I^+(p) = \{q \in M \text{ such that } p << q\},$$

$$I^-(p) = \{q \in M \text{ such that } q << p\}.$$

The **causal future** for p can be defined in a similar way as $I^+(p)$ by replacing "future-directed time-like curve" with "future-directed causal curve"(see Fig. 24). Thus,

$$J^+(p) = \{q \in M \text{ such that } p < q\},$$

$$J^-(p) = \{q \in M \text{ such that } q < p\}.$$

The above definitions imply that

$$\overline{I^+(p)} = \overline{J^+(p)} \quad and \quad \dot{I}^+(p) = \dot{J}^+(p),$$

where for a set A, $\overline{A}$ is the closure of A and $\dot{A} = \overline{A} \cap \overline{M - A}$ denotes the topological boundary.

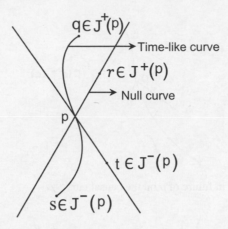

Figure 24 Causal future and past.

In general, p does not belong to $I^+(p)$. However, if $p \in I^+(p)$, then $\exists$ a closed time-like curve starts and ends at p. $p \in J^+(p)$ as the causal future of p consists of points that can be joined to p by time-like or null curves, and the constant path $\lambda(t) = p$ connecting p to p itself has vanishing tangent vector and, hence, is a null curve. One can notice that the relations $<$ *and* $<<$ are transitive. Also, if $p < q$ and $q << r$, or, $p << q$ and $q < r$, then $p << r$.

Any point p in the spacetime (M, g_{ab}), i.e., $p \in M$, actually represents an event in the spacetime.

If $q \in I^+(p)$, then, $p \in I^-(q)$. If $p \in I^-(q)$, then $I^-(p)$ is a subset of $I^-(q)$. Let γ be future directed either time-like or null curve that joins p to q, then $I^+(q) \subset I^+(p)$ (see Fig. 25).

The **chronological (causal) future of any set** $S \subset M$ is defined by (see Fig. 26)

$$I^+(S) = \bigcup_{p \in S} I^+(p),$$

$$J^+(S) = \bigcup_{p \in S} J^+(p).$$

The **chronological (causal) pasts** for subsets of spacetime are similarly defined.

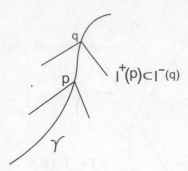

Figure 25 q lies in future of p on the causal curve γ.

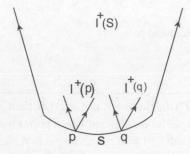

Figure 26   Chronological (causal) future set of a set.

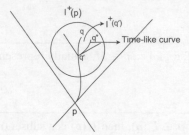

Figure 27   Every point in $I^+(p)$ is an interior point.

Exercise 8.1

Show that $I^+(p)$ is an open set.

Hint: Let $q \in I^+(p)$ and fix a future-directed time-like curve γ from p to q. Now we can choose q' on γ sufficiently near to q (just prior to q) in a convex normal neighborhood of q (i.e., an open set N with $q \in N$ such that $\forall x, y \in N$, $\exists$ a distinctive geodesic γ' connecting x, y and lying totally inside of N) such that the set of points $q'' \in I^+(q')$ forms an open neighborhood of $q \in I^+(p)$ (see Fig. 27). Hence, $I^+(p)$ is an open set.

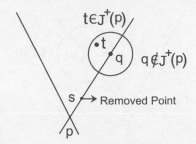

Figure 28 All the limit points of $J^+(p)$ are not contained in $J^+(p)$.

Note 8.4

$I^+(S)$ is an open set since arbitrary collection of open sets is open. Further, $I^+(I^+(S)) = I^+(S)$ and $I^+(\overline{S}) = I^+(S)$.

Exercise 8.2

Show that in general spacetime, $J^+(p)$ is not a closed set.

Hint: Let us consider Minkowski spacetime with a point s lying on the future light cone of p is removed. In this spacetime, no causal curve connects p and q such that $q \notin J^+(p)$, but $q \in \overline{J^+(p)}$ (see Fig. 28). Hence, $J^+(p)$ is not closed.

[Here obviously $s \in \overline{J^+(p)} = \overline{I^+(p)}$.]

Note 8.5

In Minkowski spacetime $J^+(p)$ is a closed set.

Exercise 8.3

Show that $J^+(p)$ is not an open set.

Hint: Suppose r is a point in $J^+(p) - I^+(p)$. Let us consider a neighborhood of r. Then, some points of this neighborhood lie outside of $J^+(p)$ and some points lie within $J^+(p)$ (see Fig. 29). Therefore, r is not an interior point. Hence, $J^+(p)$ is not an open set.

Note 8.6

In a spacetime M, if $q \in J^+(p) - I^+(p)$, i.e., $p < q$ but $p \not\ll q$, then any future-directed causal curve connecting p to q must be a null geodesic.

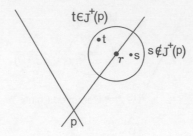

Figure 29   Every point in $J^+(p)$ is not an interior point.

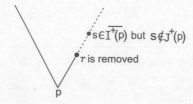

Figure 30 $J^+(p) \subset \overline{I^+(p)}$.

Exercise 8.4

For all subsets $S \subset M$, show that $J^+(p) \subset \overline{I^+(p)}$.

Hint: Let us consider a Minkowski space with a point r removed as shown in Fig. 30. The points after the deleted point r, i.e., on the dotted line are not in $J^+(p)$. However, points on the dotted line are in the closer $\overline{I^+(p)}$. Hence, $J^+(p) \subset \overline{I^+(p)}$.

Exercise 8.5

For all subsets $S \subset M$, show that $Int J^+(p) = I^+(p)$.

Hint: Obviously, $I^+(p) \subset Int J^+(p)$. Now, let $r \in Int J^+(p)$, then $p \ll r$, in other words, $r \in I^+(p)$ (see Fig. 31). Therefore, $Int J^+(p) \subset I^+(p)$. Hence, $Int J^+(p) = I^+(p)$.

Exercise 8.6

Let $J^+(x)$ be closed in M. Show that M is causal if $\dot{I}^+(x) \cap \dot{I}^-(x) = \{x\}$ for all $x \in M$.

Hint: Given that M is causal. Now let, if possible, $\exists \, y \, (\neq x)$ such that $y \in \dot{I}^+(x) \cap \dot{I}^-(x)$. We know $\dot{I}^+(x) = \dot{J}^+(x)$, therefore, $y \in \dot{J}^+(x)$. Since $J^+(x)$ is closed in M, therefore, $y \in \dot{J}^+(x)$. This means, $x < y$. Similarly, $y \in \dot{J}^-(x)$ indicates $y < x$. This suggests that there is a closed causal curve through x. Definitely, this violates the causality. Hence, $y \neq x$ is not true, i.e., $y = x$.

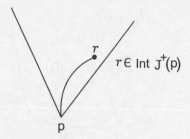

Figure 31 *r* is an interior point.

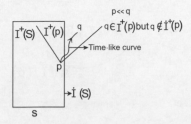

Figure 32 An event *p* lies just inside the boundary of the chronological future $\dot{I}^+(S)$.

8.1 Theorem: The boundary of the chronological future $\dot{I}^+(S)$ of a set *S* cannot be time-like.

Proof: We consider a point *p* just inside the boundary of the chronological future $\dot{I}^+(S)$ (see Fig. 32). The figure indicates that there should exist some points, say *q* in future of the point *p* lies outside of the boundary of the chronological future $\dot{I}^+(S)$, such that $p << q$. This is definitely a contradiction of the boundary of the chronological future $\dot{I}^+(S)$ of a set *S*. Hence, $\dot{I}^+(S)$ is not time-like.

A set *F* of *M* is said to be a **future set** if $\exists$ a subset *S of M* such that $F = I^+(S)$. **Past set** *P of M* is defined similarly, i.e., $P = I^-(S)$.

Obviously, $I^+(F) \subset F$. One can notice that if *F* is a future, then $M - F$ is a past set.

Exercise 8.7

Show that future sets are open set.
Hint: A future set $F = I^+(S)$ is the union of $I^+(p)$ for all $p \in S$ and $I^+(p)$ is an open set (see Fig. 33).

Exercise 8.8

Show that future sets are not closed sets.
Hint: $I^+(p)$ is not a closed set, etc.

A set *S* is said to be **achronal** if no two points of *S* are time-like related, i.e., there do not exist $p, q \in S$ such that $q \in I^+(p)$. In other words, $I^+(S) \cap S = \varphi$. This means no points in *S* can be joined by a time-like curve (see Fig. 34).

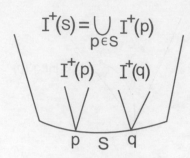

Figure 33   A future set of a set is the union of $I^+(p)$.

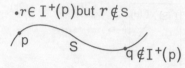

Figure 34   Achronal set.

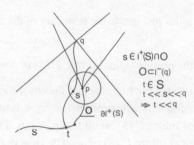

Figure 35 $I^+(p) \subset I^+(S)$.

8.2 Theorem: Let $\partial I^+(S)$ be an achronal boundary of the achronal set S. If $p \in \partial I^+(S)$, then $I^+(p) \subset I^+(S)$ and $I^-(p) \subset M - I^+(S)$.

Hint: Let $p \in \partial I^+(S)$. If $q \in I^+(p)$, then $p \in I^-(q)$. Since $I^-(q)$ is an open set, there is an open neighborhood O of p contained in $I^-(q)$. Since, p is a boundary of $I^+(S)$, therefore $O \cap I^+(S) \neq \varphi$ and hence, $q \in I^+(S)$ (see Fig. 35), i.e., $I^+(p) \subset I^+(S)$.

Similarly, we can prove $I^-(p) \subset M - I^+(S)$.

8.3 Theorem: The achronal boundary $\partial I^+(S)$ of the achronal set S is achronal.

Hint: Suppose, if possible, $\partial I^+(S)$ is not achronal, then $\exists\, p, q \in \partial I^+(S)$ such that $q \in I^+(p)$. By the above example, $q \in I^+(S)$. However, this is impossible since $I^+(S)$ is an open set and, therefore, $I^+(S) \cap \partial I^+(S) = \varphi$. Thus, $\partial I^+(S)$ is achronal.

Suppose, S is an achronal set. The **edge of S** is the collection of points $p \in S$ such that each open neighborhood O of p comprises a point $q \in I^+(p)$ as well as a point $r \in I^-p$ and there exists a time-like curve λ connecting r and q, which does not intersect S (see Fig. 36).

For any an achronal set S if *edge of S* $= \varphi$, we say that S is **edgeless**.

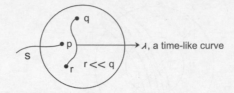

Figure 36 Edge of a set.

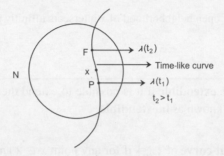

Figure 37 Convex normal neighborhood N of x.

8.4 Theorem: The achronal boundary $\partial I^+(S)$ is edgeless.

Hint: We know if $p \in \partial I^+(S)$, then $I^+(p) \subset I^+(S)$ and $I^-(p) \subset M - I^+(S)$. This implies that any time-like curve from $I^-(p)$ to $I^+(p)$ must meet $\partial I^+(S)$. Thus, $\partial I^+(S)$ is edgeless.

Causal relations in a spacetime are defined by the existence of smooth nonspace-like, i.e., either time-like or null, curves between two events.

A **continuous curve** λ is said to be a future-directed time-like (causal) if each $x \in \lambda$, there is a convex normal neighborhood N of x such that given any t_1, t_2, with $t_1 < t_2$, $\lambda(t_1)$, $\lambda(t_2) \in N$, then there is a smooth future-directed time-like (causal) curve in N from $\lambda(t_1)$ to $\lambda(t_2)$ (see Fig. 37).

The nature, time-like or causal, remains unchanged for a continuous curve for a continuous, one–one transformation.

Let $\lambda(t)$ be future-directed causal curve. Then $p \in M$ is said to be a **future end point** of λ if for every neighborhood O of p, $\exists \, t_0$ such that $\lambda(t) \in O \, \forall \, t > t_0$.

Note 8.7

Here, λ has at most one future end point because M is Hausdorff. The end point p may lie outside the curve, i.e., we cannot get a value of t such that $\lambda(t) = p$. The past end point is defined similarly. The curve λ is said to be **future or past inextendible** if it has no future or past end point, respectively in M. For inextendible causal curve λ, it is not possible to extend λ beyond p, i.e., it might be departing to infinity or it might reach in spacetime singularity.

Let $\{\lambda_n\}$ be a sequence of causal curves. A point $x \in M$ is called a **limit point** of the sequence $\{\lambda_n\}$ if every open neighborhood of x intersects infinitely many $\{\lambda_n\}$ (see Fig. 38).

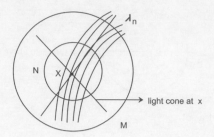

Figure 38 Every open neighborhood of x intersects infinitely many $\{\lambda_n\}$.

Note 8.8

Actually a curve is said to be **extendible** if it is possible to extend the curve to an end point in the spacetime. Else, the curve is known as **inextendible**.

A curve λ is called a **limit curve** of $\{\lambda_n\}$ if for any point $x \in \lambda$ and any open neighborhood U of x, there is a subsequence $\{\lambda_n'\}$ such that for sufficiently large n all $\{\lambda_n'\}$ intersect U, i.e., λ is a convergence curve. Thus, if λ is a limit curve, then every $x \in \lambda$ is a limit point of $\{\lambda_n\}$.

Note 8.9

If $\{\lambda_n\}$ be a sequence of future inextendible causal curves with a limit point p, then $\exists$ a future inextendible causal curve λ passing through p, may be a limit curve of the $\{\lambda_n\}$. Also if $\{\lambda_n\}$ be a sequence of time-like curves, then the limit curve λ may be only a causal curve.

Note 8.10

The study of causal relations is same in a spacetime (M, g_{ab}) and its conformal geometry. Suppose (M, g_{ab}) is the physical spacetime. Now, we assume the set of all conformal metrics $\overline{g}_{ab}$ defined by $\overline{g}_{ab} = \Omega^2 g_{ab}$, where Ω is a nonzero C^r function. Then, if two events p and q are either time-like or causally related, i.e., if $p << q$ or $p < q$ in (M, g_{ab}), then the same relation is well maintained in $(M, \overline{g}_{ab})$. Hence, causal relationships remain invariant under a conformal transformation of the metric.

8.3 Causal Relation

Local causality condition for a spacetime indicates that over small regions of spacetime, the causal structure remains the same as in special theory of relativity. It is generally believed that no closed time-like or causal curve exists in physically realistic spacetime because it would give rise to the situation of entering one's own past. However, in general relativity closed time-like or causal curve may exist. For example, Gödel universe possesses closed time-like curves through each point of M. For another example, we consider a cylinder $M = S' \times M$ obtained from the Minkowski spacetime,

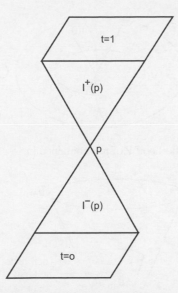

Figure 39 Identifying $t = 0$ and $t = 1$ hypersurfaces.

possessing the metric $ds^2 = dt^2 - dx^2$, by identifying $t = 0$ and $t = 1$ hypersurfaces. Here, $x =$ constant, representing circles that are closed time-like curves. Here, actually $I^+(p) = I^-(p) = M, \forall p \in M$ (see Fig. 39). Also, one can notice that wormhole spacetime, possessing the multiple connected nature of topology of space, may contain closed time-like curve.

Causality condition: The spacetime (M, g_{ab}) does not admit any closed time-like or null curves.

A spacetime M is said to be **chronological** if it possesses no closed time-like curves. Thus, here, $p \notin I^+(p), \forall p \in M$.

A spacetime M is said to be **causal** if it possesses no closed nonspace-like curves.

Consider a cylinder $M = S' \times M$ taken from the Minkowski spacetime, which possesses the metric $ds^2 = -dtdx$, then M is chronological but not causal. Here, $x =$ constant, representing circles that are null geodesics.

8.5 Theorem: Every compact spacetime possesses a closed time-like curve.

Alt: If M is chronological, M cannot be compact.

Proof: Let M be a compact spacetime and the sets $\{I^+(p) \mid p \in M\}$ cover M. Then $\exists$ a finite subcover of M, i.e., for finite set of points $p_1, p_2, \ldots, p_n$, the set $I^+(p_1) \cup I^+(p_2) \cup \ldots \cup I^+(p_n)$ covers M. These sets cover M implies that $p_1 \in I^+(p_{i_1})$, i.e., $p_1 \gg p_{i_1}$ for some i_1 in $2, \ldots, n$. Hence, we have $p_1 \gg p_{i_1} \gg p_{i_2} \gg \ldots \gg p_{i_{n-1}}$, where all n points are included. This implies that there exists i_k such that $p_{i_k} \gg p_{i_k}$, i.e., $p_{i_k} \in I^+(p_{i_k})$, which indicates that there should exit a closed time-like curve through p_{i_k}.

A spacetime (M, g_{ab}) is said to be **strongly causal** if $\forall$ *events*, $p \in M$, and each neighborhood O of p, $\exists$ a neighborhood V of p confined fully in O such that no nonspace-like or causal curve meets V more than once (see Fig. 40).

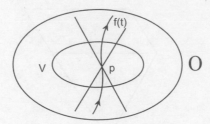

Figure 40 The neighborhood V of p is contained in O.

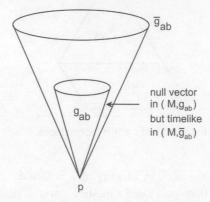

Figure 41 The light cone of $\bar{g}_{ab}$ is strictly larger than that of g_{ab}.

Thus, if a spacetime does not follow the strong causality at p, then there exist nonspace-like or causal curves from neighborhoods of p, which come arbitrarily close to intersecting themselves. In such a spacetime, a small modification of g_{ab} in an arbitrarily small neighborhood of p can produce closed causal curves.

Let g_{ab} be the spacetime metric and for any time-like vector t^a defined at a point $p \in M$, we can construct another spacetime metric $\bar{g}_{ab}$ at p by

$$\bar{g}_{ab} = g_{ab} - t_a t_b.$$

It is interesting to notice that $\bar{g}_{ab}$ has the same signature as g_{ab} at p and the light cone of $\bar{g}_{ab}$ is strictly larger than that of g_{ab} (see Fig. 41). This means that every time-like and null vector of g_{ab} is a time-like vector of $\bar{g}_{ab}$.

A spacetime (M, g_{ab}) is **stably causal** if $\exists$ continuous non-null time-like vector field t^a such that the spacetime $(M, \bar{g}_{ab})$, where $\bar{g}_{ab} = g_{ab} - t_a t_b$, admits no closed time-like curves.

Actually, stably causality indicates the existence of a global time function on the spacetime. The following theorem confirms it.

8.6 Theorem: The necessary and sufficient condition for a spacetime (M, g_{ab}) to be stably causal is that a differentiable function f exists on M so that $\nabla^a f$ is a past-directed time-like vector field.

Proof: We will prove one part only and for other part we refer Hawking and Ellis' book.

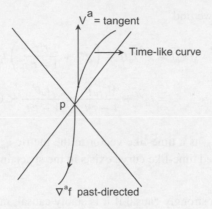

Figure 42 — $\nabla^a f$ is future-directed.

Suppose, a differentiable function f exists on the spacetime (M, g_{ab}) so that $\nabla^a f$ is a past-directed time-like vector field. We will prove that the spacetime (M, g_{ab}) is stably causal.

For a differentiable function f on (M, g_{ab}), let $\nabla^a f$ be an arbitrary past-directed time-like vector field, then, $-\nabla^a f$ is future-directed (see Fig. 42). Then, along $-\nabla^a f$, i.e., in future-directed time-like curve with tangent v^a, we have $g_{ab}v^a(-\nabla^b f) < 0$, or, $g_{ab}v^a(\nabla^b f) > 0$ (here, the signature is assumed to be $(-, +, +.+)$). This implies tangent vector field $v(f) > 0$. Hence, f is severely nondecreasing along each future-directed time-like curve. This indicates that there does not exist any closed time-like curve in (M, g_{ab}) because f never comes back to its initial position.

Now, we define $t^a = \nabla^a f$ and construct a new metric $\bar{g}_{ab}$ by

$$\bar{g}_{ab} = g_{ab} - t_a t_b.$$

The inverse metric of $\bar{g}_{ab}$ is

$$\bar{g}^{ab} = g^{ab} + \frac{t^a t^b}{1 - t^c t_c}.$$

$$[\quad \bar{g}_{ab}\bar{g}^{ab} = (g_{ab} - t_a t_b)\left(g^{ab} + \frac{t^a t^b}{1 - t^c t_c}\right)$$

$$= g_{ab}\, g^{ab} + \frac{g_{ab}\, t^a t^b}{1 - t^c t_c} - g^{ab}\, t_a t_b - \frac{t_a t_b\, t^a t^b}{1 - t^c t_c}$$

$$= 1 + \frac{t_b t^b}{1 - t^c t_c} - t_b t^b - \frac{(t_c t^c)^2}{1 - t^c t_c}$$

$$= 1 + \frac{t_c t^c}{1 - t^c t_c} - t_c t^c - \frac{(t_c t^c)^2}{1 - t^c t_c} = 1.$$

(changing the index b into c)]

Now, for the new metric $\bar{g}_{ab}$, we find

$$\bar{g}^{ab}\nabla_a f \nabla_b f = \bar{g}^{ab} t_a t_b = \left(g^{ab} + \frac{t^a t^b}{1 - t^c t_c} \right) t_a t_b$$

$$= t_a t^a + \frac{(t^a t_a)^2}{1 - t^c t_c} = \frac{t_a t^a}{1 - t^c t_c} < 0.$$

This indicates that $t^a = \bar{g}^{ab}\nabla_b f$ is a time-like vector in the metric $\bar{g}_{ab}$. Using the above argument, one can conclude that no closed time-like curve exists in the spacetime $(M, \bar{g}_{ab})$. Hence, (M, g_{ab}) is stably causal.

8.7 Theorem: A spacetime is strongly causal if it is stably causal; in other words, stable causality implies strong causality.

Proof: Let us consider a global time function f defined on the spacetime M. For any point $p \in M$, we can consider two open neighborhoods O and V of p, where V is contained in O, i.e., $V \subset O$. This specific choice is done in such a way that when a future-directed nonspace-like curve leaves V, the limiting value of f is greater than that when a future-directed nonspace-like curve enters V (see Fig. 43). This indicates that f is increasing along every future-directed causal curve and hence, no nonspace-like curve can enter V more than once.

Suppose, S is an achronal set that is closed. The set $D^+(S)$ is said to be a **future domain of dependence or future Cauchy development** of S if

$$D^+(S) = p \in M$$

such that each past inextendible nonspace-like curve through p meets S (see Fig. 44).

Note that

$$S \subset D^+(S) \subset J^+(S)$$

and also

$$D^+(S) \cap \Gamma^-(S) = \varphi$$

as S is achronal.

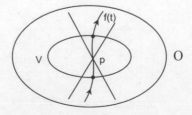

Figure 43 When a future-directed nonspace-like curve leaves V, the limiting value of f is greater than that when a future-directed nonspace-like curve enters V.

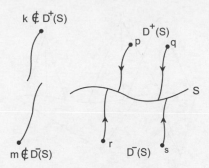

Figure 44 Future and past domain of dependence.

The above definition was proposed by Hawking and Ellis. However, Penrose and Geroch used time-like curve in place of causal curve to define future domain of dependence. Since, no particle (massive or massless) can have velocity more than light, a light signal sent to $p \in D^+(S)$ should be "recorded" on S. Therefore, the knowledge about the set $D^+(S)$ is important. Thus, if we have information about the initial condition on S, then it is conceivable to envisage what will happen at $p \in D^+(S)$. Again, if p lies in $I^+(p)$ but not in $D^+(S)$, then information of the situations on S does not help to decide the circumstances at p.

We can define the **past domain of dependence of S**, $D^-(S)$ by exchanging "future" and "past" in the above definition.

Thus, $D^-(S) = \{p \in M$: every future-directed inextendible causal curve from p intersects $S\}$ (see Fig. 44).

Similarly, as above, we are able to retrodict conditions at all $q \in D^-(S)$ from a knowledge of conditions on S.

The **full domain of dependence**, denoted by $D(S)$ is defined by

$$D(S) = D^+(S) \cup D^-(S).$$

Thus, having the knowledge of conditions on S, one can determine all conditions of the events in $D(S)$. The importance of domain of dependence lies in the fact that all points in $D^+(S)$ (respective $D^-(S)$) can be considered as effects (respective causes) of the events in $D(S)/.

A closed achronal set Σ in spacetime M is known as **Cauchy surface** if $D(\Sigma) = M$.

Thus, for any Cauchy surface $edge(\Sigma) = \varphi$ (see Fig. 45). All Cauchy surfaces Σ are embedded C^0 submanifold of M as $edge(\Sigma) = \varphi$. The Cauchy surface Σ characterizes an instant of time throughout the spacetime as Σ is achronal.

Suppose, S is an achronal as well as closed set. The **future Cauchy horizon** of S is defined by

$$H^+(S) = \{p \in D^+(S) \ such \ that \ I^+(p) \cap D^+(S) = \varphi \ (\ see \ Fig. \ 46)\}$$

$$= \overline{D^+(S)} - I^-[D^+(S)]$$

Actually, $H^+(S)$ is a portion of the boundary of the past of the set $D^+(S)$.

The **past Cauchy horizon**, $H^-(S)$, is defined in a similar manner.

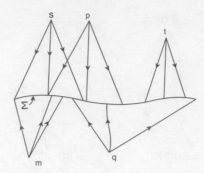

Figure 45 Cauchy surface.

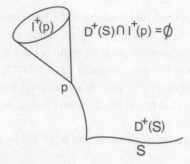

Figure 46 Future Cauchy horizon.

If $S \in M$ is not a Cauchy surface, then $N = D^+(S) \cap D^-(S) \neq M$, i.e., $N \subset M$. The boundary of N in M divides M into two regions that are $H^+(S)$ and $H^-(S)$. Cauchy horizons are the regions of the spacetime M, which are not completely determined by information on $S \in M$.

Note 8.11

The future Cauchy horizon can also be defined as $H^+(S) = \dot{D}(S)$. If $H^+(S) = \dot{D}(S) = \varphi$, then $\overline{D(S)} = int[D(S)] = D(S)$. This implies here, $D(S)$ is both open and closed.

8.8 Theorem: $H^+(S)$ is achronal and closed.

Proof: If possible, two points x and y in $H^+(S)$ are time-like related, i.e., $x, y \in H^+(S)$ with $x \ll y$. This indicates that x lies on the chronological past of y, i.e., $x \in I^-(y)$. Also, $\exists$ a neighborhood N_x of x lying entirely within $I^-(y)$, i.e., $N_x \subset I^-(y)$. Let p be any point within N_x, i.e., $p \in N_x \cap I^+(x)$. Then, every past-directed time-like curve (λ) from p must meet the achronal set S. Also, $\exists$ a time-like curve (γ), which relates p and y, i.e., $p \ll y$. Thus, we have a past-directed time-like curve from y (combining two curves λ and γ), which must meet S (see Fig. 47). Then, $p \in D^+(S)$, which is a contradiction to $D^+(S) \cap I^+(x) = \phi$. Therefore, it is not possible for two points in $H^+(S)$ to be time-like related. Hence, $H^+(S)$ is achronal.

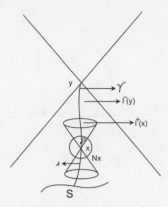

Figure 47 No two points in $H^+(S)$ are time-like related.

We know $H^+(S)$ can be defined as

$$H^+(S) = \overline{D^+(S)} - \Gamma[D^+(S)] = \overline{D^+(S)} \cap \{\Gamma^-[D^+(S)]\}^c.$$

Here, we see that $\overline{D^+(S)}$ is a closed set and

$$\{\Gamma^-[D^+(S)]\}^c = M - \Gamma^-[D^+(S)]$$

also is a closed set. Thus, $H^+(S)$ is the intersection of two closed sets $\overline{D^+(S)}$ and $\{\Gamma^-[D + (S)]\}^c$. Hence, $H^+(S)$ is closed.

The full Cauchy horizon of a closed achronal set S is expressed as

$$H(S) = H^+(S) \cup H^-(S).$$

The boundary of the domain of dependence of S is actually the Cauchy horizon.

Exercise 8.9

Let $p \in \overline{D^+(S)}$, then show that every past inextendible time-like curve from p intersects S.

Hint: Let $p \in \overline{D^+(S)}$ and if possible, there exists a past inextendible time-like curve from p that does not intersect S. Let O be a neighborhood of p. Let q be any point in O. Then, obviously, there exists a past inextendible time-like curve from q that does not intersect S. So, $O \cap D^+(S) = \varphi$. Hence, $p \notin \overline{D^+(S)}$ is a contradiction. Therefore, if $p \in \overline{D^+(S)}$, then each past inextendible time-like curve from p meets S.

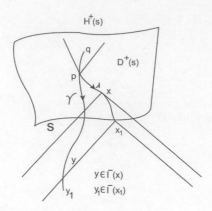

Figure 48 $p \in D^+(S) - H^+(S)$.

Exercise 8.10

If $p \in D^+(S) - H^+(S)$, then show that each past inextendible causal curve from p meets $I^-(S)$.

Hint: Let $p \in D^+(S) - H^+(S)$ and consider a past inextendible causal curve γ from p. Then, $\exists$ a point $q \in I^+(p) \cap D^+(S)$ and a past inextendible causal curve λ from q (which meets S) such that for each point $x \in \lambda$, one can obtain a point $y \in \gamma$, where y lies within $I^-(x)$ (see Fig. 48). Let λ meet S at the point x_1, then by above argument, there will be a y_1 such that $y_1 \in \gamma \cap I^-(S)$. Hence, the proof.

Exercise 8.11

If $p \in int(D^+(S))$, then show that each past inextendible causal curve from p intersects $I^-(S)$ and $I^+(S)$.

Hint:

$$int(D^+(S)) = D^+(S) - \{H^+(S) \cup H^-(S)\}.$$

A spacetime (M, g_{ab}) containing a Cauchy surface Σ is called **globally hyperbolic**. Alternatively, we can say a causal spacetime (M, g_{ab}) is globally hyperbolic if $\forall x, y \in M$, the sets $J^+(x) \cap J^-(y)$ are compact.

Globally hyperbolic spacetime is strongly and stably causal.

Exercise 8.12

Let (M, g_{ab}) be globally hyperbolic spacetime. Show that the sets $J^\pm(x)$ are closed, $\forall x \in M$.

Hint: If possible, let $J^+(x)$ is not closed. Then, $\exists$ limit point y of $J^+(x)$ such that $y \notin J^+(x)$. Suppose, $z \in J^+(y)$, then $y \in \overline{J^+(x) \cap J^-(z)}$ but $y \notin J^+(x) \cap J^-(z)$. By definition of globally hyperbolic, we know $x, z \in M$, the set $J^+(x) \cap J^-(z)$ is compact, i.e., the set $J^+(x) \cap J^-(z)$ is closed; in other words, all limit points belong to $J^+(x) \cap J^-(z)$. This contradiction implies $J^+(x)$ is closed.

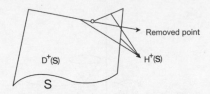

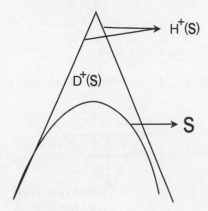

Figure 49 The future domain of dependence $D^+(S)$ and Cauchy horizon $H^+(S)$.

Figure 50 S is asymptotically null to the right and becomes exactly null to the left.

Exercise 8.13

Give an example of the future domain of dependence $D^+(S)$ and Cauchy horizon $H^+(S)$.
Hint: Let us consider the Minkowski spacetime M with a point removed and let S be a closed achronal subset of M. The diagram (Fig. 49) shows the future domain of dependence $D^+(S)$ and the Cauchy horizon $H^+(S)$.

Exercise 8.14

Show that $H^+(S)$ can intersect S even if $edge(S) = \phi$ for any closed achronal set S.
Hint: Let us consider Minkowski spacetime M with a point removed and let S be a closed achronal subset of M. Here, S is asymptotically null to the right and becomes exactly null to the left (see Fig. 50). This shows that $H^+(S)$ can intersect S even if $edge(S) = \phi$.

Exercise 8.15

Give an example of Cauchy surface.
Hint: Let M be globally hyperbolic with topology $M = S \times R$, for any $t \in R$, the space-like surface S_t of constant time is a Cauchy surface for the spacetime (see Fig. 51).

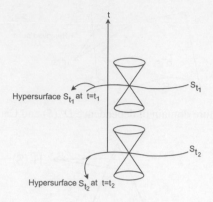

Figure 51   The surface S_t of constant time in Minkowski spacetime.

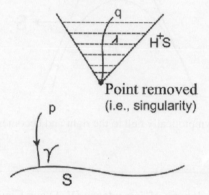

Figure 52 S is not globally hyperbolic.

Exercise 8.16

Give two examples of a spacetime that are not globally hyperbolic.

Hint: (i) Let us consider Minkowski spacetime M with a point removed and S be a closed achronal subset of M. Then, S is not globally hyperbolic. The point $q \notin D^+(S)$, as there are no causal curves like λ, intersects S in the past. The event $p \in D^+(S)$. The points in the boundary of the shaded regions that do not belong to $D^+(S)$ is the Cauchy horizon (see Fig. 52).

(ii) Let us consider Minkowski spacetime M and let p be any point in M. We consider an edgeless space-like hyper surface Σ which lies strictly within $I^-(p)$ (see Fig. 53). Here, Σ is an achronal surface, however, $D^+(\Sigma)$ ends at the light cone. Thus, it is not possible to use the information on Σ to predict what will happen throughout M.

An open set U is said to be **causally simple** if for every compact set $K \subset U$, $J^+(K)$ *and* $J^-(K)$ are closed in U.

The spacetime M is said to be **causally simple** if the causal future and past $J^{\pm}(x)$ are closed $\forall x$ *in* M.

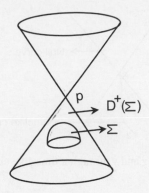

Figure 53 Example of nonglobally hyperbolic spacetime.

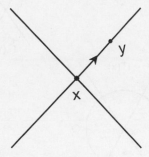

Figure 54 y lies strictly on the light cone.

Notation: We say that $x \longrightarrow y$ if $x < y$ but $x \not<< y$, i.e., $x < y$ and further $y \in J^+(x) - I^+(x)$ (see Fig. 54).

8.9 Theorem: Let M be a causally simple spacetime and x and y be distinct points of M. Then $x \longrightarrow y$ if $x \not<< y$ and $\forall z$, $z << x$ implies $z << y$.

Proof: Given: $x \longrightarrow y$ and we will show that $x \not<< y$ and $\forall z$, $z << x$ implies $z << y$.

Now, $x \longrightarrow y$ implies $x < y$ but $x \not<< y$, i.e., y lies on $J^+(x)$ but not in $I^+(x)$. Let us consider a point z lying in $I^-(x)$, then $z << x$ (see Fig. 55). Thus, we have $z << x < y$ which implies $z << y$. Hence, the proof.

Conversely, if $x \not<< y$ and $\forall z$, $z << x$ implies $z << y$, then $x \longrightarrow y$.

If possible, the above conditions hold good but $x \nrightarrow y$.

Consider a point y lying outside the causal future and past of $x \in M$, where M is a causally simple spacetime. Then

$$y \notin J^+(x) \cup J^-(x) \quad (\text{see Fig. 56})$$

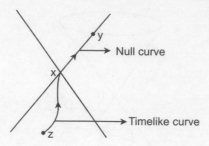

Figure 55 One can join z and y through a time-like curve.

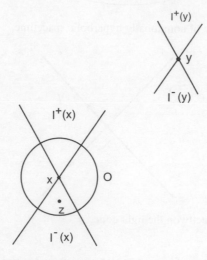

Figure 56 The point y lies outside of the causal future and past of $x \in M$.

Since, M is a causally simple spacetime, y should not be the limit point of $I^+(x)$ and $I^-(x)$, i.e.,

$$y \notin \overline{I^+(x)} \cup \overline{I^-(x)}.$$

This implies that y lies in space-like region. Hence, x is not a limit point of $I^+(y)$ and $I^-(y)$, i.e.,

$$x \notin \overline{I^+(y)} \cup \overline{I^-(y)}.$$

This indicates that x belongs to the complement of these closed sets. Let us consider an open neighborhood O of x in this complement and choose a point $z \in I^-(x)$ such that

$$z \in O \cap I^-(x).$$

Then definitely,

$$z \notin \overline{I^+(y)} \cup \overline{I^-(y)},$$

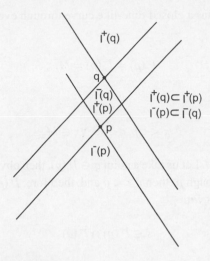

Figure 57 Reflecting spacetime.

i.e.,

$$z \notin J^+(y) \cup J^-(y).$$

This implies $z \not\ll y$. But, we have $z \ll x$. Thus,

$$x \nrightarrow y \Rightarrow z \not\ll y,$$

which is a contradiction to the given condition. Hence, $x \longrightarrow y$.

A spacetime $(M; g_{ab})$ is said to be **reflecting** if $\forall p$ *and* $q \in M$,

$$I^+(q) \subseteq I^+(p) \Leftrightarrow I^-(p) \subseteq I^-(q) \; (Fig. 57)$$

Exercise 8.17

Let M be a reflecting spacetime. If $p, q \in M$ such that $q \in \dot{I}^+(p)$, then show that $p \in \dot{I}^-(q)$.

Hint: Let $p, q \in M$ such that $q \in \dot{I}^+(p)$. Then, q is not chronologically related to p. Since, M is reflecting,

$$I^+(q) \subseteq I^+(p) \; and \; I^-(p) \subseteq I^-(q).$$

The given condition implies that p is not chronologically related to q, therefore, $p \in \dot{I}^-(q)$.

8.10 Theorem: Let, there be a point p in a reflecting spacetime M through which a closed time-like curve is passing. Then, M contains closed time-like curve through every point of M.

Proof: To prove that M contains a closed time-like curve through every point of M means we have to show for $p \in M$,

$$I^+(p) = I^-(p) = M.$$

Now, $I^+(p) = I^-(p) = M$ means

$$x << p << x, \quad \forall x \in M.$$

Suppose, if possible, $I^+(p) \neq M$. Let us take a point $q \in \dot{I}^+(p)$, then obviously, $p \in \dot{I}^-(q)$. Since, there is a closed time-like curve through p, then $p << p$ and, therefore, $I^+(p)$ is an open neighborhood of p. As a result, there is a point s with

$$s \in I^+(p) \cap I^-(q)$$

such that

$$p << s << q.$$

This implies $q \in I^+(p)$, which is a contradiction of the assumption. Hence, $I^+(p) = M$. Similarly, one can prove $I^-(p) = M$.

A spacetime M is called **future distinguishing** if $\forall x, y \in M$ such that

$$I^+(x) = I^+(y) \Rightarrow x = y.$$

A spacetime M is called **past distinguishing** if $\forall x, y \in M$ such that

$$I^-(x) = I^-(y) \Rightarrow x = y.$$

Generally, one does not imply the other, i.e., M can be future distinguishing but may not past distinguishing and vice versa. A spacetime M is called **distinguishing** if $\forall x, y \in M$ such that

$$I^+(x) = I^+(y) \text{ and } I^-(x) = I^-(y) \text{ when } x = y.$$

In a spacetime M, an event $q \in M$ is said to **causally connectible** to another event $p \in M$ if $q \in J^+(p) \cup J^-(p)$.

Thus, two events in a given spacetime are causally connected if there exists a causal signal that connects to each other.

8.4   Causal Function

Let us consider an event in the spacetime M, then one can define the size or volume of the past or future of that event. Suppose, we consider an arbitrary positive volume element dV on the spacetime

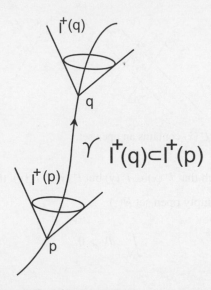

Figure 58 $I^+(q)$ is strictly contained in $I^+(p)$.

M. Then, total volume of M is equal to unity, i.e.,

$$\int_M dV = 1.$$

Let $x \in M$ and $I^+(x)$ be the chronological future of x, then the **future causal function** h^+ is defined as

$$h^+(x) = \int_{I^+(x)} dV > 0.$$

Similarly, one can define **past causal function**

$$h^-(x) = \int_{I^-(x)} dV > 0.$$

The values of the causal functions (future or past) are characterized by the size of chronological futures and pasts, respectively. Let us consider a future-directed time-like curve γ with two events on it in such a way that q lies in the future of p, i.e., $p \ll q$, then $I^+(q)$ is strictly contained in $I^+(p)$, i.e., $I^+(q) \subset I^+(p)$ (see Fig. 58). Then obviously future causal function $h^+(p)$ is decreasing function. Likewise, past causal function $h^-(p)$ is increasing function.

Note 8.12

Causal functions $h^{\pm}(x)$ are discontinuous.

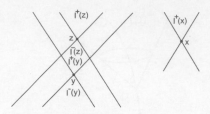

Figure 59 $I^+(y) - I^+(x)$ contains an open set.

8.11 Theorem: If $x, y \in M$ such that $I^+(x) \subset I^+(y)$ but $I^+(x) \neq I^+(y)$, then $h^+(x) < h^+(y)$.

Proof: We know for any nonempty open set $N(z)$

$$\int_{N(z)} dV > 0.$$

Now,

$$\int_{I^+(y)-I^+(x)} dV = \int_{I^+(y)} dV - \int_{I^+(x)} dV = h^+(y) - h^+(x).$$

Therefore, we have to show that $I^+(y) - I^+(x)$ has a nonempty open subset. Let z be a point in $I^+(y)$ but not in $I^+(x)$ (see Fig. 59), i.e.,

$$z \in I^+(y) - I^+(x).$$

Then, obviously

$$I^-(z) \bigcap I^+(x) = \varphi,$$

otherwise, $z >> x$. However, $I^-(z) \bigcap I^+(y)$ is nonempty set and open as we have $z >> y$. Thus, $I^-(z) \bigcap I^+(y)$ is a nonempty open set contained in $I^+(y) - I^+(x)$.

8.12 Theorem: The spacetime M is chronological if $h^\pm(x)$ is strictly monotone along every time-like curve.

Proof: Let the spacetime M be chronological and γ be a time-like curve in M. We will show that $h^+(x)$ is strictly decreasing along γ. Since γ is a time-like curve in M, which is chronological, γ is not closed. Hence, for any two points $x, y \in \gamma$ such that $x << y$ (see Fig. 60). Hence, $I^+(y) \subset I^+(x)$. Therefore, by above theorem, $h^+(y) < h^+(x)$.

Conversely, let $h^+(x)$ be a strictly decreasing function. We will show that the spacetime M is chronological. If possible, the condition holds good but M is not chronological. Then $\exists$ a closed time-like curve γ in M. Therefore, there are points $x, y \in \gamma$ such that $x << y$ and $y << x$. These give $h^+(y) < h^+(x)$ and $h^+(x) < h^+(y)$, which is impossible. Hence, the spacetime M is chronological.

8.13 Theorem: Let the spacetime M be future (past) distinguishing, then causal function h^+ (respectively h^-) is injective along all future-(past) directed causal curve.

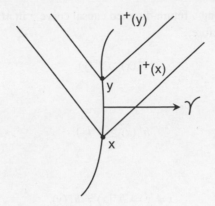

Figure 60 $I^+(y)$ is strictly contained in $I^+(x)$.

Proof: Since the spacetime M is future distinguishing, then $\forall x, y \in M$ such that for

$$x \neq y \Rightarrow I^+(x) \neq I^+(y).$$

Suppose x and y are any two distinct points on a future-directed causal curve $\gamma \in M$ such that $x < y$. Therefore,

$$I^+(x) \supseteq I^+(y).$$

As

$$I^+(x) \neq I^+(y),$$

therefore,

$$I^+(x) \supset I^+(y).$$

Since for chronological spacetime, h^+ is strictly monotonic, therefore, this yields

$$h^+(x) > h^+(y).$$

Hence, distinct points in γ yield distinct images. Therefore, h^+ is injective along future-directed causal curve.

8.14 Theorem: Let M be a spacetime such that causal function h^+ (respectively h^-) is injective along all future-(past) directed causal curve. Then, h^+ (respectively h^-) is strictly decreasing (increasing) along all future-(past) directed causal curves.

Proof: Let h^+ be injective along a future-directed causal curve γ in M. Suppose two distinct points $x, y \in \gamma$ such that $x < y$. Therefore,

$$I^+(x) \supset I^+(y).$$

This yields

$$h^+(x) \geq h^+(y).$$

Given that h^+ is injective, i.e.,

$$x \neq y \Rightarrow h^+(x) \neq h^+(y).$$

Hence,

$$h^+(x) > h^+(y).$$

In other words, h^+ is strictly decreasing along all future-directed causal curves.

A nonempty subset P of M is said to be a **past set** if $\exists$ a subset A of M so that $I^-(A) = P$.

The past set P is said to be **indecomposable past set (IP)** if it is not possible to express P as the union of two proper past subsets.

An indecomposable past set P is said to be **proper indecomposable past set or PIP** if there is some $x \in M$ such that $I^-(x) = P$, i.e., P is the past of the point x in M.

An indecomposable past set P is said to be **terminal indecomposable past set or TIP** if an IP set P is not a PIP, i.e., P is not the past of the point x in M.

The future set and future counter parts of indecomposable future sets, IF, proper indecomposable future set, PIF, and terminal indecomposable future set, TIF, are defined similarly.

Note 8.13

According to some authors, a subset P of M is an indecomposable past set if $\exists$ a future-directed time-like curve γ such that $I^-(\gamma) = P$.

Let M^* be a spacetime with boundary of M, then obviously $M \subset M^*$.

Point $x \in M$ is called a **regular point** if it is expressed by a PIP or PIF.

The points in M^* that are not regular points are presented by TIPs or TIFs. These points are known as the **ideal or causal boundary points** of M.

Result: Let γ be a future-directed time-like curve. Then, $P = I^-(\gamma)$ is a PIP if γ has a future end point. Again, $P = I^-(\gamma)$ is a TIP if γ is future inextendible without a future end point.

The above result was proved by Geroch, Kronheimer, and Penrose in 1972.

8.15 Theorem: Let γ be a future-directed time-like curve in a distinguishing spacetime M. Then, $I^-(\gamma)$ is a PIP if h^- attains its maximum value along γ.

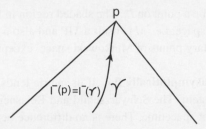

Figure 61 For future-directed time-like curve γ with a future end point p, $I^-(\gamma) = I^-(p)$.

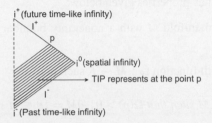

Figure 62 The shaded region in figure is the TIP representing the point p.

Proof: Let γ be a future-directed time-like curve and $I^-(\gamma)$ is a PIP. Then, by above result, γ has a future end point p (see Fig. 61). Then, obviously

$$I^-(\gamma) = I^-(p).$$

As M is distinguishing, h^- is strictly increasing along γ. This implies γ possesses a maximum value at the future end point p.

On the other hand, suppose h^- has a maximum at some $p \in \gamma$, which is not future end point. Then, $\exists\, q \in \gamma$ such that $p << q$. We know h^- is strictly increasing, therefore, this indicates

$$h^-(q) > h^-(p).$$

It contradicts the assumption that h^- has a maximum at p. Hence, $p \in \gamma$ is the future end point.

8.16 Theorem: Let γ be a future-directed time-like curve in a distinguishing spacetime M. Then $I^-(\gamma)$ is a PIP if h^+ attains its minimum value along γ.

The ideal points or boundary points in $M^* - M$ may be the points at infinity and singularities, which are characterized by TIPs and TIFs of spacetime.

A TIP is said to be **nonsingular** if $\exists$ a time-like curve of infinite length.

The ideal points that are nonsingular are known as **singularities** of spacetime.

A future (past) ideal point in M^* defined by a TIP, say $I^-(\gamma)$ (a TIF say $I^+(\lambda)$) is called a **future (past) 0-ideal point** if h^+ (respectively h^-) converges to 0 along the curve.

Also, x will be called a **future (past) k-ideal point** if h^+ (respectively h^-) converges to k with $0 < k < 1$ along γ (respectively λ).

In Minkowski space, let p be a point on I^+. The shaded region in Fig. 62 is the TIP representing the point p. In fact, the whole space, i.e., M is itself a TIP and also a TIF representing the points i^+ and i^-. Actually, all the boundary points of Minkowski space, except i, can be represented as TIPs or TIFs.

A spacetime is said to be **asymptotically flat** if its metric tends to Minkowski metric at large distance from the physical system. The Schwarzschild and Reissner–Nordström solutions are the examples of asymptotically flat spacetime. There is no difference between the conformal structure of null infinity in these spaces with Minkowski spacetime.

A spacetime (M, g) is known as **asymptotically simple and empty** if there exists a strongly causal spacetime $(\overline{M}, \overline{g}_{ab})$ and has properties given below:

(1) $\overline{M}$ has an open submanifold M with a nonempty boundary ∂M (one can assume $\overline{M} = M \cup \partial M$).

(2) There exists a smooth function $\Omega : \overline{M} \to R$ such that

$$M = \{p \in \overline{M} \text{ such that } \Omega(p) > 0\}, \partial M = \{p \in \overline{M} \text{ such that } \Omega(p) = 0\},$$

$$d\Omega \neq 0 \text{ everywhere on } \partial M \text{ and } \overline{g} = \Omega^2 g \text{ on } M.$$

(3) All null geodesics in M contain two end points (past and future) on ∂M.

(4) There is an open neighborhood U of ∂M such that $R_{ab} = 0$ on $U \cap M$.

Asymptotic simplicity is the base of the infinity in asymptotically flat spacetimes as well as spacetimes with other asymptotic structures. Asymptotically simple and empty spacetime has a boundary that has similar properties of infinity in Minkowski spacetime. The boundary ∂M of asymptotically simple and empty space is a null surface and consists of two connected components J^+ and J^- that are null. Each null geodesic in (M, g) has past and future end point on I^- and I^+, respectively.

Existence of Geodesics

For a complete Riemannian manifold possessing positive definite metric, any two points can be linked by a geodesic of minimum length. This geodesic may not be unique. The same result can be stated for Lorentzian metrics:

8.17 Theorem: Let (M, g_{ab}) be globally hyperbolic spacetime. Suppose two events are causally related $p, q \in M$ such that $p < q$. Then $\exists$ a nonspace-like (causal) geodesic from p to q whose length is greater than or equal to that of any other future-directed causal curve from p to q.

In a globally hyperbolic spacetime with a Cauchy surface S, let a point $p \in I^+(S)$. Then $\exists$ a time-like geodesics-like γ (see Fig. 63) orthogonal to S, which is greater than the lengths of all nonspace-like(causal) curves from p to S.

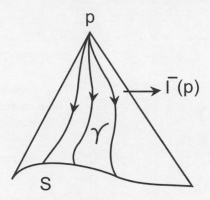

Figure 63 The time-like geodesics γ from p.

Note 8.13

For all complete Riemannian manifolds possessing positive definite metrics, the minimal length result holds.

Exact Solutions of Einstein Equations and Their Causal Structures

9.1 Minkowski Spacetime

Exact solutions of Einstein equations mean that the spacetime metric, which satisfies the Einstein field equations with stress-energy tensor T_{ab}

$$R_{ab} - \frac{1}{2}g_{ab}R + \Lambda g_{ab} = \frac{8\pi G}{c^4}T_{ab}.$$

Now, we will study the causal structures of some exact solutions of Einstein field equations.

The most simple empty spacetime in general theory of relativity is **Minkowski spacetime**. This is actually the spacetime in special theory of relativity. Using the natural coordinates $\left(x^1, x^2, x^3, x^4\right)$ on $R^4 = M$ (M = manifold), one can express the metric in the form

$$ds^2 = (dx^4)^2 - (dx^1)^2 - (dx^2)^2 - (dx^3)^2, \tag{9.1}$$

with the range of coordinates as $-\infty < x^1, x^2, x^3, x^4 < \infty$. In this spacetime, all the components of Riemann tensor $R^i_{jkl} = 0$, therefore, it is a flat spacetime. The vector $\frac{\partial}{\partial x^4}$ offers a time orientation of this spacetime.

For the choice of spherical polar coordinates (t, r, θ, ϕ), where

$$x^4 = t, x^3 = r\cos\theta, x^2 = r\sin\theta\cos\phi, x^1 = r\sin\theta\sin\phi,$$

the metric assumes the following form,

$$ds^2 = dt^2 - dr^2 - r^2\left(d\theta^2 + \sin^2\theta d\phi^2\right). \tag{9.2}$$

In these new coordinate system the ranges are

$$-\infty < t < \infty, 0 < r < \infty, 0 < \theta < \pi \ and \ 0 < \phi < 2\pi.$$

Here all the Christoffel symbols Γ^i_{jk} will not all vanish. However, due to flat spacetime, all the Riemann curvature components will vanish.

To know the structure of infinity in Minkowski spacetime is our next target. For this, we use the interesting representation of this spacetime proposed by Roger Penrose.

A two-dimensional diagram of the spacetime where the causal relations and infinity structure are depicted through the use of conformal transformations is known as the **Penrose diagram**. They are also called the **conformal diagram**.

Actually conformal is a methodological word related to the rescaling of size. The light cones could not be changed under conformal transformation. The metric on a Penrose diagram and the actual metric in spacetime are conformally equivalent. The conformal factor is selected in such a way that the whole infinite spacetime is converted into a finite size in the Penrose diagram.

Result: If any two metrics G and g are chosen from the same manifold M such that they are connected by a positive definite conformal factor $\Omega^2(x)$ as

$$G_{ij}dx^i dx^j = \Omega^2(x)g_{ij}dx^i dx^j,$$

then the null geodesics with respect to the metric G are the same to the null geodesics with respect to the metric g and vice-versa.

In the metric (9.2), we choose an alternative coordinate system on v, w, θ, ϕ defined by the advanced and retarded null coordinate given by

$$v = t + r, w = t - r, \theta = \theta, \phi = \phi.$$

This actually provides a reference frame based on null cones. Here definitely $v \geq w$.

Using this transformations, the metric (9.2) converts into,

$$ds^2 = dvdw - \frac{1}{4}(v - w)^2 \left(d\theta^2 + \sin^2\theta d\phi^2\right), \tag{9.3}$$

where $-\infty < w < \infty$ *and* $-\infty < v < \infty$.

Here, the coefficients of dv^2 *and* dw^2 are zero. This indicates that the surfaces ($v = $ constant), ($w = $ constant) are null. Here obviously

$$w_{;a}w_{;b}\, g^{ab} = 0 = v_{;a}v_{;b}\, g^{ab}.$$

One can consider the null coordinate $v(w)$ as an incoming (outgoing) spherical waves, which moves with the same speed of light. They are actually advanced (retarded) time coordinate. The intersection between two surfaces ($v = $ constant and $w = $ constant) is a two sphere.

The future null infinity corresponds to $v \to \infty$ can be attained by moving along $w =$ constant light cones (see Fig. 64). Likewise, past null infinity corresponds to $w \to \infty$ can be attained by moving along $v =$ constant light cones.

If we plot (t, r) and (v, w) in a single origin, then v, w axes are new axes due to rotation (see Fig. 65).

Our target was to bring infinity to a finite coordinate value. For this, we define new null coordinate in which the infinities of v, w have been transformed to finite values. Now, to confine the Minkowski spacetime into a finite region, we take a conformal transformation

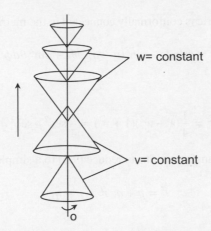

Figure 64 Null coordinate $v(w)$ can be regarded as an incoming (outgoing) spherical wave.

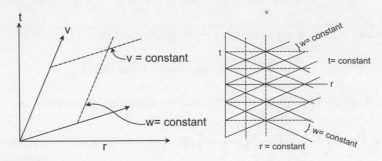

Figure 65 (t, r) and (v, w) in a single origin.

$$\Omega^2 = \frac{1}{4}(1 + w^2)(1 + v^2).$$

Or, equivalently, we take new null coordinates (p, q) to transform the infinities of v, w to finite values as

$$\tan p = v, \tan q = w.$$

Now, the ranges of these new coordinates are

$$-\frac{\pi}{2} < p < \frac{\pi}{2}, \ -\frac{\pi}{2} < q < \frac{\pi}{2} \ and \ (p \ge q).$$

Now the metric (9.3) transforms to the following form

$$ds^2 = \sec^2 p \ \sec^2 q \left[dpdq - \frac{1}{4} \sin^2(p - q) \left(d\theta^2 + \sin^2 \theta d\phi^2 \right) \right]. \tag{9.4}$$

Therefore, the Minkowski metric is conformally connected to the metric $d\bar{s}^2$ ($ds^2 = \Omega^2 d\bar{s}^2$) given by

$$d\bar{s}^2 = 4dpdq - \sin^2(p - q)(d\theta^2 + \sin^2\theta d\phi^2). \tag{9.5}$$

Here, the conformal factor is

$$\Omega^2 = \frac{1}{4}(1 + w^2)(1 + v^2) = \frac{1}{4}\sec^2 p \sec^2 q.$$

Now, we define another transformations to reduce (9.5) to a simple form

$$t' = p + q, \ r' = p - q,$$

where the coordinate ranges

$$-\pi < t' + r' < \pi, \ -\pi < t' - r' < \pi, \ r' \geq 0. \tag{9.6}$$

Then (9.5) will take the following form in (t', r', θ, ϕ) coordinates

$$d\bar{s}^2 = dt'^2 - dr'^2 - \sin^2 r'(d\theta^2 + \sin^2\theta d\phi^2). \tag{9.7}$$

Thus, in (t', r', θ, ϕ) coordinates the whole Minkowski spacetime is expressed as

$$ds^2 = \frac{1}{4}\sec^2\left(\frac{t' + r'}{2}\right)\sec^2\left(\frac{t' - r'}{2}\right)[dt'^2 - dr'^2 - \sin^2 r'(d\theta^2 + \sin^2\theta d\phi^2)]. \tag{9.8}$$

Thus, ultimately, the coordinates t, r are connected to t', r' by

$$2t = \tan\left(\frac{t' + r'}{2}\right) + \tan\left(\frac{t' - r'}{2}\right), \tag{9.9}$$

$$2r = \tan\left(\frac{t' + r'}{2}\right) - \tan\left(\frac{t' - r'}{2}\right). \tag{9.10}$$

The metric (9.7), i.e., $d\bar{s}^2$ metric on four-dimensional cylinder $S^3 \times R^1$, which is the Einstein static universe. However, coordinates range are restricted by Eq. (9.6). Thus, the whole Minkowski spacetime is conformal to the region of Einstein static universe restricted by Eq. (9.6). Hence, the boundary of this restricted region (on the cylinder) characterizes the conformal structure of infinity of Minkowski spacetime. The consequence of this result indicates that we can describe infinity of Minkowski spacetime. Figure 66 indicates that the restricted Einstein static universe, i.e., the restricted cylinder can be decomposed into various components: (i) lower vertex point i^-, known as **past time-like infinity** and specified by the coordinates,

$$r' = 0, \ t' = -\pi \ or \ p = -\frac{\pi}{2}, \ q = -\frac{\pi}{2},$$

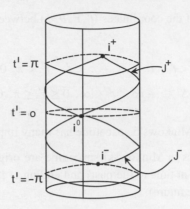

Figure 66 Einstein static cylinder can be decomposed into various components.

(ii) the three-dimensional surface **J⁻**, known as **past null infinity** and specified by the coordinates,

$$t' = -\pi + r' \; for \; 0 < r' < \pi \; or \; q = -\frac{\pi}{2},$$

(iii) the point i^0, known as **spatial infinity** and specified by the coordinates,

$$r' = \pi, \; t' = 0 \; or \; p = \frac{\pi}{2}, \; q = -\frac{\pi}{2},$$

(iv) the three-dimensional surface **J⁺**, known as **future null infinity** and specified by the coordinates,

$$t' = \pi - r' \; for \; 0 < r' < \pi \; or \; p = \frac{\pi}{2},$$

(v) upper vertex point i^+, known as **future time-like infinity** and specified by the coordinates,

$$r' = 0, \; t' = \pi \; or \; p = \frac{\pi}{2}, \; q = \frac{\pi}{2}.$$

Note 9.1
_____

The coordinate transformations (9.9) and (9.10) indicate that one can have a mapping

$$h : M_{Mink} \longrightarrow M_{ESU} \cong R \times S^3,$$

that inserts the whole of Minkowski space into a finite volume region of the Einstein Static universe, which is isomorphic to $R \times S^3$. Actually, the image of Minkowski space through the *h*-mapping:

$$h(M_{Mink}) \subset R \times S^3.$$

The comparison of the ranges of the coordinates (t', r', θ, ϕ) between Minkowski space and $R \times S^3$ are as follows:

$$\text{Minkowski space}: \quad -\pi < t' + r' < \pi, \ -\pi < t' - r' < \pi, \ 0 \le \theta \le \pi, \ 0 \le \phi \le 2\pi.$$

$$R \times S^3: \quad -\infty < t' < \infty, \ 0 \le r' \le \pi, \ 0 \le \theta \le \pi, \ 0 \le \phi \le 2\pi.$$

The conformal diagram for Minkowski spacetime has many important features:

(a) All time-like geodesics in Minkowski spacetime are originating from i^- (past time-like infinity, i.e., far away in time to the past) and ending at i^+ (future time-like infinity, i.e., far away in time to the future).

(b) All space-like geodesics in Minkowski spacetime originate and end at i^0 (spatial infinity, i.e., far away in distance).

(c) All null geodesics in Minkowski spacetime originate from $\mathbf{J}^-$ (past null infinity) and end at $\mathbf{J}^+$ (future null infinity).

Clearly Minkowski spacetime, M is bounded by the infinities $\mathbf{J}^+$ *and* $\mathbf{J}^-$ and the points i^-, i^+, and i^0. Thus, the boundary of the Minkowski spacetime ∂M is comprised into the following parts

$$\partial M = i^0 \cup i^- \cup i^+ \cup \mathbf{J}^+ \cup \mathbf{J}^-$$

The Minkowski spacetime representing the restricted cylinder could be dissected at the point i^0 and will be developed on a plane. This typical form of the plane can be used to represent the conformal Minkowski world. This is well-known Penrose diagram. Now, we describe the conformal structure of infinity through the diagram in (t', r') plane (see Fig. 67).

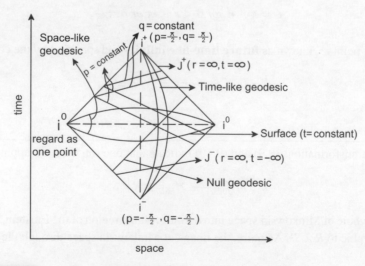

Figure 67   Diagram of Minkowski spacetime in (t', r') plane.

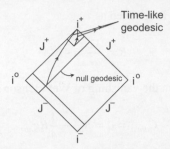

Figure 68 Any point can be causally connected with future time-like infinity.

Note 9.2

In the Penrose diagram, one can see that any point can be causally connected with future time-like infinity (see Fig. 68).

Finally, we express whole Minkowski spacetime in a small region.

9.2 de Sitter Spacetime

Similar to Minkowski space, a de Sitter space is the spacetime of a sphere in ordinary Euclidean space. It is maximally symmetric and simply connected and has constant positive curvature. Willem de Sitter (1872–1934) discovered this spacetime and, therefore, it is named after him. de Sitter space is demarcated as a submanifold of a Minkowski space of one extra dimension and described by the hyperboloid of one sheet,

$$v^2 - u^2 - x^2 - y^2 - z^2 = \alpha^2.$$

Here, the nonzero constant α has the same dimension of length.

The isometry group of four-dimensional de Sitter space is the Lorentz group $O(1, 3)$ and the metric has 10 independent Killing vector fields. As de Sitter space is maximally symmetric, therefore, it has constant curvature. The de Sitter metric of constant curvature is locally described as

$$R_{abcd} = \frac{1}{\alpha^2} \left[g_{ac}\, g_{bd} - g_{ad}\, g_{bc} \right].$$

In de Sitter space, the Ricci tensor and the given metric are proportional to each other, i.e.,

$$R_{ab} = \frac{3}{\alpha^2}\, g_{ab}.$$

This indicates that the de Sitter space is nothing but a vacuum solution of Einstein's field equation in presence of cosmological constant where

$$\Lambda = \frac{3}{\alpha^2}.$$

The Ricci scalar in de Sitter space is obtained as

$$R = \frac{12}{\alpha^2} = 4\Lambda.$$

The above result can be found from the vanishing of Weyl tensor as

$$C_{abcd} = 0 = R_{ab} - \frac{1}{4}Rg_{ab}.$$

This indicates that the Riemann tensor can be found from the Ricci scalar R alone. After contracting Bianchi identities, one can find that R is a constant through spacetime. This spacetime is actually homogeneous. From the above equation, one can write the Einstein tensor as

$$G_{ab} = R_{ab} - \frac{1}{2}Rg_{ab} = -\frac{1}{4}Rg_{ab}.$$

This yields the spacetime as solutions of the field equations for an empty space with $\Lambda = \frac{1}{4}R$.

The space of constant curvature with $R = 0$ corresponds to Minkowski spacetime.

The space with $R > 0$ is known as **de Sitter spacetime** possessing the topology $R^1 \times S^3$. The de Sitter spacetime is represented by hyperboloid imbedded in a five-dimensional flat space with metric

$$ds^2 = dv^2 - dw^2 - dx^2 - dy^2 - dz^2.$$

By introducing new coordinates (t, χ, θ, ϕ) on the hyperboloid by the relations

$$\alpha \sinh\left(\frac{t}{\alpha}\right) = v$$

$$\alpha \cosh\left(\frac{t}{\alpha}\right)\cos\chi = w,$$

$$\alpha \cosh\left(\frac{t}{\alpha}\right)\sin\chi\cos\theta = x,$$

$$\alpha \cosh\left(\frac{t}{\alpha}\right)\sin\chi\sin\theta\cos\phi = y,$$

$$\alpha \cosh\left(\frac{t}{\alpha}\right)\sin\chi\sin\theta\sin\phi = z,$$

the above metric transforms to

$$ds^2 = dt^2 - \alpha^2\cosh^2\left(\frac{t}{\alpha}\right)\left\{d\chi^2 + \sin^2\chi\left(d\theta^2 + \sin^2\theta d\phi^2\right)\right\}.$$

This spacetime has some trivial singularities namely, at $\chi = 0$, $\chi = \pi$, and $\theta = 0, \theta = \pi$. Except these trivial singularities, the ranges of the coordinates that cover the entire space are

$$-\infty < t < \infty, \ 0 \leq \chi \leq \pi, \ 0 \leq \theta \leq \pi, \ 0 \leq \phi \leq 2\pi.$$

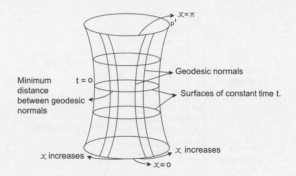

Figure 69 The image of de Sitter spacetime. This is a hyperboloid embedded in a flat five-dimensional spacetime given by general coordinates (t, χ, θ, ϕ).

For the above coordinates, de Sitter spacetime looks like a three-sphere (S^3) of constant positive curvature for the spatial sections $t = $ constant. The surfaces $t = $ constant are actually Cauchy surfaces. There geodesic normal lines start out infinitely large at $t = -\infty$ and shrink gradually to a lowest spatial separation, i.e., to a smallest fixed size at $t = 0$, after that expands again to infinite size as $t \to \infty$ (see Fig. 69).

It is interesting to study the infinity in de Sitter spacetime and for that we have to define a new time coordinate t'.

Now, one can change the time variable with the conformal time as

$$t' = 2 \tan^{-1}\left(exp\left(\frac{t}{\alpha}\right)\right) - \frac{1}{2}\pi$$

or

$$\tan\left(\frac{t'}{2}\right) = \tanh\left(\frac{t}{2\alpha}\right),$$

where

$$-\frac{\pi}{2} < t' < \frac{\pi}{2}.$$

Finally, one can get a metric conformally equivalent to Einstein static universe as

$$ds^2 = \alpha^2 cosh^2\frac{t}{\alpha}\overline{ds}^2,$$

where

$$\overline{ds}^2 = dt'^2 - dr'^2 - \sin^2 r'\left(d\theta^2 + \sin^2\theta d\phi^2\right).$$

Thus the de Sitter spacetime is conformal to part of the Einstein static universe (see Fig. 70) with the conformal factor

$$\Omega^2(t) = \alpha^2 cosh^2\frac{t}{\alpha}.$$

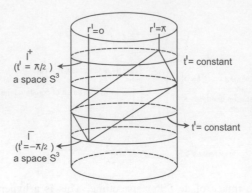

Figure 70   Einstein static universe.

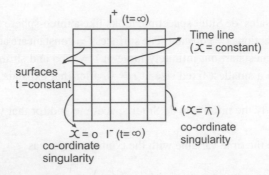

Figure 71   Past infinity (I^-) and future infinity (I^+), which are S^3 sphere.

Here, we identify $r' = \chi$. Thus, the de Sitter space is conformally related with the Einstein static universe to the restricted region bounded by $-\frac{\pi}{2} < t' < \frac{\pi}{2}$.

The de Sitter space has past infinity (I^-) and future infinity (I^+), which are S^3 spheres (see Fig. 71) and in contrast to Minkowski space, these infinities have space-like character. This new property is responsible for the existence of both particle and event horizons for geodesic of observers in de Sitter spacetime.

Note that the normal vector defined as

$$n^\mu = g^{\mu\nu}\Omega_{,\nu},$$

is time-like (where $g_{\mu\nu}$ is the fundamental tensor of de Sitter space).

In de Sitter spacetime, all the time-like geodesics must begin from the space-like infinity I^- and end at the space-like infinity I^+.

Let O be an observer whose world line is γ as shown in Fig. 72. Let p be any point on γ. Then the observer O can observe the set of events, which lie on $I^-(p)$, i.e., events in the past null cone of p at that time. Thus, the observer O can see the world lines of some particles, which intersect this null cone $I^-(p)$ but he cannot see the particles whose world lines don't intersect this null cone. Hence,

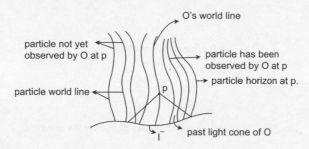

Figure 72 Particle horizon.

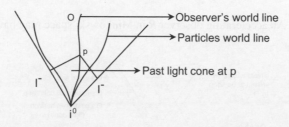

Figure 73 In Minkowski spacetime all the particles are seen at any event p on $O's$ world.

there is a sharp partition into those seen by O at p and those are not observed by O at p. This partition is said to be **particle horizon** for the observer O at the event p (see Fig. 72).

Actually, particle horizon is the limit of $O's$ vision, i.e., it represents the boundary between the observable and the unobservable particles. However, for Minkowski space I^- is null and all the particles are seen at any event p on $O's$ world line if they move on time-like geodesics (see Fig. 73).

Now we consider the observer O at F on I^+, then the past light cone $I^-(F)$ of O at F is called the **future event horizon** of O. This is actually a boundary between events, which can be seen in some time by O and those that will never be seen by O. Hence, events outside this boundary cannot be observable by O. Future event horizon is the borderline of the past of the observer world line. But for Minkowski space (here, I^- is null), the limiting null cone of any observer moving on a time-like geodesic will see eventually the whole spacetime, i.e., no events to be unseen. In other words, the observer O does not possess an event horizon. On the other hand, for an accelerating observer R in Minkowski space may have future event horizon (see Fig. 74). Note that these types of event horizon in Minkowski space depend on the observers, i.e., it is observer-dependent event horizon.

As we see that there is a limit to $O's$ world line on I^+, similarly, there is a limit to $O's$ world line on I^-. In de Sitter space, the future null cone of the point S is a boundary between events, which can be seen in some time by O and those that will never be observed by O. This surface is known as **past event horizon**.

Let world line of O begin at S on I^- and end at F on I^+. We draw light cones at S and F, which represent past and future event horizons, respectively (see Fig. 75).

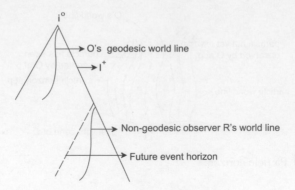

Figure 74   An accelerating observer R in Minkowski space may have future event horizon.

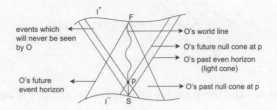

Figure 75   Future and past event horizons.

9.3   Anti-de Sitter Space

Anti-de Sitter space is suitably determined as a quadric in a five-dimensional flat spacetime with signature $(2, 3)$, i.e., the coordinate points (x, y, z, v, w) follow the relation

$$-x^2 - y^2 - z^2 + v^2 + w^2 = 1. \tag{9.11}$$

It has the topology $S^1 \times R^3$ and the Lorentzian metric induced from the metric on the five-dimensional flat spacetime is given by

$$ds^2 = dv^2 + dw^2 - dx^2 - dy^2 - dz^2.$$

It is evident that anti-de Sitter spacetime is conformally flat and in this spacetime, the Ricci scalar is a negative constant throughout the spacetime. Let us take the following transformation as

$$v = R \cos t, \quad w = R \sin t,$$

then Eq. (9.11) takes the form as

$$-x^2 - y^2 - z^2 + R^2 = 1.$$

The metric on the five-dimensional spacetime becomes

$$ds^2 = -dx^2 - dy^2 - dz^2 + dR^2 + R^2 dt^2. \tag{9.12}$$

Now, we consider another transformation by

$$R = \sqrt{1 + \rho^2}.$$

Again substitute the coordinates (x, y, z) by

$$x = \rho \cos \theta, \ y = \rho \sin \theta \cos \phi, \ z = \rho \sin \theta \sin \phi.$$

The induced metric assumes the following form as

$$ds^2 = -d\rho^2 - \rho^2 \left(d\theta^2 + \sin^2 \theta d\phi^2\right) + \frac{\rho^2 d\rho^2}{1 + \rho^2} + (1 + \rho^2)dt^2$$

$$= (1 + \rho^2)dt^2 - \frac{d\rho^2}{1 + \rho^2} - \rho^2 \left(d\theta^2 + \sin^2 \theta d\phi^2\right)$$

$$= (1 + \rho^2)dt^2 - \frac{d\rho^2}{1 + \rho^2} - \rho^2 d\Omega^2.$$

The transformation $\rho = \sinh r$ yields the line element as

$$ds^2 = \cosh^2 r dt^2 - dr^2 - \sinh^2 r \left(d\theta^2 + \sin^2 \theta d\phi^2\right). \tag{9.13}$$

The whole space can be covered by the surfaces $t = $ constant, which have nongeodesic normals.

Note 9.3

The metric

$$ds^2 = l^2 \cosh^2(r/l)dt^2 - dr^2 - l^2 \sinh^2(r/l)\left(d\theta^2 + \sin^2 \theta d\phi^2\right),$$

is solution to the Einstein equations with cosmological constant, where

$$\Lambda = -\frac{3}{l^2}.$$

Here the time-like coordinate t is an angular coordinate, which is restricted by the interval $t \in \ ]-\pi, \pi\ [$. This indicates that anti-de Sitter spacetime is a spacetime possessing closed time-like curves. Here, half of the hyperboloid is covered by these coordinates, therefore, to cover all of anti-de Sitter spacetime we should consider two portions $r > 0$ and $r < 0$. Therefore, the time-like coordinate t is restricted by the interval $t \in \ ]-\frac{\pi}{2}, \frac{\pi}{2}\ [$ to cover only half of the hyperboloid.

To know the structure at infinity, one can use the following transformation

$$r' = 2\tan^{-1}(e^r) - \frac{\pi}{2}$$

or

$$\tan\frac{r'}{2} = \tanh\frac{r}{2}, \quad 0 \le r' < \frac{\pi}{2}. \tag{9.14}$$

Then one can get,

$$ds^2 = \cosh^2 r d\bar{s}^2,$$

where $d\bar{s}^2$ is given by

$$d\bar{s}^2 = dt^2 - dr'^2 - \sin^2 r'\left(d\theta^2 + \sin^2\theta d\phi^2\right). \tag{9.15}$$

This implies once again that the whole anti-de Sitter space is expected to be conform to the part of Einstein static universe. The region of Einstein static universe is restricted by $0 \le r' \le \frac{\pi}{2}$. The Penrose diagram of universal anti-de Sitter space is actually the one half of the Einstein static universe. Here, null and space-like infinity can be considered as a time-like surface possessing the topology $R^1 \times S^2$. Usually it is not possible to get conformal transformation that transform the time-like infinity to finite without pocketing the Einstein static universe to a point, therefore, one can symbolize the time-like infinity by the two distinct points i^+, i^-. Thus, the infinity comprises of the time-like surface I with two distinct points i^+, i^-. The projection of some time-like and null geodesics is shown in Fig. 76. All the geodesic normals from $t = 0$ converge at p and q. All the time-like geodesics start from p and end at q without reaching the time-like surface I, however, future null geodesics from p will reach I.

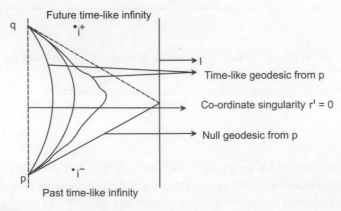

Figure 76   Penrose diagram in anti-de Sitter space.

9.4 Robertson–Walker Spaces

The universe appears to be homogeneous and isotropic (i.e., matter content is uniformly distributed and looks qualitatively the same in all direction) around us at sufficiently large scales (more than a 100 million light years or so). That is on this large scale the density of galaxies is roughly the same and all directions from us seem to be alike. Walker has shown that if all points and all directions of the universe are same (i.e., for exact spherically symmetry about all point), then the universe is spatially homogeneous and admits six isometries whose surfaces of transitivity are three-spaces of constant curvature, which are space-like. This space is known as **Robertson–Walker space**.

(A displacement of the type for which the displaced space is indistinguishable from its original states known as **isometry**.)

Under certain conditions Robertson–Walker space can be transformed into Minkowski space, de Sitter, anti-de Sitter spaces.

Robertson–Walker spacetimes are foliated by the three-dimensional hypersurfaces Σ of constant curvature. The one-parameter family of constant curvature three-spaces S are characterized by the time coordinate $t = $ constant. For this structure, the metric of a general Robertson–Walker spacetime can be written in the form

$$ds^2 = dt^2 - S^2(t)d\sigma^2, \tag{9.16}$$

where $d\sigma^2$ is the metric of a three space of constant curvature is independent of time. In Chapter Eleven, the deduction of the Robertson–Walker metric is given.

The geometry of these three spaces may have only three types and characterized by a parameter k, which is the sign of their curvature and actually they are three spaces of constant positive, zero, or negative curvature. The three-dimensional space is flat spacetime when $k = 0$, a three-sphere S^3 when $k = +1$, and a hyperbolic three-space H^3 when $k = -1$.

Also, general Robertson–Walker spacetime can be written (by alternative coordinate parametrizations of the three-spaces of constant curvature) as

$$ds^2 = dt^2 - S^2(t)\left[\frac{dr^2}{1 - kr^2} + r^2(d\theta^2 + \sin^2\theta d\phi^2)\right], \tag{9.17}$$

Now, after rescaling the function S, we can normalize this curvature k to be $1, 0, or -1$. According to k, we can categorize three possibilities:

$k = 0$: This case corresponds to a flat space and replacing r by χ in (9.17), metric ($d\sigma^2$) of the three-spaces of constant curvature takes the form

$$d\sigma^2 = d\chi^2 + \chi^2(d\theta^2 + \sin^2\theta\, d\phi^2),$$

$k = +1$: This represents a three-space of constant positive curvature, which is a three-sphere S^3 and using $r = \sin\chi$ in (9.17), the metric $d\sigma^2$ takes the form

$$d\sigma^2 = d\chi^2 + \sin^2\chi(d\theta^2 + \sin^2\theta\, d\phi^2).$$

$k = -1$: This characterizes a hyperbolic three-space H^3 and known as Lobatchevski space of constant negative curvature and using $r = \sinh \chi$ in (9.17), the metric $d\sigma^2$ assumes the form

$$d\sigma^2 = d\chi^2 + \sinh^2 \chi (d\theta^2 + \sin^2 \theta \, d\phi^2).$$

The above spherically symmetric forms of constant curvature three-spaces appear remarkably similar and simple and is written in a more compacted form as

$$d\sigma^2 = d\chi^2 + f^2(\chi)(d\theta^2 + \sin^2 \theta \, d\phi^2),$$

where

$$f(\chi) = \sin \chi, \; if \, k = 1$$

$$= \chi, \; if \, k = 0$$

$$= \sinh \chi, \; if \, k = -1$$

Here, the ranges of χ are different for different k

$$\chi \in [0, \infty) \; for \, k = 0, \, -1,$$

$$\chi \in [0, \pi] \; for \, k = 1.$$

To develop a homogeneous and isotropic model of the universe, i.e., to get the Robertson–Walker solutions one needs the perfect fluid energy-momentum tensor whose density μ and pressure p depend on time coordinate 't' only. Here the flow lines are the curves with $(\chi, \theta, \phi) = constant$ (i.e., coordinates are comoving).

The flow lines of fluids or history of photons are represented by a families of time-like or null curves. These families produce some influence on the spacetime curvature. Here, the perfect fluid energy-momentum tensor is taken as

$$T^{ab} = (\mu + p)V^a V^b - pg^{ab},$$

where, V^a is the four-velocity of the fluid as measured in some observer's local inertial frame.

The Raychaudhuri equation is

$$\frac{d\theta}{ds} = -R_{ab}V^a V^b + 2w^2 - 2\sigma^2 - \frac{1}{3}\theta^2 + \dot{V}^a_{;a},$$

where volume expansion,

$$\theta = V^a_{;a} = \theta_{ab}h^{ab}, \; \theta_{ab} = expansion \, tensor; \; \sigma_{ab} = \theta_{ab} - \frac{1}{3}h_{ab}\theta = shear \, tensor,$$

$$\omega_{ab} = vorticity \, tensor, \; h^a_b = \delta^a_b + V^a V_b, \; s = affine \, parameter,$$

$$\dot{V}^a = V^a_{;b}V^b \; is \, the \, acceleration \, of \, the \, flow \, lines.$$

Now the conservation equation in these spaces yields

$$\dot{\mu} = -3(\mu + p)\frac{\dot{S}}{S}. \tag{9.18}$$

The Raychaudhuri equation takes the form

$$9\pi(\mu + 3p) - \Lambda = -3\frac{\ddot{S}}{S}. \tag{9.19}$$

The Einstein field equation for the metric (9.17) gives

$$3\dot{S}^2 = \frac{8\pi(\mu S^3)}{S} + \Lambda S^2 - 3k.$$

We see that if $\Lambda = 0$, then (9.19) implies that S is not constant. In other words, the universe is either expanding or contracting. (Hubble confirmed that universe is expanding in present time). Also (9.18) implies that the density is decreasing with the expansion of the universe and as a consequence, we can predict that the density was greater in the earlier and growing indeterminately as $S \to 0$. This singularity (known as big-bang singularity) is the utmost remarkable feature of Robertson–Walker solutions. For the dust case (i.e. $p = 0$ and $\Lambda = 0$), the solutions of the above equations can be written in terms of τ where $\frac{d\tau}{dt} = \frac{1}{S(t)}$ as [details is given in Chapter Eleven]

$$S = \frac{E}{3}(\cosh \tau - 1), \quad t = \frac{E}{3}(\sinh \tau - \tau) \ for \ k = -1,$$

$$S = \tau^2, \quad t = \frac{1}{3}\tau^3 \ for \ k = 0,$$

$$S = \frac{-E}{3}(1 - \cos \tau), \quad t = \frac{-E}{3}(\tau - \sin \tau) \ for \ k = 1.$$

Here, E is a constant.

9.5 Penrose Diagrams of Robertson–Walker Spacetime for the Dust Case

Let us define a new coordinate τ using the transformation as

$$\frac{d\tau}{dt} = \frac{1}{S(t)},$$

then Robertson–Walker metric is expressed as

$$ds^2 = S^2(\tau)\left[d\tau^2 - d\chi^2 - f^2(\chi)(d\theta^2 + \sin^2\theta \, d\phi^2)\right]. \tag{9.20}$$

Case I: $k = 1, f(\chi) = \sin \chi$

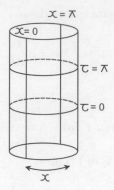

Figure 77 Robertson–Walker spacetime for $k = 1$ is mapped into the region in the Einstein static universe.

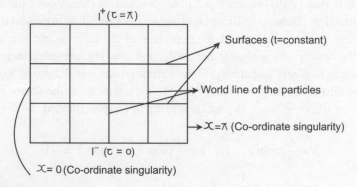

Figure 78 Past infinity (I^-) and future infinity (I^+) in Robertson–Walker spacetime for $k = 1$.

(9.20) implies $ds^2 = S^2(\tau)d\bar{s}^2$ where

$$d\bar{s}^2 = dt^{1^2} - dr^{1^2} - \sin^2 r^1(d\theta^2 + \sin^2 \theta \, d\phi^2),$$

which is (putting $\tau = t^1$, $\chi = r^1$) **Einstein static space**.

Here, τ lies in the range $0 < \tau < \pi$, so the whole space is mapped into this region in the Einstein static universe (see Figs. 77 and 78).

Case II: $k = 0$, $f(\chi) = \chi$

Here, the Robertson–Walker space is conformal to the Minkowski flat space as,

$$ds^2 = dt^{1^2} - dr^2 - r^2(d\theta^2 + \sin^2 \theta d\phi^2),$$

$$[here, \; \tau = t^1; \; r = \chi]$$

The actual space can be fixed by the values taken by τ. Here $0 < \tau < \infty$. So, Robertson–Walker space is conformal to the half ($t^1 > 0$) of the Minkowski spacetime.

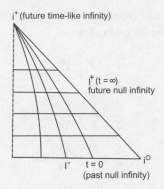

Figure 79 Penrose diagram of Robertson–Walker spacetime for $k = -1$.

Case III: $k = -1$, $f(\chi) = \sinh \chi$

In this case, Robertson–Walker space is conformal to the part of the region of the Einstein static space. Let us assume the following transformations

$$t^1 = \tan^{-1}\left[\tanh\frac{1}{2}(\tau + \chi)\right] + \tan^{-1}\left[\tanh\frac{1}{2}(\tau - \chi)\right],$$

$$r^1 = \tan^{-1}\left[\tanh\frac{1}{2}(\tau + \chi)\right] - \tan^{-1}\left[\tanh\frac{1}{2}(\tau - \chi)\right].$$

Here the region of the Einstein static space is bounded by

$$-\frac{1}{2}\pi \le t^1 + r^1 \le \frac{\pi}{2},$$

$$-\frac{1}{2}\pi \le t^1 - r^1 \le \frac{\pi}{2}.$$

Depending on the range of τ, this is the part of diamond-shaped region. Here, the space is mapped into the upper half (see Fig. 79).

Thus, we have seen that Robertson–Walker space with three different three-spaces of different constant curvatures are conformally related to some regions of Einstein static space.

9.6 Spatially Homogeneous Cosmological Models

The spatially homogeneous spacetime possesses group of isometries and weakest are those in which the group of isometrics is Abelian. According to Bianchi classifications, this spacetime is known as **Bianchi – I space**.

The metric for the spatially homogeneous spacetime possessing abelian isometry assumes the following form for the comoving coordinate (t, x, y, z) is

$$ds^2 = dt^2 - X^2(t)dx^2 - Y^2(t)dy^2 - Z^2(t)dz^2.$$

Here, X, Y, Z are scale factors along x, y, and z directions.

Let us assume, for simplicity, that the matter content is a pressureless perfect fluid, i.e., dust. The Einstein field equations take the following forms

$$\frac{\ddot{Y}}{Y} + \frac{\ddot{Z}}{Z} + \frac{\dot{Y}\dot{Z}}{YZ} = 0,$$

$$\frac{\ddot{Z}}{Z} + \frac{\ddot{X}}{X} + \frac{\dot{X}\dot{Z}}{XZ} = 0,$$

$$\frac{\ddot{X}}{X} + \frac{\ddot{Y}}{Y} + \frac{\dot{X}\dot{Y}}{XY} = 0,$$

$$\frac{\dot{X}\dot{Y}}{XY} + \frac{\dot{Y}\dot{Z}}{YZ} + \frac{\dot{Z}\dot{X}}{ZX} = \rho.$$

Conservation equation yields

$$\dot{\rho} + 3\frac{\dot{S}}{S}\rho = 0,$$

where, we define $S = (XYZ)^{\frac{1}{3}}$ as average scale factor. Conservation equation yields the density of matter as

$$\frac{4}{3}\pi\rho = \frac{M}{S^3},$$

where M is a suitable constant.

The solutions of the above field equations can be obtained as,

$$X = S\left(\frac{t^{\frac{2}{3}}}{S}\right)^{2\sin\alpha} \quad ; \quad Y = S\left(\frac{t^{\frac{2}{3}}}{S}\right)^{2\sin(\alpha+\frac{2}{3}\pi)},$$

$$Z = S\left(\frac{t^{\frac{2}{3}}}{S}\right)^{2\sin(\alpha+\frac{4}{3}\pi)} \quad ; \quad S^3 = \frac{9}{2}Mt(t+a),$$

where α and a are constants.

Here $a > 0$ determines the amount of the anisotropy and $\alpha(-\frac{1}{6}\pi < \alpha \le \frac{1}{2}\pi)$ decides the direction in which the most rapid expansion takes place. For $a = 0$, the space will be isotropic, which is Einstein–de Sitter universe.

The average rate of expansion can be obtained as

$$\frac{\dot{S}}{S} = \frac{2}{3t}\frac{t+\frac{a}{2}}{t+a}.$$

The expansion in x, y, and z directions are given by

$$\frac{\dot{X}}{X} = \frac{2}{3t} \frac{t + \frac{a(1+2\sin\alpha)}{2}}{t + a},$$

$$\frac{\dot{Y}}{Y} = \frac{2}{3t} \frac{t + \frac{a\{1+2\sin(\alpha+\frac{2}{3}\pi)\}}{2}}{t + a},$$

$$\frac{\dot{Z}}{Z} = \frac{2}{3t} \frac{t + \frac{a\{1+2\sin(\alpha+\frac{4}{3}\pi)\}}{2}}{t + a}.$$

If $\alpha \neq \frac{\pi}{2}$ then $-1 + 2\sin(\alpha + \frac{4}{3}\pi)$ will be negative, then universe collapses along z direction, whereas in x and y directions the universe expands monotonically at all times.

Now consider the case $\alpha = \frac{\pi}{2}$. Then metric coefficients assume the following forms as

$$X(t) = t\left\{\frac{9}{2}M(t+a)\right\}^{-\frac{1}{3}}; \quad Y(t) = Z(t) = \left\{\frac{9}{2}M(t+a)\right\}^{\frac{2}{3}}.$$

Following Hawking and Ellis, one can choose new coordinates τ, η such that

$$\tanh\left(\frac{2X}{9Ma}\right) = \frac{\eta}{\tau}; \quad exp\left[\frac{4}{9M}\int_0^t \frac{dt}{X(t)}\right] = \tau^2 - \eta^2.$$

Here, the whole space $(t > 0)$ is mapped into region Ω defined by

$$\tau > 0, \quad \tau^2 - \eta^2 > 0.$$

Here $t(\tau, \eta) > 0$ will be known from

$$\tau^2 - \eta^2 = \frac{9}{2}Mt^2 exp\frac{2(t+a)}{a}.$$

The (τ, η) plane is shown in region Ω, which is bounded by the surfaces $t = 0$ (see Fig. 80). For this space, the world lines of particles are straight lines starting from origin and then diverge.

$$t = 0 \Rightarrow \tau^2 = \eta^2 \Rightarrow \eta = \pm\tau,$$

$$t = constant \Rightarrow \tau^2 - \eta^2 = constant = hyperbola,$$

$$X = constant \Rightarrow \frac{\eta}{\tau} = constant,$$

$$\Rightarrow \eta = constant.\tau.$$

$\Rightarrow$ straight line passing through origin.

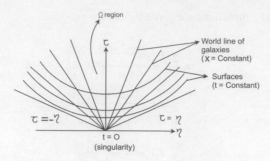

Figure 80   Dust-filled Bianchi-I model in $\tau - \eta$ plane.

9.7   Schwarzschild Solutions

The most interesting general spherically symmetric static vacuum solution of the Einstein field equations is the Schwarzschild solution. The Schwarzschild solution, i.e., the Schwarzschild metric describes the Schwarzschild black hole. It defines the gravitational field in the outer region of a spherical mass. The Schwarzschild line element is written as

$$ds^2 = \left(1 - \frac{2m}{r}\right) dt^2 - \frac{dr^2}{1 - \frac{2m}{r}} - r^2(d\theta^2 + \sin^2\theta \, d\phi^2), \tag{9.21}$$

where $m = \frac{GM}{c^2}$.

Karl Schwarzschild found this exact static spherically symmetric vacuum solutions of the Einstein field equations in 1915 while fighting during World War I in favor of Germany. To show honor, this solution is dubbed as The Schwarzschild solution. The coordinate r is a radial parameter, which has the property that the surface area of the two-sphere (t = constant, r = constant) is $4\pi r^2$.

There are two values of coordinates for which the solution has singularities. A singularity at $r = 0$ is an essential singularity, whereas singularity at

$$r = 2m = \frac{2GM}{c^2} \; (known \; as \; Schwarzschild \; radius)$$

is dubbed as coordinate singularity. Here, the Kretschmann scalar

$$R_{abcd}R^{abcd} = \frac{48m^2}{r^6},$$

which is finite at $r = 2m$ but at $r = 0$ it blows up. Thus, singularity at $r = 0$ is irremovable, and thus, it is an essential singularity.

[Coordinate singularity is a place where geometry cannot be described properly and it is not essential, i.e., it can be uninvolved by a suitable choice of coordinate system.]

Note that coefficient of dt^2, i.e., $g_{tt} = 0$ yields **infinite redshift**. Here g_{tt} vanishes at $r = 2m$, therefore, the surface $r = 2m$ is the surface of infinite redshift. Also note that $r = 2m$ is a null hypersurface, which splits the spacetime into two disconnected regions:

$$\text{I. } 2m < r < \infty, \quad \text{II. } 0 < r < 2m$$

The region for $r > 2m$ represents the external field, whereas usual t is time-like and r is space-like, however, in the region $0 < r < 2m$, the role of r and t will be reversed, i.e., here, r is time-like and t is space-like. Thus, topological behavior of Schwarzschild solution is not Euclidean.

9.8 Null Curves in Schwarzschild Spacetime

In Schwarzschild geometry, we see that $r = 2m$ is a problematic radius. Here, the metric becomes singular at $r = 2m$. Therefore, it is expected that Schwarzschild solution is not appropriate for investigating the physics in the region $r \leq 2m$. However, this singular behavior has been occurred due to choice of bad coordinates. To know better the characteristic of the Schwarzschild geometry, it is essential to look after its casual structure, i.e., the light cones.

Now we consider radial null curves ($ds^2 = 0$) in the planes $\theta = constant$ and $\phi = constant$. Hence, we have

$$ds^2 = 0 = \left(1 - \frac{2m}{r}\right) dt^2 - \left(1 - \frac{2m}{r}\right)^{-1} dr^2.$$

This implies,

$$\frac{dt}{dr} = \pm \left(1 - \frac{2m}{r}\right)^{-1}. \tag{9.22}$$

Integrating the above integral, we get (taking positive sign)

$$t = r + 2m \ln |r - 2m| + constant. \tag{9.23}$$

We note that for $r > 2m$, $\frac{dt}{dr} > 0$. This indicates r is increasing with t. This radial null geodesic is outgoing (see Fig. 81).

For negative sign, the above integral yields

$$t = -(r + 2m \ln |r - 2m| + constant). \tag{9.24}$$

This radial null geodesic is ingoing (see Fig. 81).

We consider the light cones in (r, t) plane. Note that $\frac{dt}{dr}$ signifies the slope of the light cones at a given value of r. For $r \to \infty$, $\frac{dt}{dr} = \pm 1$, i.e., slope is ± 1 as in Minkowski space or flat space. When one approaches to $r = 2m$, one will get

$$\frac{dt}{dr} \to \pm \infty.$$

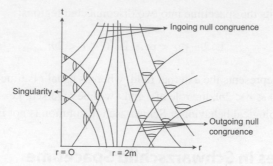

Figure 81 Outgoing and ingoing radial null geodesics.

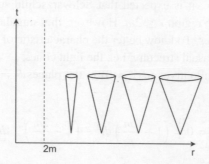

Figure 82 If we go toward $r = 2m$, the light cones become thinner and thinner and ultimately collapse entirely.

This indicates when we approach to Schwarzschild radius the light cones close up. Here coordinate velocity approaches to zero at $r = 2m$, i.e., coordinate velocity in the r direction is gradually diminished with respect to the coordinate time t. As a result, the casual structure of the Schwarzschild geometry in Schwarzschild coordinate (r, t) indicates that as we are approaching $r = 2m$, the light cones become thinner and thinner and ultimately collapse totally (see Fig. 82).

9.9 Time-like Geodesics in Schwarzschild Spacetime

Now, we consider the geodesic of radial time-like particles in the planes $\theta = constant$ and $\phi = constant$. Hence, we have the Lagrangian

$$L = \left(1 - \frac{2m}{r}\right)\left(\frac{dt}{d\tau}\right)^2 - \left(1 - \frac{2m}{r}\right)^{-1}\left(\frac{dr}{d\tau}\right)^2 - r^2\left(\frac{d\theta}{d\tau}\right)^2 - r^2\sin^2\theta\left(\frac{d\phi}{d\tau}\right)^2.$$

For time-like particle, we know $L = 1$, the above equation takes the form

$$1 = \left(1 - \frac{2m}{r}\right)\left(\frac{dt}{d\tau}\right)^2 - \left(1 - \frac{2m}{r}\right)^{-1}\left(\frac{dr}{d\tau}\right)^2. \tag{9.25}$$

Also Euler–Lagrangian equation yields

$$\left(1 - \frac{2m}{r}\right)\left(\frac{dt}{d\tau}\right) = E$$

where E is a constant and it assumes different values for different initial conditions.

The above two equations imply

$$\left(\frac{dr}{d\tau}\right)^2 = (E^2 - 1) + \frac{2m}{r}. \tag{9.26}$$

If the initial velocity of the particle is zero, which starts from infinite distance, then, *at $r \to \infty$*, $\left(\frac{dr}{d\tau}\right) = 0$. Putting these conditions in the above equation, we get $E^2 = 1$. Now Eq. (9.25) takes the form

$$\left(\frac{dr}{d\tau}\right) = -\sqrt{\frac{2m}{r}}.$$

[We have taken negative sign as particle is moving towards decreasing r.

Note that this radial free particle from infinity has the four velocity $u^\mu = (\frac{dt}{d\tau}, \frac{dr}{d\tau}, \frac{d\theta}{d\tau}, \frac{d\phi}{d\tau}) = ([1 - \frac{2m}{r}]^{-1}, -\sqrt{\frac{2m}{r}}, 0, 0).]$

Solving this equation, we get the proper time of fall as

$$\tau - \tau_0 = \frac{2}{3\sqrt{2m}}\left(r_0^{\frac{3}{2}} - r^{\frac{3}{2}}\right), \tag{9.27}$$

where initial position of the particle is at r_0 at the proper time τ_0. Interestingly, we note that there are no singularities at $r = 2m$ (Schwarzschild radius) and $r = 0$ (origin). Also a body takes finite proper time to reach from $r = 2m$ to $r = 0$, which is given by (see left panel of Fig. 83)

$$\tau = \frac{4m}{3}.$$

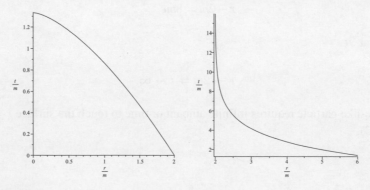

Figure 83 (Left) A body takes finite proper time to reach from $r = 2m$ to $r = 0$. (Right) Any time-like particle requires infinite amount of time to touch the surface $r = 2m$.

Now, to search how a viewer sees this phenomenon. Here, the viewer's time is t (Schwarzschild time coordinate) instead of proper time τ. Using as before $E = 1$, we get

$$\frac{dt}{dr} = \frac{\left(\frac{dt}{d\tau}\right)}{\left(\frac{dr}{d\tau}\right)} = -\frac{\sqrt{\frac{r}{2m}}}{\left(1 - \frac{2m}{r}\right)}.$$

Solving this equation, we get

$$t - t_0 = -\frac{2}{3\sqrt{2m}}\left[\left(r^{\frac{3}{2}} - r_0^{\frac{3}{2}}\right) + 6m\left(r^{\frac{1}{2}} - r_0^{\frac{1}{2}}\right)\right]$$

$$+ 2m \ln \frac{\left[r^{\frac{1}{2}} + (2m)^{\frac{1}{2}}\right]\left[r_0^{\frac{1}{2}} - (2m)^{\frac{1}{2}}\right]}{\left[r_0^{\frac{1}{2}} + (2m)^{\frac{1}{2}}\right]\left[r^{\frac{1}{2}} - (2m)^{\frac{1}{2}}\right]}. \tag{9.28}$$

Note that this result has changed significantly from (9.27). However, for $r, r_0 \gg 2m$ Eqs. (9.27) and (9.28) are almost identical. Also we note that for large r_0 but r is very closed to $2m$, the expression (9.28) can be written approximately as

$$t - t_0 \approx 2m \ln \frac{\left[r^{\frac{1}{2}} + (2m)^{\frac{1}{2}}\right]}{\left[r^{\frac{1}{2}} - (2m)^{\frac{1}{2}}\right]}$$

or

$$\frac{t - t_0}{2m} = \ln \frac{8m}{r - 2m} \qquad (using \ r \approx 2m),$$

or

$$r - 2m = 8me^{-\frac{t-t_0}{2m}}. \tag{9.29}$$

It is obvious that

$$r \to 2m \Rightarrow t \to \infty.$$

Hence any time-like particle requires infinite amount of time to touch the surface $r = 2m$ (see right panel of Fig. 83).

9.10 Tortoise Coordinates

Now we use new coordinate system to describe the region around $r = r_s = 2m$. We consider the Schwarzschild metric in the following form,

$$ds^2 = \left(1 - \frac{2m}{r}\right)\left[dt^2 - \left(1 - \frac{2m}{r}\right)^{-2} dr^2\right] - r^2 d\Omega^2. \tag{9.30}$$

Now, we introduce new coordinate r^* such that

$$dr^* = \left(1 - \frac{2m}{r}\right)^{-1} dr. \tag{9.31}$$

Solving this, we get,

$$r^* = r + 2m \ln\left(\frac{r}{2m} - 1\right). \tag{9.32}$$

The above new radial co-ordinate is said to be **Tortoise coordinate**.

From (9.30), we can get

$$dt = \pm\left(1 - \frac{2m}{r}\right)^{-1} dr \Rightarrow dt = \pm dr^*.$$

This implies

$$t = \pm r^* + c,$$

which is characterizing the light cones. In terms of r^*, the Schwarzschild metric takes the form as

$$ds^2 = \left(1 - \frac{2m}{r}\right)[dt^2 - dr^{*2}] - r^2 d\Omega^2, \tag{9.33}$$

where r be a function of r^*.

Here, the light cones are defined as $dt^2 = dr^{*2}$, i.e., light cones comprise with the constant slope $\frac{dt}{dr^*} = \pm 1$ (see Fig. 84).

The appearance of these light cones are the same as the light cones in Minkowski space and no longer fold up as $r \to 2m$ (for which $r^* \to -\infty$). Here, surface $r = 2m$ lies infinitely far away. Thus, the Tortoise coordinate system pushes the singularity at finite distance, i.e., at $r = 2m$ to infinitely far away ($r^* \to -\infty$).

9.11 Eddington–Finkelstein Coordinates

Till now t is not workable for sightseeing the region beyond $r = r_s = 2m$. Also, time-like geodesic takes finite proper time to reach the surface $r = r_s = 2m$. Therefore, it is reasonable to express

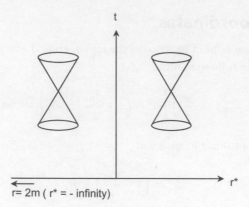

r= 2m (r* = - infinity)

Figure 84 The light cones in Schwarzschild geometry for the Tortoise coordinate (r^*, t).

coordinates, which are logically adjusted to the null geodesics as

$$u = t + r^*, \quad v = t - r^*, \tag{9.34}$$

where r^* is the Tortoise coordinate. Here u *and* v are known as **advanced time** parameter and **retarded time** parameter, respectively.

The new coordinate system (u, r, θ, ϕ) or (v, r, θ, ϕ) is known as **Eddington–Finkelstein coordinates**.

Here,

$$dt = du - dr^* = du - \left(1 - \frac{2m}{r}\right)^{-1} dr,$$

or

$$dt^2 = du^2 - \left(1 - \frac{2m}{r}\right)^{-1} 2drdu + \left(1 - \frac{2m}{r}\right)^{-2} dr^2.$$

Keeping the original radial coordinate r and substituting the time-like coordinate t with the new coordinate $u(or\ v)$ the Schwarzschild metric takes the form as

$$ds^2 = \left(1 - \frac{2m}{r}\right) du^2 - 2dudr - r^2 d\Omega^2. \tag{9.35}$$

Note that the value of metric coefficient g_{uu} or g_{vv} will be zero at $r = 2m$ but the determinant of the metric is

$$g = -r^4 \sin^2 \theta,$$

which is completely regular at $r = 2m$. As a result, the metric is nonsingular and one can note that $r = 2m$ is just a coordinate singularity in initial Schwarzschild coordinate (t, r, θ, ϕ) system. Note also that $r = 0$ is only the physical singularity.

Now, we calculate the radial null geodesics as before to define the light cones in Eddington–Finkelstein coordinate system. Using $(ds = 0, d\theta = d\phi = 0)$, we get

$$\left(1 - \frac{2m}{r}\right) du^2 = 2dudr,$$

$\Rightarrow$ either,

$$\frac{du}{dr} = 0, \quad i.e., \quad u = constant,$$

which defines incoming null geodesics, or,

$$\frac{du}{dr} = 2 \left(1 - \frac{2m}{r}\right)^{-1},$$

which characterizes the outgoing null geodesics. The following equation provides the path of radial light

$$u = 2(r + 2m \ln | r - 2m |) + constant.$$

Note that for $r > 2m$, u will increase with the increase of r. This indicates radial light rays are outgoing. Also the light cones are well performed at $r = 2m$, i.e., do not abolish rather do tilt over. Therefore, we can expect that the null or time-like geodesics can be well defined beyond $r = 2m$. We notice that for $r < 2m$, u will decrease with the decrease of r. This indicates radial light rays are ingoing.

If we draw the light cones in (u, r) plane, then one can note that one side of the light cone permanently leftovers horizontal (at $u = constant$), the other side keeps vertical at $r = 2m$ ($\frac{du}{dr} = \infty$) and then inclines to the other side such that for $r < 2m$ all future-directed paths are in the direction of decreasing r (see Fig. 85). Thus, once a null or time-like particle enters the surface $r = 2m$, it could not be able to come out. This surface is termed as **event horizon**. This is a null surface. There is no correspondence between the events in $r < 2m$ and $r > 2m$.

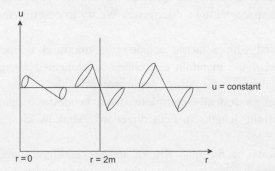

Figure 85   The behavior of the light cone in Eddington–Finkelstein coordinate.

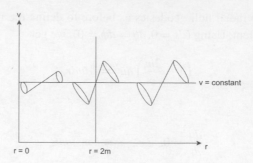

Figure 86 The behavior of the light cone in Eddington–Finkelstein coordinate.

Note that no particle (light or any massive) can return from the event horizon and, therefore, it is not possible to see what happens inside the event horizon. Thus, this configuration is dubbed as **black hole**, an object that is entirely undetectable.

Actually, in (u, r) coordinate system, the event horizon can be crossed only by future-directed paths, not on past-directed ones. This behavior will be changed, if we use (v, r) coordinate system. Here the Schwarzschild line element becomes

$$ds^2 = \left(1 - \frac{2m}{r}\right) dv^2 + 2dvdr - r^2 d\Omega^2. \tag{9.36}$$

As before, in this case one can able to cross the event horizon ($r = 2m$) but this time the geodesic paths will be only past-directed (see Fig. 86). Actually, this is a past extensions of the original Schwarzschild geometry. In this case, the region behind $r = 2m$ plays the role opposite to a black hole and known as **white hole**. Thus, white holes are flare-ups of light and nothing can get inside them. This is a hypothetical and a mathematical peculiarity than a real thing. Actually if radius of a star increases unavoidably through its Schwarzschild radius, then white hole may create.

9.12 Kruskal–Szekeres Coordinates

We now give our attention in describing space-like geodesics instead of null geodesics to find region of Schwarzschild space yet to be discovered. We try to describe the maximal extension of spacetime.

(A manifold endowed with a metric geometry is **maximal** if each geodesic originating from an arbitrary point of the manifold can either be elongated along the geodesic in both directions to infinite values of the affine parameter or ends on an intrinsic singularity. A manifold is said to be **geodesically complete** if all geodesics originating from any point can be elongated to infinite lengths in both directions. Minkowski spacetime is geodesically complete.)

Schwarzschild solution is not maximal. Kruskal and Szekeres have found the maximal extension of the Schwarzschild solution. They have started the following transformations,

$$u = t + r^*, \; v = t - r^*, \tag{9.37}$$

where r^* is the Tortoise coordinate defined by

$$r^* = r + 2m \ln\left(\frac{r}{2m} - 1\right).$$

For this transformations, the Schwarzschild metric (9.33) takes the form as

$$ds^2 = \left(1 - \frac{2m}{r}\right) du\,dv - r^2 d\Omega^2, \tag{9.38}$$

where r is given by,

$$\frac{1}{2}(u - v) = r + 2m \ln\left(\frac{r}{2m} - 1\right). \tag{9.39}$$

However, coordinate $r = 2m$ is infinitely far away at either $u = -\infty$ or $v = \infty$.

Now to pull these infinite points into finite coordinate values, we make the following transformations

$$u^1 = e^{\frac{u}{4m}}, \quad v^1 = e^{-\frac{v}{4m}}. \tag{9.40}$$

(here the horizon is at either $u^1 = 0$ or $v^1 = 0$.)

These transformations can be expressed in terms of original Schwarzschild (t, r) system as,

$$u^1 = \left(\frac{r}{2m} - 1\right)^{\frac{1}{2}} e^{\frac{r+t}{4m}}, \quad v^1 = \left(\frac{r}{2m} - 1\right)^{\frac{1}{2}} e^{\frac{r-t}{4m}}. \tag{9.41}$$

The transformations (9.40) yield the Schwarzschild metric in u^1, v^1, θ, ϕ coordinates as

$$ds^2 = \frac{32m^3}{r} e^{-\frac{r}{2m}} (du^1 dv^1) - r^2 d\Omega^2. \tag{9.42}$$

This metric is absolutely nonsingular and regular except at $r = 0$. This is actually a physical singularity, which cannot be avoided by any coordinate transformation. Since coefficients of both du^1 and dv^1 are zero, therefore, both u^1 and v^1 are null coordinates and their partial derivatives $\frac{\partial}{\partial u^1}$ and $\frac{\partial}{\partial v^1}$ are null vectors. Now changing these null coordinate u^1, v^1 to more readable coordinates (T, X) of which first is time-like and later is space-like as,

$$X = \frac{1}{2}(u^1 + v^1) = \left(\frac{r}{2m} - 1\right)^{\frac{1}{2}} e^{\frac{r}{4m}} \cosh\left(\frac{t}{4m}\right), \tag{9.43}$$

$$T = \frac{1}{2}(u^1 - v^1) = \left(\frac{r}{2m} - 1\right)^{\frac{1}{2}} e^{\frac{r}{4m}} \sinh\left(\frac{t}{4m}\right). \tag{9.44}$$

These transformations yield the Schwarzschild metric (9.42) as

$$ds^2 = \frac{32m^3}{r} e^{-\frac{r}{2m}} (dT^2 - dX^2) - r^2 d\Omega^2, \tag{9.45}$$

where r is defined by,

$$X^2 - T^2 = \left(\frac{r}{2m} - 1\right) e^{\frac{r}{2m}}. \tag{9.46}$$

The coordinates (T, X, θ, ϕ) are known as **Kruskal–Szekeres** coordinates.

The radial null curves are given by (putting, $d\theta = d\phi = 0$)

$$ds^2 = 0 = \frac{32m^3}{r} e^{-\frac{r}{2m}} (dT^2 - dX^2),$$

which gives

$$\frac{dX}{dT} = \pm 1 \Rightarrow X = \pm T + constant. \tag{9.47}$$

This is very similar to Minkowski spacetime and light cones are same.

Contrasting to the Tortoise coordinates (t, r^*), the singularity $r = 2m$ is not located at infinite distance rather the singularity is now at the null surfaces

$$X = \pm T. \tag{9.48}$$

The surfaces $r = constant$ are given by

$$X^2 - T^2 = constant. \tag{9.49}$$

Thus, the surfaces $r = constant$ represent hyperbolae in (T, X) plane.

Again $t = $ constant, surfaces are given by

$$\frac{T}{X} = \tanh\left(\frac{t}{4m}\right). \tag{9.50}$$

These straight lines are passing through the origin with slope $\left(\tanh \frac{t}{4GM}\right)$.

We see that as $t \to \pm\infty$, $T = \pm X$, therefore, these surfaces coincide with $r = 2m$.

Thus, one can note that Kruskal–Szekeres coordinates (T, X, θ, ϕ) are everywhere finite without possessing the singularity at $r = 2m$. The range of these coordinates are given by

$$-\infty \leq X \leq \infty \quad and \quad T^2 < X^2 + 1. \tag{9.51}$$

The only real singularity at $r = 0$, is described by two sheets of the hyperboloid.

$$r = 0 \Leftrightarrow T^2 - X^2 = 1. \tag{9.52}$$

Hence, the metric is defined for $r > 0$ and

$$r > 0 \Rightarrow T^2 - X^2 < 1. \tag{9.53}$$

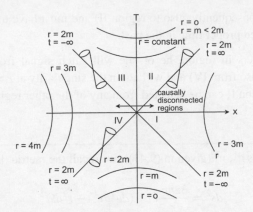

Figure 87 Kruskal–Szekeres diagram.

By suppressing θ, ϕ coordinates one can draw a spacetime diagram in $T-X$ plane. This diagram is known as **Kruskal–Szekeres diagram** and it characterizes the complete spacetime of Schwarzschild metric (see Fig. 87).

To get a clear view of Kruskal–Szekeres diagram, one can divide the diagram into four regions. Now, we will show some significant characteristics of Kruskal–Szekeres diagram.

1. X is a global radial coordinate and T is a global time coordinate.

2. The radial null curves are diagonals, $X = \pm T + constant$, which is very similar to Minkowski spacetime and light cones are same.

3. The lines $T = \pm X$, coincide with the horizon surface $r = 2m$.

4. The surfaces $r = constant$ represent hyperbolae in (T, X) plane. In a more specific, the hyperbolae lie in quadrants I and III for $r > 2m$ and the hyperbolae lie in quadrants II and IV for $r < 2m$.

5. The only real singularity at $r = 0$, is described by $T^2 - X^2 = 1$ (two sheets of the hyperbola).

6. The world lines with $r = constant$ but $r < 2m$ lie in quadrants II and IV are space-like.

7. The regions I and II are concealed by the Eddington–Finkelstein coordinates (u, r) and regions I and IV by (v, r).

8. Region I (curves with $r > 2m$) is covered by the original Schwarzschild coordinates and is divided by the horizons from regions II and IV.

9. The region in quadrant III is totally different and is also filled with curves, $r > 2GM$. It is just a copy of region I. This region is detached from region I by a space-like distance. Thus, there is no causal relation between regions I and III, i.e., they are casually disconnected.

10. An observer sitting either in region I or III can collects signals from region IV and send signals to region II. A spectator sitting in region IV can transmit signals into both region

I and III (and consequently, also to region II) and must have transpired from singularity at $r = 0$ at a fixed proper time in the past.

11. If a viewer enters in region, he or she will catch signal from region I and III (and consequently, also from IV) and will attain the singularity at $r = 0$ in a fixed future time. No event in region II can be detected from any of the other regions.

Note 9.4

In Kruskal null coordinates u^1, v^1 (given in (9.41)), we recall the metric (9.42) as

$$ds^2 = \frac{32m^3}{r} e^{-\frac{r}{2m}} (du^1 dv^1) - r^2 d\Omega^2,$$

with

$$u^1 v^1 = \left(\frac{r}{2m} - 1 \right) e^{\frac{r}{2m}}.$$

This metric is absolutely nonsingular and regular except at $r = 0$. Let us take other coordinate transformations (u^{11}, v^{11}) to bring infinity into finite coordinate values

$$u^{11} = \tan^{-1} \left[\frac{u^1}{\sqrt{2m}} \right], \quad v^{11} = \tan^{-1} \left[\frac{v^1}{\sqrt{2m}} \right].$$

The range of the coordinates

$$-\frac{\pi}{2} < u^{11} < \frac{\pi}{2}, \quad -\frac{\pi}{2} < v^{11} < \frac{\pi}{2}.$$

In these coordinates $(u^{11}, v^{11}, \theta, \phi)$, the metric is conformally related to Minkowski spacetime. It shows all possible regions of the complete analytically extended manifold. Here, the null geodesics (black hole horizon $r = 2m$) are the straight lines that make 45^0, i.e., slope ± 1 in the Penrose diagram for the maximally extended Schwarzschild solution (see Fig. 88). The time-like infinity in one asymptotic region to time-like infinity in the other are joined by the straight lines, which are characterized by $r = 0$ singularity in the new coordinates. Outside the horizons, we have two causally separated static asymptotically Minkowski-like regions $r > 2m$. All the surfaces of constant r intersect at $i^{\pm}$ (i.e., far away in time, either to the future or past). Interestingly, we note that the points i^+ and i^- are separated from the point $r = 0$ because there are many timelike curves that do not touch the singularity.

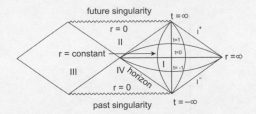

Figure 88 Penrose diagram of Schwarzschild solution in Kruskal coordinates.

9.13 Reissner–Nordström Solution

The Reissner–Nordström metric is the unique spherically symmetric and asymptotically flat solution of the coupled Einstein–Maxwell equations describing the field around an isolated spherical object with mass m and charge e, i.e., it describes the geometry of the region outside of a spherically symmetric electrically charged star or black hole. It is given by the following metric

$$ds^2 = g_{tt}dt^2 - g_{rr}dr^2 - r^2(d\theta^2 + \sin^2\theta d\phi^2), \tag{i}$$

where,

$$g_{tt} = (g_{rr})^{-1} = \Delta = \left(1 - \frac{2m}{r} + \frac{e^2}{r^2}\right) = \frac{1}{r^2}(r - r_+)(r - r_-).$$

Here, the outer and inner horizons $(r_\pm)$ are read as

$$r_\pm = m \pm \sqrt{m^2 - e^2}.$$

Thus, we have three cases, $m^2 - e^2 < 0, = 0, > 0$, which depend on the relative size of the gravitational mass m and charge e.

Case I: $m^2 - e^2 < 0$: Here, $m < e$, and so $\Delta > 0$ for all $r > 0$. Therefore, no horizon exists and we have a **naked curvature singularity** at $r = 0$ (i.e., it is a gravitational singularity with no event-horizon, in other words, event-horizon does not cover it). It is the case of an "overcharged" star. This case has a little astrophysical relevance. Here, the coordinate r is valid very near to $r = 0$, i.e., one can go to very close to the singularity and come back, this is actually a tragedy. Since the coordinate r is valid very near to $r = 0$, then the Penrose diagram of this spacetime is shown in Fig. 89. Here, the vertical line $r = 0$ now signifies a singularity, observable all the way to I^+.

Case II: $m^2 - e^2 > 0$: Here one can have two real roots, $r_+ > r_- > 0$. The metric is regular within all of the three distinct areas

$$r > r_+, \quad r_- < r < r_+ \quad and \quad r < r_-.$$

We note that in the region (I) for $r > r_+$, r is a space-like and t is time-like coordinate. However, the region (II) between two roots, i.e., $r_- < r < r_+$, r is timelike and t spacelike. Note again that the

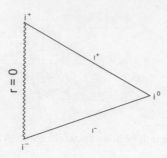

Figure 89   Penrose diagram of naked singularity in Reissner–Nordström solution.

innermost region (III), $0 < r < r_-$ the coordinate r turns into spacelike again and t is timelike. The roots $r = r_+$ and $r = r_-$ are called the **outer horizon** and the **inner horizon**, respectively.

However, these singularities are coordinate singularities and can be uninvolved by some appropriate coordinate transformations as done earlier. Let us use the coordinate transformation r^* where

$$dr^* = \frac{dr}{\left(1 - \frac{2m}{r} + \frac{e^2}{r^2}\right)},$$

$$or, \quad r^* = r + \frac{r_+^2}{r_+ - r_-}\ln(r - r_+) - \frac{r_-^2}{r_+ - r_-}\ln(r - r_-).$$

Now, we introduce Eddington–Finkelstein type coordinates similar to Schwarzschild solution as

$$u = t - r^*, \quad v = t + r^*.$$

In the advanced null coordinate, the Reissner–Nordström metric can be rewritten as

$$ds^2 = -2dvdr + \left(1 - \frac{2m}{r} + \frac{e^2}{r^2}\right)dv^2 - r^2(d\theta^2 + \sin^2\theta d\phi^2).$$

Note that this metric is regular at both horizons $r = r_\pm$. Incoming and outgoing null geodesics can now be easily identified and an Eddington–Finkelstein-like diagram can be constructed, as shown in Fig. 90. To see the incoming and outgoing null geodesics in the spacetime, one can draw the diagram for $\bar{t} = t + r^* - r$ with respect to radial coordinate r (see Fig. 90). Note that light signal cannot go from regions II to I. Hence $r = r_+$ is an event horizon. Particles entering region II move toward region III and may cross $r = r_-$ or reach asymptotically. The light cones in region III indicates that particles need not reach the $r = 0$ singularity. Hence, no point within the region $r < r_+$ can send a signal to the region $r > r_+$, so this region corresponds to a black hole.

In the retarded null coordinate, the Reissner–Nordström metric can be rewritten as

$$ds^2 = 2dudr + \left(1 - \frac{2m}{r} + \frac{e^2}{r^2}\right)du^2 - r^2(d\theta^2 + \sin^2\theta d\phi^2).$$

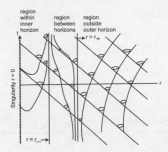

Figure 90 Light cones in Reissner–Nordström sacetime.

In this case one can able to cross the event horizon ($r = r_+$) but this time the geodesic paths will be only past-directed, hence the region $r < r_+$ plays the role opposite to a black hole and known as white hole.

Now, we try to remove coordinate singularities by using Kruskal–Szekeres-like coordinates. Note that one cannot remove both coordinate singularities simultaneously, however, it can be removed separately.

Now, let us assume the double null coordinates as

$$v = \bar{t} + r, \quad u = 2t - v,$$

where,

$$\bar{t} = t + \frac{r_+^2}{r_+ - r_-} \ln(r - r_+) - \frac{r_-^2}{r_+ - r_-} \ln(r - r_-).$$

For the above new null coordinates, the Reissner–Nordström line element takes the following form

$$ds^2 = \left(1 - \frac{2m}{r} + \frac{e^2}{r^2} \right) dv du - r^2 (d\theta^2 + \sin^2 \theta d\phi^2).$$

Let us first consider the case for outer horizon $r = r_+$ and take the following transformations,

$$U_+ = -\frac{2r_+^2}{(r_+ - r_-)} exp\left(\frac{(r_+ - r_-)}{2r_+^2} \right) u, \quad V_+ = \frac{2r_+^2}{(r_+ - r_-)} exp\left(\frac{(r_+ - r_-)}{2r_+^2} \right) v,$$

which converts the above equation as

$$ds^2 = \frac{r_- r_+}{r^2} \left(\frac{(r - r_-)}{r_-} \right)^{1 + \frac{r_-^2}{r_+^2}} exp\left(-\frac{(r_+ - r_-)}{r_+^2} r \right) dU_+ dV_+$$

$$- r^2 (d\theta^2 + \sin^2 \theta d\phi^2).$$

Now, in Kruskal–Szekeres coordinates,

$$T_+ = \frac{1}{2}(V_+ + U_+), \quad R_+ = \frac{1}{2}(V_+ - U_+),$$

the metric becomes,

$$ds^2 = \frac{r_- r_+}{r^2}\left(\frac{(r - r_-)}{r_-}\right)^{1+\frac{r_-^2}{r_+^2}} exp\left(-\frac{(r_+ - r_-)}{r_+^2}r\right)(dT_+^2 - dR_+^2)$$
$$- r^2(d\theta^2 + \sin^2\theta d\phi^2).$$

Here, $r = r_+$ is a regular point, however, $r = r_-$ is not regular. Actually, the metric is regular from $r = r_-$ to $r = r_+$ and out to ∞. The outer horizon $r = r_+$ corresponds to both null hyper surfaces $U_+ = 0, V_+ = 0$.

For the region $0 < r \le r_-$, we introduce another transformations as

$$U_- = \frac{2r_-^2}{(r_+ - r_-)}exp\left(\frac{(r_+ - r_-)}{2r_-^2}\right)u, \quad V_- = -\frac{2r_-^2}{(r_+ - r_-)}exp\left(-\frac{(r_+ - r_-)}{2r_-^2}\right)v,$$

which converts the above equation as

$$ds^2 = \frac{r_- r_+}{r^2}\left(\frac{(r_+ - r)}{r_+}\right)^{1+\frac{r_+^2}{r_-^2}} exp\left(-\frac{(r_+ - r_-)}{r_-^2}r\right)dU_- dV_- - r^2(d\theta^2 + \sin^2\theta d\phi^2).$$

Now, in Kruskal–Szekeres coordinates,

$$T_- = \frac{1}{2}(V_- + U_-), \quad R_- = \frac{1}{2}(V_- - U_-),$$

the metric becomes,

$$ds^2 = \frac{r_- r_+}{r^2}\left(\frac{(r_+ - r)}{r_+}\right)^{1+\frac{r_+^2}{r_-^2}} exp\left(-\frac{(r_+ - r_-)}{r_-^2}r\right)(dT_-^2 - dR_-^2)$$
$$- r^2(d\theta^2 + \sin^2\theta d\phi^2).$$

Here, $r = r_-$ is a regular point. In this case, the metric is regular between the region $0 < r \le r_-$. The inner horizon $r = r_-$ corresponds to both null hyper surfaces $U_- = 0, V_- = 0$. The Penrose diagram for Reissner–Nordström spacetime is shown in Fig. 91. Note that there simultaneously exists another identical asymptotically flat region I (in fact infinite number of regions) outside the horizons, i.e., for $r > r_+$. These regions are connected by the two regions II for $r_- < r < r_+$, i.e., region between two horizons and III for $0 < r < r_-$, i.e., the region between the center of the black hole and the inner event horizon. Figure 91 also indicates that the events inside the outer event horizon (at $r = r_+$)

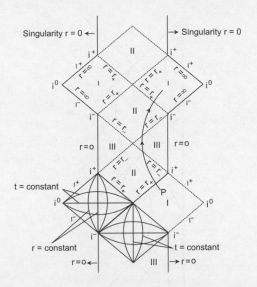

Figure 91 Light cones in Reissner–Nordström spacetime.

cannot be seen at I^+ or i^+. The region inside the inner horizon ($0 < r < r_-$) is moderately restricted by the naked singularity at $r = 0$, which is time-like in character.

Case III: $m^2 - e^2 = 0$:

Now, we will consider the extreme case for which $m^2 = e^2$. The word "extreme" signifies actually that this black hole is very similar to a black body with zero temperature. Here the metric function $g_{tt}(r)$ has a double zero at $r = m$ and metric takes the form as

$$ds^2 = \left(1 - \frac{m}{r}\right)^2 dt^2 - \left(1 - \frac{m}{r}\right)^{-2} dr^2 - r^2(d\theta^2 + \sin^2\theta d\phi^2).$$

Note that this case is totally different from Schwarzschild case because if one passes the event horizon, time and space are never interchanged. In Schwarzschild case, the time and space are interchanged as $g_{tt} = \left(1 - \frac{2m}{r}\right)$ changes sign. However, for extreme Reissner–Nordström spacetime, when r reach the horizon, g_{tt} becomes zero but does not change sign. Here, $r = m$ represents the event horizon but the r coordinate is never time-like rather it is space-like on either side. It becomes null at $r = m$.

To know the spacetime structure of extreme Reissner–Nordström spacetime, we introduce the following coordinate transformation in the region $r > m$ as

$$dr^* = g_{rr}dr = \left(1 - \frac{m}{r}\right)^{-2} dr,$$

$$or.\ r^* = \frac{r(r - 2m)}{r - m} + 2m \ln\left|\frac{r}{m} - 1\right|.$$

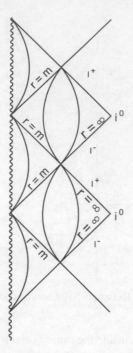

Figure 92 Light cones in extreme Reissner–Nordström spacetime.

Here, r^* is known as tortoise coordinate via

$$ds^2 = \left(1 - \frac{m}{r}\right)^2 \left[dt^2 - dr^{*2}\right] - r^2(r^*)(d\theta^2 + \sin^2\theta d\phi^2).$$

Using the retarded and advanced time-coordinates u and v by

$$u = t - r^*, \quad v = t + r^*,$$

we get Eddington–Finkelstein-like forms of the metric. The infalling and outgoing radial null geodesics ($\frac{dr^*}{dt} = -1$) are characterized by $v = constant$, and $v = constant$ respectively. In the ingoing Eddington–Finkelstein coordinates (v, r, θ, ϕ), the metric assumes the following form as

$$ds^2 = \left(1 - \frac{m}{r}\right)^2 dv^2 - 2dvdr - r^2(d\theta^2 + \sin^2\theta d\phi^2),$$

and in the outgoing Eddington–Finkelstein coordinates (u, r, θ, ϕ), the metric takes the form as

$$ds^2 = \left(1 - \frac{m}{r}\right)^2 du^2 + 2dudr - r^2(d\theta^2 + \sin^2\theta d\phi^2).$$

One can also express extreme Reissner–Nordström spacetime in terms of double-null coordinates (u, v, θ, ϕ) as

$$ds^2 = \left(1 - \frac{m}{r}\right)^2 dudv - r^2(d\theta^2 + \sin^2\theta d\phi^2), \quad where \ r = r(u, v).$$

All the above forms of metric are regular across the horizon $r = m$ and we can also go beyond $r = m$, which includes all positive values for r. It is to be noted that after crossing the horizon into the interior region, further extensions to different exterior regions could be possible. The singularity $r = 0$ is actually a time-like line. In case of extreme Reissner–Nordström spacetime one can evade the singularity and remain to travel to the future copies of asymptotically flat regions preserving singularity to the left. The Penrose diagram is depicted in Fig. 92. Here one requires to add an infinite sequence of domains to get a geodesically complete spacetime.

Rotating Black Holes

10.1   Null Tetrad

The tetrad formalism is a transformed coordinate approach to general relativity (GR) that replaces the coordinate basis by the local basis for the tangent bundle, which is less restricted. It is constituted by a set of four linearly independent vector fields known as a **tetrad** or **vierbein**. Usually, specific tetrad basis, which is a set of four independent vector fields, is used to represent a tensor in tetrad formalism. This basis spans the four-dimensional (4D) vector tangent space at each point in spacetime. Therefore, at any point P on the curve, we can present an orthogonal frame of three unit space-like vectors

$$e_i^\alpha = (e_1^\alpha, \ e_2^\alpha, \ e_3^\alpha),$$

which are all orthogonal to v^α (which is unit tangent vector), a time-like vector and we define

$$e_0^\alpha \equiv v^\alpha.$$

These four vectors $e_i^\alpha (i = 0, 1, 2, 3)$ are known to form a **frame or tetrad** at P.

Treating e_i^α as a 4×4 matrix at P, we can define its inverse (called the **dual basis or dual tetrad**) e_α^j, which follows

$$e_i^\alpha e_\alpha^j = \delta_i^j. \tag{10.1}$$

Actually, e_i^α are four linearly independent vector fields.

We introduce a new matrix g_{ij} defined by

$$g_{ij} = g_{\alpha\beta} e_i^\alpha e_j^\beta = e_{\beta i} e_j^\beta, \tag{10.2}$$

which is known as **frame metric**.

Here, e_i^α are linearly independent and the global metric, $g_{\alpha\beta}$, is nonsingular. As a result the matrix g_{ij} is nonsingular and hence invertible, i.e., g^{ij} exists, which is **contravariant frame metric**. Note that

$$g_{ij} g^{ik} = \delta_j^k. \tag{10.3}$$

Exactly the same way as metric tensor, we can raise and lower the tensor indices with the frame metric g_{ij}.

Eq. (10.2) yields the inverse relation as

$$g_{\alpha\beta} = g_{ij}e^i_\alpha e^j_\beta. \tag{10.4}$$

Note that tetrad vectors determine the linear differential forms

$$e^a = e^a_\mu dx^\mu,$$

and from this the metric takes the form

$$ds^2 = e^a e_a = g_{\mu\nu} dx^\mu dx^\nu.$$

For a simple physical interpretation of the frame, we can think $e^\alpha_0 = v^\alpha$ as the four velocity of an observer whose world line is C and three space-like vectors $e^\alpha_i (i = 1, 2, 3)$ are rectangular coordinate vectors (such as usual cartesian basis $\hat{i}, \hat{j}, \hat{k}$) at P. For different tetrad, one gets different frame metric.

Exercise 10.1

Let the tetrad consist of one unit time-like vector, v^α, and three unit space-like vectors, i^α, j^α and k^α. Construct the frame metric, g_{ij}.

Hints: Here $e^\alpha_m = (v^\alpha, i^\alpha, j^\alpha, k^\alpha)$ $(\alpha = 0, 1, 2, 3)$ are the tetrad that follow

$$v^\alpha v_\alpha = e^\alpha_0 e_{0\alpha} = 1 \ \ and \ \ i^\alpha i_\alpha = e^\alpha_1 e_{1\alpha} = j^\alpha j_\alpha = e^\alpha_2 e_{2\alpha} = k^\alpha k_\alpha = e^\alpha_3 e_{3\alpha} = -1,$$

$$and \ e^\alpha_m e_{n\alpha} = 0, m \neq n, \ \ m, n = 0, 1, 2, 3.$$

[v^α is time-like and i^α, etc., are space-like. Also v^α, and i^α etc., are mutually perpendicular to each other.]

Hence,

$$e^\alpha_m e_{n\alpha} = diag(1, -1, -1, -1).$$

Here, the frame metric $g_{mn} = g_{\alpha\beta}e^\alpha_m e^\beta_n = e^\beta_m e_{n\beta} = \eta_{mn} = diag(1, -1, -1, -1)$ is the Minkowski metric.

Exercise 10.2

Let the tetrad consist of $l^\alpha, n^\alpha, j^\alpha$ and k^α, where

$$l^\alpha = \frac{1}{\sqrt{2}}(v^\alpha + i^\alpha), \ n^\alpha = \frac{1}{\sqrt{2}}(v^\alpha - i^\alpha),$$

v^α is a unit time-like vector and i^α, j^α and k^α are three unit space-like vectors. Show that l^α *and* n^α are null vectors and also construct the frame metric, g_{ij}.

Hints: It is easy to show that

$$l^{\alpha} l_{\alpha} = n^{\alpha} n_{\alpha} = 0 \ \ since \ \ v^{\alpha} v_{\alpha} = 1, \ v^{\alpha} i_{\alpha} = v_{\alpha} i^{\alpha} = 0, \ and \ i^{\alpha} i_{\alpha} = -1.$$

Thus, l^{α} and n^{α} are null vectors. Note that l^{α} and n^{α} satisfy the normalization condition as

$$l^{\alpha} n_{\alpha} = 1,$$

which is very similar to light cone.

Here,

$$e_0^{\alpha} = l^{\alpha} = \frac{1}{\sqrt{2}} (v^{\alpha} + i^{\alpha}), \ \ e_1^{\alpha} = n^{\alpha} = \frac{1}{\sqrt{2}} (v^{\alpha} - i^{\alpha}),$$

and assume

$$e_2^{\alpha} = j^{\alpha} \ \ and \ \ e_3^{\alpha} = k^{\alpha},$$

then the frame metric will take the following form:

$$g_{ij} = \begin{pmatrix} 0 & 1 & 0 & 0 \\ 1 & 0 & 0 & 0 \\ 0 & 0 & -1 & 0 \\ 0 & 0 & 0 & -1 \end{pmatrix}.$$

$[g_{00} = g_{\alpha\beta} e_0^{\alpha} e_0^{\beta} = e_0^{\alpha} e_{\alpha 0} = l^{\alpha} l_{\alpha} = 0; \ g_{01} = g_{\alpha\beta} e_0^{\alpha} e_1^{\beta} = e_0^{\alpha} e_{\alpha 1} = l^{\alpha} n_{\alpha} = 1, \ etc.]$

Exercise 10.3

Let the tetrad consist of $l^{\alpha}, n^{\alpha}, m^{\alpha}$ and $\overline{m}^{\alpha}$, where

$$l^{\alpha} = \frac{1}{\sqrt{2}} (v^{\alpha} + i^{\alpha}), \ n^{\alpha} = \frac{1}{\sqrt{2}} (v^{\alpha} - i^{\alpha}), m^{\alpha} = \frac{1}{\sqrt{2}} (j^{\alpha} + ik^{\alpha}), \overline{m}^{\alpha} = \frac{1}{\sqrt{2}} (j^{\alpha} - ik^{\alpha}),$$

where v^{α} is a unit time-like vector and i^{α}, j^{α} and k^{α} are three unit space-like vectors. Show that l^{α}, n^{α}, m^{α}, and $\overline{m}^{\alpha}$ are null vectors and also construct the frame metric, g_{ij}.

Hints: It is easy to show that

$$l^{\alpha} l_{\alpha} = n^{\alpha} n_{\alpha} = 0, \ m^{\alpha} m_{\alpha} = 0 \ and \ \overline{m}^{\alpha} \overline{m}_{\alpha} = 0,$$

$$since \ v^{\alpha} v_{\alpha} = 1, \ i^{\alpha} i_{\alpha} = -1 \ etc, \ v^{\alpha} i_{\alpha} = v_{\alpha} i^{\alpha} = 0, \ etc.$$

Thus, $l^{\alpha}, n^{\alpha}, m^{\alpha}$, and $\overline{m}^{\alpha}$ are null vectors.

Note that m^α *and* $\overline{m}^\alpha$ satisfy the normalization condition as

$$m^\alpha \overline{m}_\alpha = -1.$$

Here,

$$e_0^\alpha = l^\alpha = \frac{1}{\sqrt{2}}(v^\alpha + i^\alpha), \ \ e_1^\alpha = n^\alpha = \frac{1}{\sqrt{2}}(v^\alpha - i^\alpha),$$

$$e_2^\alpha = m^\alpha = \frac{1}{\sqrt{2}}(j^\alpha + ik^\alpha), \ \ e_3^\alpha = \overline{m}^\alpha = \frac{1}{\sqrt{2}}(j^\alpha - ik^\alpha),$$

are the **null tetrad**.

For the event on two-sphere, l^α *and* n^α are the ingoing and outgoing null normals, respectively, whereas m^α and $\overline{m}^\alpha$ are tangential null vectors.

Finally, one can get the frame metric as

$$g_{ij} = \begin{pmatrix} 0 & 1 & 0 & 0 \\ 1 & 0 & 0 & 0 \\ 0 & 0 & 0 & -1 \\ 0 & 0 & -1 & 0 \end{pmatrix}.$$

The null tetrad is used above in special cases. However, in general case, considering the definitions of above null tetrad, the global metric can be written as

$$g_{\alpha\beta} = l_\alpha n_\beta + l_\beta n_\alpha - m_\alpha \overline{m}_\beta - m_\beta \overline{m}_\alpha.$$

The contravariant form of this equation is

$$g^{\alpha\beta} = l^\alpha n^\beta + l^\beta n^\alpha - m^\alpha \overline{m}^\beta - m^\beta \overline{m}^\alpha.$$

Newman and Penrose established a set of notations for GR, which is known as **Newman–Penrose (NP) formalism**. In NP formalism, usually the vector basis is chosen as null tetrad given above (two real, l^α *and* n^α, and a complex conjugate pair, m^α and $\overline{m}^\alpha$).

Typically two types of signature and normalization conventions are used in literature. The signature $(+, -, -, -)$ yields

$$l^\alpha n_\alpha = 1, \ m^\alpha \overline{m}_\alpha = -1, \ l_\alpha m^a = l_\alpha \overline{m}^\alpha = n_\alpha m^a = n_\alpha \overline{m}^\alpha = 0,$$

and signature $(-, +, +, +)$ yields

$$l^\alpha n_\alpha = -1, \ m^\alpha \overline{m}_\alpha = 1, \ l_\alpha m^a = l_\alpha \overline{m}^\alpha = n_\alpha m^a = n_\alpha \overline{m}^\alpha = 0.$$

Normally the first type is used in various other areas of astrophysics such as black hole physics, gravitational waves as different areas in GR, and the second type is used in modern research of black holes from quasilocal viewpoints such as isolated horizons, dynamical horizons, etc.

For the signature $(-, +, +, +)$, the global metric can be written as

$$g_{\alpha\beta} = -l_\alpha n_\beta - l_\beta n_\alpha + m_\alpha \overline{m}_\beta + m_\beta \overline{m}_\alpha.$$

The contravariant form of this equation is

$$g^{\alpha\beta} = -l^\alpha n^\beta - l^\beta n^\alpha + m^\alpha \overline{m}^\beta + m^\beta \overline{m}^\alpha.$$

In NP formalism, the metric itself is represented by a null tetrad. The first derivatives of the metric are described in three groups with twelve complex spin coefficients that define the variation in the tetrad from point to point. The second derivatives of the metric are expressed by five complex functions encoding Weyl tensors and ten other functions encrypting Ricci tensors in the tetrad basis.

Let us write the **four directional covariant derivatives** alongside with all tetrad vector as

$$D := \nabla_\ell = l^\alpha \nabla_\alpha, \quad \Delta := \nabla_\mathbf{n} = n^\alpha \nabla_\alpha, \quad \delta := \nabla_\mathbf{m} = m^\alpha \nabla_\alpha, \quad \bar\delta := \nabla_{\overline{\mathbf{m}}} = \overline{m}^\alpha \nabla_\alpha.$$

The **twelve spin coefficients** (complex numbers in three groups) are assigned as contractions of covariant derivatives of the null tetrad:

$$\kappa := -m^a D l_a = -m^a l^b \nabla_b l_a, \quad \tau := -m^a \Delta l_a = -m^a n^b \nabla_b l_a$$

$$\sigma := -m^a \delta l_a = -m^a m^b \nabla_b l_a, \quad \rho := -m^a \bar\delta l_a = -m^a \overline{m}^b \nabla_b l_a$$

$$\pi := \overline{m}^a D n_a = \overline{m}^a l^b \nabla_b n_a, \quad \nu := \overline{m}^a \Delta n_a = \overline{m}^a n^b \nabla_b n_a$$

$$\mu := \overline{m}^a \delta n_a = \overline{m}^a m^b \nabla_b n_a, \quad \lambda := \overline{m}^a \bar\delta n_a = \overline{m}^a \overline{m}^b \nabla_b n_a$$

$$\varepsilon := -\frac{1}{2}\left(n^a D l_a - \overline{m}^a D m_a \right) = -\frac{1}{2}\left(n^a l^b \nabla_b l_a - \overline{m}^a l^b \nabla_b m_a \right)$$

$$\gamma := -\frac{1}{2}\left(n^a \Delta l_a - \overline{m}^a \Delta m_a \right) = -\frac{1}{2}\left(n^a n^b \nabla_b l_a - \overline{m}^a n^b \nabla_b m_a \right)$$

$$\beta := -\frac{1}{2}\left(n^a \delta l_a - \overline{m}^a \delta m_a \right) = -\frac{1}{2}\left(n^a m^b \nabla_b l_a - \overline{m}^a m^b \nabla_b m_a \right)$$

$$\alpha := -\frac{1}{2}\left(n^a \bar\delta l_a - \overline{m}^a \bar\delta m_a \right) = -\frac{1}{2}\left(n^a \overline{m}^b \nabla_b l_a - \overline{m}^a \overline{m}^b \nabla_b m_a \right).$$

In NP formalism, the five complex self dual components (**Weyl scalars** Ψ_n) defined by certain contractions of the Weyl tensor with tetrad vectors are written as

$$\Psi_0 := C_{abcd} l^a m^b l^c m^d, \quad \Psi_1 := C_{abcd} l^a n^b l^c m^d, \quad \Psi_2 := C_{abcd} l^a m^b \overline{m}^c n^d$$

$$\Psi_3 := C_{abcd} l^a n^b \overline{m}^c n^d, \quad \Psi_4 := C_{abcd} n^a \overline{m}^b n^c \overline{m}^d.$$

Rotating Black Holes

[Actually five complex Weyl-NP scalars represent the ten independent components of the Weyl tensor.]

These Weyl scalars have some physical implications: Ψ_0 represents the transverse radiation along n^a, Ψ_1 represents the longitudinal radiation along n^a, Ψ_2 represents the Coulomb field and spin, Ψ_3 represents the longitudinal radiation along l^a, and Ψ_4 represents the transverse radiation along l^a.

The **Ricci scalars** Λ and Φ_{nm} are the ten independent components of the Ricci tensor, which are expressed in terms of three complex scalars and four real scalars with their complex conjugates as

$$\Lambda = \frac{1}{24}R, \ \Phi_{00} = -\frac{1}{2}R_{ab}l^a l^b, \ \Phi_{11} = -\frac{1}{4}R_{ab}(l^a n^b + m^a \overline{m}^b), \ \Phi_{22} = -\frac{1}{2}R_{ab}n^a n^b,$$

$$\Phi_{01} = -\frac{1}{2}R_{ab}l^a m^b, \ \Phi_{10} = -\frac{1}{2}R_{ab}l^a \overline{m}^b,$$

$$\Phi_{12} = -\frac{1}{2}R_{ab}n^a m^b, \ \Phi_{21} = -\frac{1}{2}R_{ab}n^a \overline{m}^b,$$

$$\Phi_{02} = -\frac{1}{2}R_{ab}m^a m^b, \ \Phi_{20} = -\frac{1}{2}R_{ab}\overline{m}^a \overline{m}^b.$$

After defining NP formalism, we should stop the discussion here and suggest the reader to consult the famous book of S. Chandrasekhar, *The Mathematical Theory of Black Holes*, for its applications.

10.2 Null Tetrad of Some Black Holes

Here, we wish to write the metric of some black holes in terms of null tetrad.

(i) **Schwarzschild metric:** The Schwarzschild metric is written in standard coordinates as

$$ds^2 = f(r)dt^2 - f(r)^{-1}dr^2 - r^2(d\theta^2 + \sin^2\theta d\phi^2), \ where \ f(r) = 1 - \frac{2m}{r}.$$

Using the succeeding transformation, we can write it in advanced Eddington–Finkelstein coordinates as

$$dt = du + f(r)^{-1}dr.$$

The metric takes the following form in the new coordinates as

$$ds^2 = f(r)du^2 + 2dudr - r^2(d\theta^2 + \sin^2\theta d\phi^2).$$

This form of Schwarzschild metric is known as the advanced Eddington–Finkelstein form. Here, $u = constant$ surface is a spherically symmetric null surface. Now, this metric is expressed in terms of complex null tetrad,

$$Z_i^\alpha = (l^\alpha, n^\alpha, m^\alpha, \overline{m}^\alpha), \ i = 1, 2, 3, 4,$$

which are given by

$$l^\alpha = \delta_r^\alpha = \frac{\partial}{\partial r},$$

$$n^\alpha = \left[\delta_u^\alpha - \frac{1}{2}f(r)\delta_r^\alpha\right] = \left[\frac{\partial}{\partial u} - \frac{1}{2}\left(1 - \frac{2m}{r}\right)\frac{\partial}{\partial r}\right],$$

$$m^\alpha = \frac{1}{\sqrt{2}r}\left[\delta_\theta^\alpha + \frac{i}{\sin\theta}\delta_\phi^\alpha\right] = \frac{1}{\sqrt{2}r}\left[\frac{\partial}{\partial\theta} + \frac{i}{\sin\theta}\frac{\partial}{\partial\phi}\right],$$

$$\overline{m}^\alpha = \frac{1}{\sqrt{2}r}\left[\delta_\theta^\alpha - \frac{i}{\sin\theta}\delta_\phi^\alpha\right] = \frac{1}{\sqrt{2}r}\left[\frac{\partial}{\partial\theta} - \frac{i}{\sin\theta}\frac{\partial}{\partial\phi}\right]. \tag{10.5}$$

$$[l^\mu = \partial_r x^\mu = \frac{\partial x^\mu}{\partial r}, \ x^\mu = (u, r, \theta, \phi), \ i.e. \ l^\mu = (0, 1, 0, 0)]$$

The contravariant metric

$$g^{\alpha\beta} = l^\alpha n^\beta + l^\beta n^\alpha - m^\alpha \overline{m}^\beta - m^\beta \overline{m}^\alpha,$$

yields

$$g^{uu} = 0, g^{rr} = -f(r) = -\left(1 - \frac{2m}{r}\right), g^{ru} = 1,$$

$$g^{\theta\theta} = -\frac{1}{r^2}, g^{\phi\phi} = -\frac{1}{\sin^2\theta r^2}.$$

Thus, the contravariant Schwarzschild metric in null coordinates is given by

$$g^{\mu\nu} = \begin{pmatrix} 0 & 1 & 0 & 0 \\ 1 & \left\{-\left(1 - \frac{2m}{r}\right)\right\} & 0 & 0 \\ 0 & 0 & -\frac{1}{r^2} & 0 \\ 0 & 0 & 0 & -\frac{1}{r^2\sin^2\theta} \end{pmatrix}.$$

One can also find the duals of the null tetrad as

$$l_\mu dx^\mu = du,$$

$$n_\mu dx^\mu = \frac{1}{2}\left(1 - \frac{2m}{r}\right)du + dr,$$

$$m_\mu dx^\mu = \frac{r}{\sqrt{2}}(d\theta + i\sin\theta d\phi).$$

From this complex null tetrad system, one can derive the Kerr metric, which will be discussed later.

(ii) Reissner–Nordström metric: The Reissner–Nordström metric is written in standard coordinates as

$$ds^2 = f(r)dt^2 - f(r)^{-1}dr^2 - r^2(d\theta^2 + \sin^2\theta d\phi^2), \text{ where } f(r) = 1 - \frac{2m}{r} + \frac{e^2}{r^2}.$$

Using the succeeding transformation as above, we can write it in advanced Eddington–Finkelstein coordinates as

$$dt = du + f(r)^{-1}dr.$$

The metric takes the following form in the new coordinates as

$$ds^2 = f(r)du^2 + 2dudr - r^2(d\theta^2 + \sin^2\theta d\phi^2).$$

This form of Reissner–Nordström metric is known as the advanced Eddington–Finkelstein form. Here, $u = constant$ surface is a spherically symmetric null surface. Now this metric is expressed in terms of complex null tetrad,

$$Z_i^\alpha = (l^\alpha, n^\alpha, m^\alpha, \overline{m}^\alpha), \quad i = 1, 2, 3, 4,$$

which are given by

$$l^\alpha = \delta_r^\alpha = \frac{\partial}{\partial r},$$

$$n^\alpha = \left[\delta_u^\alpha - \frac{1}{2}f(r)\delta_r^\alpha\right] = \left[\frac{\partial}{\partial u} - \frac{1}{2}\left(1 - \frac{2m}{r} + \frac{e^2}{r^2}\right)\frac{\partial}{\partial r}\right],$$

$$m^\alpha = \frac{1}{\sqrt{2}r}\left[\delta_\theta^\alpha + \frac{i}{\sin\theta}\delta_\phi^\alpha\right] = \frac{1}{\sqrt{2}r}\left[\frac{\partial}{\partial\theta} + \frac{i}{\sin\theta}\frac{\partial}{\partial\phi}\right],$$

$$\overline{m}^\alpha = \frac{1}{\sqrt{2}r}\left[\delta_\theta^\alpha - \frac{i}{\sin\theta}\delta_\phi^\alpha\right] = \frac{1}{\sqrt{2}r}\left[\frac{\partial}{\partial\theta} - \frac{i}{\sin\theta}\frac{\partial}{\partial\phi}\right]. \tag{10.6}$$

The contravariant metric

$$g^{\alpha\beta} = l^\alpha n^\beta + l^\beta n^\alpha - m^\alpha \overline{m}^\beta - m^\beta \overline{m}^\alpha,$$

yields

$$g^{uu} = 0, \ g^{rr} = -f(r) = -\left(1 - \frac{2m}{r} + \frac{e^2}{r^2}\right),$$

$$g^{ru} = 1, \ g^{\theta\theta} = -\frac{1}{r^2}, \ g^{\phi\phi} = -\frac{1}{\sin^2\theta r^2}.$$

Thus, the contravariant Reissner–Nordström metric in null coordinates is given by

$$
g^{\mu\nu} = \begin{pmatrix} 0 & 1 & 0 & 0 \\ 1 & \left\{-\left(1 - \frac{2m}{r} + \frac{e^2}{r^2}\right)\right\} & 0 & 0 \\ 0 & 0 & -\frac{1}{r^2} & 0 \\ 0 & 0 & 0 & -\frac{1}{r^2 \sin^2\theta} \end{pmatrix}.
$$

One can also find the duals of the null tetrad as

$$
l_\mu dx^\mu = du,
$$

$$
n_\mu dx^\mu = \frac{1}{2}\left(1 - \frac{2m}{r} + \frac{e^2}{r^2}\right) du + dr,
$$

$$
m_\mu dx^\mu = \frac{r}{\sqrt{2}}(d\theta + i\sin\theta d\phi).
$$

From this complex null tetrad system, one can derive the Kerr–Newmann metric, which will be discussed later.

(iii) Higher dimensional black hole metric: The higher dimensional black hole metric is written in standard coordinates as

$$
ds^2 = f(r)dt^2 - \frac{dr^2}{f(r)} - r^2 d\Omega_D^2,
$$

where $d\Omega_D^2$ is the line element on the D unit sphere, i.e.,

$$
d\Omega_D^2 = d\theta_1^2 + \sin^2\theta_1 d\theta_2^2 + \ldots\ldots + \prod_{n=1}^{D-1} \sin^2\theta_n d\theta_D^2.
$$

The volume of the D unit sphere is given by

$$
\Omega_D = 2\frac{\pi^{\frac{D+1}{2}}}{\Gamma\left(\frac{D+1}{2}\right)}.
$$

For different higher dimensional black holes, $f(r)$ has different expressions.

For higher dimensional Schwarzschild black hole,

$$
f(r) = 1 - \frac{2\mu}{r^{D-1}}. \tag{10.7}
$$

For higher dimensional Reissner–Nordström black hole,

$$
f(r) = 1 - \frac{2\mu}{r^{D-1}} + \frac{q^2}{r^{2(D-1)}}. \tag{10.8}
$$

Here, μ and q are related to the mass M and charge e of the black hole as

$$\mu = \frac{8\pi GM}{D\Omega_D}, \quad q^2 = \frac{8\pi Ge}{D(D-1)}.$$

Using the succeeding transformation as above, we can write it in advanced Eddington–Finkelstein coordinates as

$$dt = du + f(r)^{-1}dr.$$

The metric takes the following form in the new coordinates as

$$ds^2 = f(r)du^2 + 2dudr - r^2d\Omega_D^2.$$

This form of higher dimensional black hole metric is known as the advanced Eddington–Finkelstein form. Here, $u = constant$ surface is a spherically symmetric null surface. Now, this metric is expressed in terms of complex **null veiltrad**,

$$Z_i^\alpha = (l^\alpha, n^\alpha, m_1^\alpha, \overline{m}_1^\alpha, m_2^\alpha, \overline{m}_2^\alpha, ..., m_{\frac{D}{2}}^\alpha, \overline{m}_{\frac{D}{2}}^\alpha), \quad i = 1, 2, 3 .... D + 2,$$

which are given by

$$l^\alpha = \delta_r^\alpha = \frac{\partial}{\partial r},$$

$$n^\alpha = \left[\delta_u^\alpha - \frac{1}{2}f(r)\delta_r^\alpha \right] = \left[\frac{\partial}{\partial u} - \frac{1}{2}f(r)\frac{\partial}{\partial r} \right], [Here, \ one \ can \ use \ f(r) \ as \ give \ in \ Eq. \ (10.7) \ or \ Eq.(10.8)],$$

$$m_k^\alpha = \frac{1}{\sqrt{2}r \sin\theta_1 \sin\theta_2 .... \sin\theta_{(k-1)}} \left[\delta_{\theta_k}^\alpha + \frac{i}{\sin\theta_k}\delta_{\theta_{k+1}}^\alpha \right]$$

$$= \frac{1}{\sqrt{2}r \sin\theta_1 \sin\theta_2 .... \sin\theta_{(k-1)}} \left[\frac{\partial}{\partial\theta_k} + \frac{i}{\sin\theta_k}\frac{\partial}{\partial\theta_{k+1}} \right],$$

$$\overline{m}_k^\alpha = \frac{1}{\sqrt{2}r \sin\theta_1 \sin\theta_2 .... \sin\theta_{(k-1)}} \left[\delta_{\theta_k}^\alpha - \frac{i}{\sin\theta_k}\delta_{\theta_{k+1}}^\alpha \right]$$

$$= \frac{1}{\sqrt{2}r \sin\theta_1 \sin\theta_2 .... \sin\theta_{(k-1)}} \left[\frac{\partial}{\partial\theta_k} - \frac{i}{\sin\theta_k}\frac{\partial}{\partial\theta_{k+1}} \right],$$

$$k = 1, 2, 3, ..., \frac{D}{2}, \quad (D \ should \ be \ an \ even \ integer) \tag{10.9}$$

The contravariant metric

$$g^{\alpha\beta} = l^\alpha n^\beta + l^\beta n^\alpha - m_1^\alpha \overline{m}_1^\beta - m_1^\beta \overline{m}_1^\alpha - m_2^\alpha \overline{m}_2^\beta - m_2^\beta \overline{m}_2^\alpha - .... - m_{\frac{D}{2}}^\alpha \overline{m}_{\frac{D}{2}}^\beta - m_{\frac{D}{2}}^\beta \overline{m}_{\frac{D}{2}}^\alpha,$$

yields

$$g^{uu} = 0, g^{rr} = -f(r), \quad g^{ru} = 1, g^{\theta_1\theta_1} = -\frac{1}{r^2},$$

$$g^{\theta_2\theta_2} = -\frac{1}{\sin^2\theta_1 r^2}, ..., g^{\theta_D\theta_D} = -\frac{1}{\sin^2\theta_1 \sin^2\theta_2 .... \sin^2\theta_{D-1} r^2}. \tag{10.10}$$

Thus, the contravariant Reissner–Nordström metric in null coordinates is given by

$$g^{\mu\nu} = \begin{pmatrix} 0 & 1 & 0 & 0 & & .....0 \\ 1 & -f(r) & 0 & 0 & & .....0 \\ 0 & 0 & -\frac{1}{r^2} & 0 & & .....0 \\ 0 & 0 & 0 & -\frac{1}{r^2\sin^2\theta_1} & & .....0 \\ .. & .. & .. & .. & & .... \\ 0 & 0 & 0 & 0 & .... & -\frac{1}{r^2\sin^2\theta_1\sin^2\theta_2...\sin^2\theta_{D-1}} \end{pmatrix}.$$

From this complex null tetrad system, one can derive the rotating black hole metric in higher dimension for uncharged or charged case.

Note that l^α and n^α are real, m_1^α and $\overline{m}_i^\alpha$ are mutually complex conjugate. This vieltrad is orthonormal and obeys the following conditions:

$$l^\alpha l_\alpha = n^\alpha n_\alpha = m_i^\alpha (m_i)_\alpha = \overline{m}_i^\alpha (\overline{m}_i)_\alpha = 0,$$

$$l_\alpha m_i^\alpha = l_\alpha \overline{m}_i^\alpha = n_\alpha m_i^\alpha = n_\alpha \overline{m}_i^\alpha = 0,$$

$$l^\alpha n_\alpha = 1, \quad (m_i)_\alpha \overline{m}_i^\alpha = -1.$$

10.3 The Kerr Solution

Kerr solution is actually the solution of the vacuum Einstein's equations which describe a rotating black hole in 4D spacetime, in other words, it is the metric outside of a rotating axially symmetric body. It was discovered in 1963 by Roy Kerr. This black hole has a curvature singularity as in the case of Schwarzschild spacetime, which is covered by horizon. This black hole does not vary with time explicitly, therefore, it is **stationary**. Also it is independent of ϕ, therefore, it is **axisymmetric**. It remains unchanged under this specified transformation $(t, \phi) \to (-t, -\phi)$. This means time reversal reverses the angular velocity, i.e., the time inversion of a rotating object suggests the rotation of an object in the reverse direction. It is **not static** as it will vary when a time reversal transformation $t \to -t$ is applied. The Kerr metric becomes Minkowski's flat metric for large r, i.e., in the limit $r \to \infty$. This means that **the Kerr spacetime is asymptotically flat**. Kerr metric is actually a two parameters-dependent solution. Those parameters are m (**mass**) and ma (**angular momentum**). Importantly, one can note that **Birkhoff theorem** does not hold good for rotating spacetimes. In other words, the spacetime geometry outside of a rotating axially symmetric body is not a part of the Schwarzschild geometry.

10.4   The Kerr Solution from the Schwarzschild Solution

The Kerr black hole solution can be derived from the Schwarzschild solution by applying a complex coordinate transformation as suggested by Newman and Janis on null tetrad given in Eq. (10.5). For this method, one will have to use the contravariant components of the advanced Eddington–Finkelstein form of the Schwarzschild metric.

The Schwarzschild line element in Eddington–Finkelstein coordinate is

$$ds^2 = \left(1 - \frac{2m}{r}\right) du^2 + 2dudr - r^2(d\theta^2 + \sin^2 \theta d\phi^2).$$

Now we use complex null tetrad to express this metric as

$$l^\alpha = \delta_r^\alpha = \frac{\partial}{\partial r},$$

$$n^\alpha = \left[\delta_u^\alpha - \frac{1}{2}f(r)\delta_r^\alpha\right] = \left[\frac{\partial}{\partial u} - \frac{1}{2}\left(1 - \frac{2m}{r}\right)\frac{\partial}{\partial r}\right],$$

$$m^\alpha = \frac{1}{\sqrt{2}r}\left[\delta_\theta^\alpha + \frac{i}{\sin\theta}\delta_\phi^\alpha\right] = \frac{1}{\sqrt{2}r}\left[\frac{\partial}{\partial\theta} + \frac{i}{\sin\theta}\frac{\partial}{\partial\phi}\right],$$

$$\overline{m}^\alpha = \frac{1}{\sqrt{2}r}\left[\delta_\theta^\alpha - \frac{i}{\sin\theta}\delta_\phi^\alpha\right] = \frac{1}{\sqrt{2}r}\left[\frac{\partial}{\partial\theta} - \frac{i}{\sin\theta}\frac{\partial}{\partial\phi}\right].$$

Following Newman and Janis method, we first complexify the null coordinate system by permitting u and r to assume values in complex space. Now, the function $f(r)$ be endorsed to a function $F(r,\overline{r})$, which is complex and reduces to $f(r)$ on the real slice. Let the radial coordinate r be replaced by its complex conjugate $\overline{r}$, and one can rewrite the tetrad in the following form:

$$l^\alpha = \frac{\partial}{\partial r},$$

$$n^\alpha = \left[\frac{\partial}{\partial u} - \frac{1}{2}\left(1 - \frac{m}{r} - \frac{m}{\overline{r}}\right)\frac{\partial}{\partial r}\right],$$

$$m^\alpha = \frac{1}{\sqrt{2}\overline{r}}\left[\frac{\partial}{\partial\theta} + \frac{i}{\sin\theta}\frac{\partial}{\partial\phi}\right],$$

$$\overline{m}^\alpha = \frac{1}{\sqrt{2}r}\left[\frac{\partial}{\partial\theta} - \frac{i}{\sin\theta}\frac{\partial}{\partial\phi}\right].$$

(Note that for this complexified tetrad, we keep l^μ and n^μ real and m^α *and* $\overline{m}^\alpha$ the complex conjugates of each other.)

In this complexified tetrad, following Newman and Janis, we make a complex coordinate transformation by describing a fresh set of coordinates as

$$x^{\mu\prime} = x^\mu + ia(\delta_r^\mu - \delta_u^\mu)\cos\theta,$$

i.e.,

$$r \to r' = r + ia\cos\theta; \quad u \to u' = u - ia\cos\theta; \quad \theta \to \theta'; \quad \phi \to \phi'.$$

Here, a is a real number known as the **Kerr parameter**. Though the unprimed coordinate system is observed as complex, we will infer the new coordinates as real. Using this ordinary coordinate transformation, the above complexified null tetrad describes a real spacetime. The new null tetrad assumes the following form (new coordinates are used without prime notations):

$$l'^\alpha = \frac{\partial}{\partial r},$$

$$n'^\alpha = \left[\frac{\partial}{\partial u} - \frac{1}{2}\left(1 - \frac{2mr}{R^2}\right)\frac{\partial}{\partial r} \right],$$

$$m'^\alpha = \frac{1}{\sqrt{2}(r + ai\cos\theta)}\left[ia\sin\theta\frac{\partial}{\partial u} - ia\sin\theta\frac{\partial}{\partial r} + \frac{\partial}{\partial\theta} + \frac{i}{\sin\theta}\frac{\partial}{\partial\phi} \right],$$

where

$$R^2 = r^2 + a^2\cos^2\theta.$$

$$\left[\frac{\partial}{\partial u} = \frac{\partial}{\partial u'}\frac{\partial u'}{\partial u} + \frac{\partial}{\partial r'}\frac{\partial r'}{\partial u} + \frac{\partial}{\partial\theta'}\frac{\partial\theta'}{\partial u} + \frac{\partial}{\partial\phi'}\frac{\partial\phi'}{\partial u} = \frac{\partial}{\partial u'}, \quad etc. \right]$$

This null tetrad yields the corresponding contravariant form of Kerr metric via

$$g^{\alpha\beta} = l'^\alpha n'^\beta + l'^\beta n'^\alpha - m'^\alpha \overline{m'}^\beta - m'^\beta \overline{m'}^\alpha$$

$$g^{\mu\nu} = \begin{pmatrix} -\frac{a^2\sin^2\theta}{R^2} & \frac{a^2+r^2}{R^2} & 0 & -\frac{a}{R^2} \\ .. & -\frac{\triangle}{R^2} & 0 & \frac{a}{R^2} \\ .. & .. & -\frac{1}{R^2} & 0 \\ .. & .. & .. & -\frac{1}{R^2\sin^2\theta} \end{pmatrix},$$

where

$$\triangle = r^2 + a^2 - 2mr.$$

One can note that in the $a \to 0$ limit, $g^{\mu\nu}$ becomes the Schwarzschild metric.

The covariant form of the metric is given by

$$g_{\mu\nu} = \begin{pmatrix} 1 - \frac{2mr}{R^2} & 1 & 0 & \frac{2mra\sin^2\theta}{R^2} \\ .. & 0 & 0 & -a\sin^2\theta \\ .. & .. & -R^2 & 0 \\ .. & .. & .. & \frac{\sin^2\theta}{R^2}\{\triangle a^2\sin^2\theta - (a^2+r^2)^2\} \end{pmatrix}.$$

Now, the line element of the Kerr metric in null coordinates is given by

$$ds^2 = \left(1 - \frac{2mr}{R^2}\right) du^2 + 2dudr + \frac{4mra\sin^2\theta}{R^2}dud\phi - 2a\sin^2\theta drd\phi$$

$$- R^2 d\theta^2 + \frac{\sin^2\theta}{R^2}\{\triangle a^2 \sin^2\theta - (a^2 + r^2)^2\}d\phi^2.$$

Also, the line element of the Kerr metric in (t, r, θ, ϕ) coordinates can be written using the succeeding transformations as

$$dt = du + \frac{r^2 + a^2}{\triangle}dr, \quad d\Phi = d\phi + \frac{a}{\triangle}dr,$$

$$ds^2 = \frac{\triangle}{R^2}(dt - a\sin^2\theta d\Phi)^2 - \frac{\sin^2\theta}{R^2}[(r^2 + a^2)d\Phi - adt]^2 - \frac{R^2}{\triangle}dr^2 - R^2 d\theta^2.$$

Thus, we started Schwarzschild metric and by carrying out a single complex transformation on it, we have found the Kerr metric with only one rotation parameter.

Note that the Kerr metric depends on two physical parameters **mass** (m) and **angular momentum** (ma). One can observe that for $a = 0$, we get the Schwarzschild solution. Actually, $a = \frac{J}{m}$ = angular momentum per unit mass.

10.5 The Kerr–Newmann Solution from the Reissner–Nordström Solution

The Kerr–Newmann black hole solution can be derived from the Reissner–Nordström solution by applying a complex coordinate transformation as suggested by Newman and Janis on null tetrad given in Eq. (10.6). For this method, one will have to use the contravariant components of the advanced Eddington–Finkelstein form of the Reissner–Nordström metric.

The Reissner–Nordström line element in Eddington–Finkelstein coordinate is

$$ds^2 = \left(1 - \frac{2mr - e^2}{r^2}\right) du^2 + 2dudr - r^2(d\theta^2 + \sin^2\theta d\phi^2).$$

Now this metric can be expressed in terms of complex null tetrad,

$$l^\alpha = \delta_r^\alpha = \frac{\partial}{\partial r},$$

$$n^\alpha = \left[\delta_u^\alpha - \frac{1}{2}f(r)\delta_r^\alpha\right] = \left[\frac{\partial}{\partial u} - \frac{1}{2}\left(1 - \frac{2mr - e^2}{r^2}\right)\frac{\partial}{\partial r}\right],$$

$$m^\alpha = \frac{1}{\sqrt{2}r}\left[\delta_\theta^\alpha + \frac{i}{\sin\theta}\delta_\phi^\alpha\right] = \frac{1}{\sqrt{2}r}\left[\frac{\partial}{\partial\theta} + \frac{i}{\sin\theta}\frac{\partial}{\partial\phi}\right],$$

$$\overline{m}^\alpha = \frac{1}{\sqrt{2}r}\left[\delta_\theta^\alpha - \frac{i}{\sin\theta}\delta_\phi^\alpha\right] = \frac{1}{\sqrt{2}r}\left[\frac{\partial}{\partial\theta} - \frac{i}{\sin\theta}\frac{\partial}{\partial\phi}\right].$$

Following Newman and Janis method, we first complexify the null coordinate system by permitting u and r to assume values in complex space. Now, the function $f(r)$ can be endorsed to a function $F(r, \bar{r})$, which is complex and reduces to $f(r)$ on the real slice. Let the radial coordinate r be replaced by its complex conjugate $\bar{r}$, and the tetrad be rewritten in the form

$$l^{\alpha} = \frac{\partial}{\partial r},$$

$$n^{\alpha} = \left[\frac{\partial}{\partial u} - \frac{1}{2} \left(1 - \frac{m}{r} - \frac{m}{\bar{r}} + \frac{e^2}{r\bar{r}} \right) \frac{\partial}{\partial r} \right],$$

$$m^{\alpha} = \frac{1}{\sqrt{2}\bar{r}} \left[\frac{\partial}{\partial \theta} + \frac{i}{\sin \theta} \frac{\partial}{\partial \phi} \right],$$

$$\overline{m}^{\alpha} = \frac{1}{\sqrt{2}r} \left[\frac{\partial}{\partial \theta} - \frac{i}{\sin \theta} \frac{\partial}{\partial \phi} \right].$$

(Note that for this complexified tetrad, we keep l^{μ} and n^{μ} real and m^{α} and $\overline{m}^{\alpha}$ the complex conjugates of each other.)

In this complexified tetrad, we make a complex coordinate transformation as used by Newman and Janis by defining a new set of coordinates as

$$x^{\mu\prime} = x^{\mu} + ia(\delta_r^{\mu} - \delta_u^{\mu}) \cos \theta,$$

i.e.,

$$r \to r' = r + ia \cos \theta; \quad u \to u' = u - ia \cos \theta; \quad \theta \to \theta'; \quad \phi \to \phi'.$$

Here, a is a real number known as the **Kerr parameter**. Though the unprimed coordinate system is observed as complex, we will infer the new coordinates as real. Using this ordinary coordinate transformation, the above complexified null tetrad describes a real spacetime. The new null tetrad assumes the following form (new coordinates are used without prime notations)

$$l'^{\alpha} = \frac{\partial}{\partial r},$$

$$n'^{\alpha} = \left[\frac{\partial}{\partial u} - \frac{1}{2} \left(1 - \frac{2mr - e^2}{R^2} \right) \frac{\partial}{\partial r} \right],$$

$$m'^{\alpha} = \frac{1}{\sqrt{2}(r + ai \cos \theta)} \left[ia \sin \theta \frac{\partial}{\partial u} - ia \sin \theta \frac{\partial}{\partial r} + \frac{\partial}{\partial \theta} + \frac{i}{\sin \theta} \frac{\partial}{\partial \phi} \right],$$

where

$$R^2 = r^2 + a^2 \cos^2 \theta.$$

This null tetrad yields the corresponding contravariant form of Kerr–Newmann metric via

$$g^{\alpha\beta} = l'^{\alpha}n'^{\beta} + l'^{\beta}n'^{\alpha} - m'^{\alpha}\overline{m'^{\beta}} - m'^{\beta}\overline{m'^{\alpha}},$$

$$g^{\mu\nu} = \begin{pmatrix} -\frac{a^2\sin^2\theta}{R^2} & \frac{a^2+r^2}{R^2} & 0 & -\frac{a}{R^2} \\ .. & -\frac{\triangle}{R^2} & 0 & \frac{a}{R^2} \\ .. & .. & -\frac{1}{R^2} & 0 \\ .. & .. & .. & -\frac{1}{R^2\sin^2\theta} \end{pmatrix},$$

where

$$\triangle = r^2 + a^2 - 2mr + e^2.$$

One can note that in the $a \to 0$ limit, $g^{\mu\nu}$ becomes the Reissner–Nordström metric.

The covariant form of the metric is given by

$$g_{\mu\nu} = \begin{pmatrix} 1 - \frac{2mr-e^2}{R^2} & 1 & 0 & \frac{(2mr-e^2)a\sin^2\theta}{R^2} \\ .. & 0 & 0 & -a\sin^2\theta \\ .. & .. & -R^2 & 0 \\ .. & .. & .. & \frac{\sin^2\theta}{R^2}\{\triangle a^2\sin^2\theta - (a^2+r^2)^2\} \end{pmatrix}.$$

Now, the line element of the Kerr metric in null coordinates is given by

$$ds^2 = \left(1 - \frac{2mr-e^2}{R^2}\right)du^2 + 2dudr + \frac{2(2mr-e^2)a\sin^2\theta}{R^2}dud\phi - 2a\sin^2\theta drd\phi$$

$$- R^2d\theta^2 + \frac{\sin^2\theta}{R^2}\{\triangle a^2\sin^2\theta - (a^2+r^2)^2\}d\phi^2.$$

Also, line element of the Kerr–Newmann metric in (t, r, θ, ϕ) coordinates can be written in (t, r, θ, ϕ) coordinates by using the succeeding transformations

$$dt = du + \frac{r^2+a^2}{\triangle}dr, \quad d\Phi = d\phi + \frac{a}{\triangle}dr,$$

$$ds^2 = \frac{\triangle}{R^2}(dt - a\sin^2\theta d\Phi)^2 - \frac{\sin^2\theta}{R^2}[(r^2+a^2)d\Phi - adt]^2 - \frac{R^2}{\triangle}dr^2 - R^2d\theta^2.$$

10.6 The Higher Dimensional Rotating Black Hole Solution

The higher dimensional rotating black hole solution can be derived from the higher dimensional black hole solution by applying a complex coordinate transformation as suggested by Newman and Janis on null veiltrad given in Eq. (10.9),

$$Z_i^{\alpha} = (l^{\alpha}, n^{\alpha}, m_1^{\alpha}, \overline{m}_1^{\alpha}, m_2^{\alpha}, \overline{m}_2^{\alpha}, \ldots\ldots, m_{\frac{D}{2}}^{\alpha}, \overline{m}_{\frac{D}{2}}^{\alpha}), \quad i = 1, 2, 3\ldots\ldots D+2,$$

which are given by

$$l^\alpha = \delta^\alpha_r = \frac{\partial}{\partial r},$$

$$n^\alpha = \left[\delta^\alpha_u - \frac{1}{2}f(r)\delta^\alpha_r\right] = \left[\frac{\partial}{\partial u} - \frac{1}{2}f(r)\frac{\partial}{\partial r}\right], [\textit{Here, one can use } f(r) \textit{ as give in Eq. (10.7) or Eq. (10.8)}],$$

$$m^\alpha_k = \frac{1}{\sqrt{2}r\sin\theta_1\sin\theta_2........\sin\theta_{(k-1)}}\left[\frac{\partial}{\partial\theta_k} + \frac{i}{\sin\theta_k}\frac{\partial}{\partial\theta_{k+1}}\right],$$

$$\overline{m}^\alpha_k = \frac{1}{\sqrt{2}r\sin\theta_1\sin\theta_2........\sin\theta_{(k-1)}}\left[\frac{\partial}{\partial\theta_k} - \frac{i}{\sin\theta_k}\frac{\partial}{\partial\theta_{k+1}}\right],$$

$$k = 1, 2, 3, ..., \frac{D}{2}. \quad (D \textit{ should be an even integer})$$

Following Newman and Janis method, we first complexify the null coordinate system by permitting u and r to assume values in complex space. Now, the function $f(r)$ (given in (10.8)) can be endorsed to a function $F(r, \bar{r})$, which is complex and reduces to $f(r)$ on the real slice. Let the radial coordinate r be replaced by its complex conjugate $\bar{r}$, and one can rewrite the tetrad in the following form:

$$l^\alpha = \frac{\partial}{\partial r},$$

$$n^\alpha = \left[\frac{\partial}{\partial u} - \frac{1}{2}\left\{1 - \left(\frac{\mu}{r^{D-2}}\right)\left(\frac{1}{r} + \frac{1}{\bar{r}}\right) + \frac{q^2}{(r\bar{r})r^{D-2}}\right\}\frac{\partial}{\partial r}\right],$$

$$m^\alpha_k = \frac{1}{\sqrt{2}\bar{r}\sin\theta_1\sin\theta_2........\sin\theta_{(k-1)}}\left[\frac{\partial}{\partial\theta_k} + \frac{i}{\sin\theta_k}\frac{\partial}{\partial\theta_{k+1}}\right],$$

$$\overline{m}^\alpha_k = \frac{1}{\sqrt{2}r\sin\theta_1\sin\theta_2........\sin\theta_{(k-1)}}\left[\frac{\partial}{\partial\theta_k} - \frac{i}{\sin\theta_k}\frac{\partial}{\partial\theta_{k+1}}\right],$$

(Note that for this complexified tetrad we keep l^μ and n^μ real and m^α $and\,\overline{m}^\alpha$ the complex conjugates of one another and D is an even number.)

In this complexified tetrad, we make a complex coordinate transformation as used by Newman and Janis by defining a new set of coordinates as

$$x^{\mu\prime} = x^\mu + ia(\delta^\mu_r - \delta^\mu_u)\cos\theta,$$

i.e.,

$$r \to r' = r + ia\cos\theta_1; \quad u \to u' = u - ia\cos\theta_1; \quad \theta_i \to \theta'_i\,.$$

Here, a is a real number known as the **Kerr parameter**. Though the unprimed coordinate system is observed as complex, we will infer the new coordinates as real. Using this ordinary coordinate transformation, the above complexified null tetrad describes a real spacetime. The new null tetrad assumes the following form (new coordinates are used without prime notations):

$$l'^{\alpha} = \frac{\partial}{\partial r},$$

$$n'^{\alpha} = \left[\frac{\partial}{\partial u} - \frac{1}{2} \left(1 - \frac{2\mu}{r^{D-3}R^2} + \frac{q^2}{R^2 r^{2D-4}} \right) \frac{\partial}{\partial r} \right],$$

$$m'^{\alpha} = \frac{1}{\sqrt{2}(r + ai\cos\theta)\sin\theta_1 \ldots \sin\theta_{(k-1)}} \left[ia\sin\theta_1 \left(\frac{\partial}{\partial u} - \frac{\partial}{\partial r} \right) + \delta^{\alpha}_{\theta_k} + \frac{i}{\sin\theta_k} \delta^{\alpha}_{\theta_{k+1}} \right],$$

where

$$R^2 = r^2 + a^2 \cos^2\theta_1.$$

Now, one can write the line element of the **higher dimensional rotating Kerr charged black hole metric** in terms of null coordinates as

$$ds^2 = \left(1 - \frac{2\mu}{r^{D-3}R^2} + \frac{q^2}{R^2 r^{2D-4}} \right) du^2 + 2dudr - 2a\sin^2\theta_1 dr d\theta_2$$

$$- R^2 d\theta_1^2 - \left[(r^2 + a^2) + \left(\frac{2\mu}{r^{D-3}R^2} - \frac{q^2}{R^{2D-2}} \right) a^2 \sin^2\theta_1 \right] \sin^2\theta_1 d\theta_2^2$$

$$- 2a \left(\frac{2\mu}{r^{D-3}R^2} - \frac{q^2}{R^{2D-2}} \right) \sin^2\theta_1 d\theta_2 du - r^2 \sin^2\theta_1 \sin^2\theta_2 d\Omega_{D-2}^2.$$

For $q = 0$, one will get **higher dimensional rotating Kerr uncharged black hole solution** as

$$ds^2 = \left(1 - \frac{2\mu}{r^{D-3}R^2} \right) du^2 + 2dudr - 2a\sin^2\theta_1 dr d\theta_2$$

$$- R^2 d\theta_1^2 - \left[(r^2 + a^2) + \left(\frac{2\mu}{r^{D-3}R^2} \right) a^2 \sin^2\theta_1 \right] \sin^2\theta_1 d\theta_2^2$$

$$- 2a \left(\frac{2\mu}{r^{D-3}R^2} \right) \sin^2\theta_1 d\theta_2 du - r^2 \sin^2\theta_1 \sin^2\theta_2 d\Omega_{D-2}^2.$$

For $a = 0$, one will get **higher dimensional nonrotating black hole solution** with charge or without charge.

In $(t, r, \theta_1, \theta_2, ...)$ coordinates (by using $dt = du + f(r)^{-1}dr$), the above higher dimensional rotating Kerr charged black hole metric assumes the following form

$$ds^2 = \left(\frac{\Delta - a^2 \sin^2 \theta_1}{R^2}\right) dt^2 - \frac{R^2}{\Delta} dr^2 + 2a \left\{ 1 - \left(\frac{\Delta - a^2 \sin^2 \theta_1}{R^2}\right) \right\} dt d\theta_2$$

$$- R^2 d\theta_1^2 - \left[R^2 + a^2 \sin^2 \theta_1 \left\{ 2 - \left(\frac{\Delta - a^2 \sin^2 \theta_1}{R^2}\right) \right\} \right]$$

$$\times \sin^2 \theta_1 d\theta_2^2 - r^2 \sin^2 \theta_1 \sin^2 \theta_2 d\Omega_{D-2}^2, \tag{10.11}$$

where,

$$\Delta = r^2 + a^2 - \frac{2\mu}{r^{D-3}} + \frac{q^2}{R^{2D-4}}.$$

10.7 Different Forms of Kerr Solution

(i) Eddington–Finkelstein form of Kerr solution

In the study of black hole geometry, a pair of coordinate systems are used that are adapted to radial null geodesics. These new coordinates are known as Eddington–Finkelstein coordinates in which outward (inward) traveling radial light rays define the surfaces of constant time. The most important advantage of this coordinate system is that the singularity in the Schwarzschild metric will disappear. Here, Eddington–Finkelstein form of Kerr solution can be obtained by using $dt = du + f(r)^{-1}dr$ as

$$ds^2 = \left(1 - \frac{2mr}{R^2}\right) du^2 + 2dudr + \frac{4mra \sin^2 \theta}{R^2} dud\phi - 2a \sin^2 \theta dr d\phi$$

$$- R^2 d\theta^2 + \frac{\sin^2 \theta}{R^2} \{\triangle a^2 \sin^2 \theta - (a^2 + r^2)^2\} d\phi^2. \tag{10.12}$$

(ii) Boyer–Lindquist form of Kerr solution

Boyer–Lindquist coordinates are most extensive studies in literature. Boyer–Lindquist form of Kerr solution is marginally dissimilar but entirely comparable with the same metric form which can be comprehended from Kerr's original advanced Eddington–Finkelstein configuration. The above Eddington–Finkelstein Kerr metric (10.12) is expressed in (t, r, θ, Φ) coordinates by using the following transformations:

$$dt = du + \frac{r^2 + a^2}{\triangle} dr, \quad d\Phi = d\phi + \frac{a}{\triangle} dr,$$

$$ds^2 = \frac{\triangle}{R^2} (dt - a \sin^2 \theta d\Phi)^2 - \frac{\sin^2 \theta}{R^2} [(r^2 + a^2)d\Phi - adt]^2 - \frac{R^2}{\triangle} dr^2 - R^2 d\theta^2 \tag{10.13}$$

Note that

$$\triangle = r^2 + a^2 - 2mr + e^2 \ \text{ or } \ \triangle = r^2 + a^2 - 2mr,$$

for charged or uncharged case, respectively, with $R^2 = r^2 + a^2 \cos^2 \theta$.

These Boyer–Lindquist coordinates are interesting as this form minimizes the number of off-diagonal components of the metric. Here, the only off-diagonal component of the metric is $g_{t\Phi}$.

One can easily check when $a \to 0$, the above metric reduces to Schwarzschild solution or Reissner–Nordström solution ($\triangle \to r^2 - 2mr$ or $r^2 - 2mr + e^2$ and $R^2 = r^2$).

In the limit $m \to 0$ (with $a \neq 0$), one can retrieve flat spacetime, however, not in ordinary polar coordinate. Here, the metric takes the form

$$ds^2 = dt^2 - \frac{(r^2 + a^2 \cos^2 \theta)}{r^2 + a^2} dr^2 - (r^2 + a^2 \cos^2 \theta) d\theta^2 - (r^2 + a^2) \sin^2 \theta d\Phi^2. \tag{10.14}$$

The spatial part of this metric represents flat space in spheroidal coordinates. They are connected to the Cartesian coordinate in Euclidean three space. Actually, the above metric has the form

$$ds^2 = dt^2 - dx^2 - dy^2 - dz^2,$$

where

$$x = \sqrt{r^2 + a^2} \sin \theta \cos \Phi,$$

$$y = \sqrt{r^2 + a^2} \sin \theta \sin \Phi,$$

$$z = r \cos \theta.$$

The Kerr black hole is spinning but not static. It is spinning in exactly same way at all times, so, it is stationary.

Expanding the Boyer–Lindquist form of Kerr–Newman metric for large r to yield

$$ds^2 = \left[1 - \frac{2m}{r} + \frac{e^2}{r^2} + \mathcal{O}\left(\frac{1}{r^3}\right) \right] dt^2 - \left[\frac{4am \sin^2 \theta}{r} + \mathcal{O}\left(\frac{1}{r^2}\right) \right] dt d\Phi$$

$$- \left[1 + \frac{2m}{r} - \frac{e^2}{r^2} + \mathcal{O}\left(\frac{1}{r^3}\right) \right] (dr^2 + r^2 d\Omega_2^2). \tag{10.15}$$

One can observe that asymptotic expansion exhibits the asymptotic flatness of the Kerr spacetime and confirm through the physical identification of the parameters m and e that these are indeed mass and charge of the spinning body, respectively. Note that $e = 0$ for uncharged case.

In general, the above metric can be written as

$$ds^2 = \left[f(r) + \mathcal{O}\left(\frac{1}{r^3}\right) \right] dt^2 - \left[\frac{4am \sin^2 \theta}{r} + \mathcal{O}\left(\frac{1}{r^2}\right) \right] dt d\Phi$$

$$- \left[[f(r)]^{-1} + \mathcal{O}\left(\frac{1}{r^3}\right) \right] (dr^2 + r^2 d\Omega_2^2),$$

where

$$f(r) = 1 - \frac{2m}{r} \ or \ f(r) = 1 - \frac{2m}{r} + \frac{e^2}{r^2},$$

for Kerr solution or Kerr–Newman solution, respectively.

(iii) Rational polynomial coordinate form of Kerr solution

Let us choose $\chi = \cos\theta$ in Boyer–Lindquist form of Kerr metric so that $\chi \in [-1, 1]$. This implies

$$R^2 = r^2 + a^2\chi^2 \ and \ d\theta = \frac{d\chi}{\sqrt{1 - \chi^2}}.$$

The above Eddington–Finkelstein Kerr metric can be written in (t, r, χ, ϕ) coordinates

$$ds^2 = \frac{\triangle}{(r^2 + a^2\chi^2)}[dt - a(1 - \chi^2)d\Phi]^2 - \frac{(1 - \chi^2)}{(r^2 + a^2\chi^2)}[(r^2 + a^2)d\Phi - adt]^2$$

$$- \frac{(r^2 + a^2\chi^2)}{\triangle}dr^2 - (r^2 + a^2\chi^2)d\theta^2. \tag{10.16}$$

This form of the Kerr metric deserves special mention as all the metric components can be expressed in terms of simple rational polynomials of the coordinates.

(iv) Doran coordinates

After the discovery of Kerr metric, the first explicit form of this metric was written in Eddington–Finkelstein coordinates as

$$ds^2 = \left[1 - \frac{2mr}{r^2 + a^2\cos^2\theta}\right](du + a\sin^2\theta d\phi)^2$$

$$- 2(du + a\sin^2\theta d\phi)(dr + a\sin^2\theta d\phi) - (r^2 + a^2\cos^2\theta)(d\theta^2 + \sin^2\theta d\phi^2). \tag{10.17}$$

Recently, Chris Doran introduced a new coordinate transformation to get a new form of Kerr metric as

$$du = dt + \frac{dr}{1 + \sqrt{\frac{2mr}{r^2 + a^2}}}, \ d\Phi = d\phi - \frac{adr}{r^2 + a^2 + \sqrt{2mr(r^2 + a^2)}}.$$

Using these transformations, one can express the Kerr metric in Doran coordinates (t, r, θ, Φ) as

$$ds^2 = dt^2 - (r^2 + a^2\cos^2\theta)d\theta^2 - (r^2 + a^2)\sin^2\theta d\Phi^2$$

$$- \left[\frac{r^2 + a^2\cos^2\theta}{r^2 + a^2}\right]\left[dr + \frac{\sqrt{2mr(r^2 + a^2)}}{r^2 + a^2\cos^2\theta}(dt - a\sin^2\theta d\Phi)\right]^2. \tag{10.18}$$

Note that for $a = 0$, one gets

$$ds^2 = dt^2 - \left[dr + \sqrt{\frac{2m}{r}} dt \right]^2 - r^2(d\theta^2 + \sin^2\theta d\Phi^2). \tag{10.19}$$

This is the famous **Painlevé–Gullstard form** of Schwarzschild metric.

(v) Kerr–Schild Cartesian coordinate form of Kerr solution

None of the above forms were discovered by Kerr. He used Cartesian coordinate (t, x, y, z) to describe the spacetime outside of a rotating body. One can obtain the Kerr spacetime in Cartesian coordinate (t, x, y, z) by using suitable coordinate transformations from Eddington–Finkelstein form (10.12) of Kerr solution. The coordinate transformations are

$$t = u - r, x = \sin\theta(r\cos\phi + a\sin\phi), y = \sin\theta(r\sin\phi - a\cos\phi), z = r\cos\theta. \tag{10.20}$$

Note that here r is related to x, y, z as

$$\frac{x^2 + y^2}{r^2 + a^2} + \frac{z^2}{r^2} = 1. \tag{10.21}$$

The above transformations yield

$$ds^2 = dt^2 - dx^2 - dy^2 - dz^2$$
$$- \frac{2mr^3}{r^4 + a^2z^2} \left[dt + \frac{r(xdx + ydy)}{a^2 + r^2} + \frac{a(ydx - xdy)}{r^2 + a^2} + \frac{z}{r}dz \right]^2. \tag{10.22}$$

For $m = 0$, we get the Minkowski spacetime

$$ds^2 = dt^2 - dx^2 - dy^2 - dz^2.$$

The line element (10.22) can be written in the following general form:

$$g_{ij} = \eta_{ij} + \lambda l_i l_j, \tag{10.23}$$

where l_i is the null vector with respect to Minkowski metric η_{ij}, i.e., $\eta^{ij} l_i l_j = 0$.
Here,

$$\lambda = \frac{2mr^3}{r^4 + a^2z^2}$$

and

$$l_i = \left(1, \frac{rs + ay}{a^2 + y^2}, \frac{ry - ax}{a^2 + y^2}, \frac{z}{r} \right). \tag{10.24}$$

We can also find the contravariant component as

$$g^{ij} = \eta^{ij} - \lambda l^i l^j, \tag{10.25}$$

with

$$l^i = \left(-1, \frac{rs + ay}{a^2 + y^2}, \frac{ry - ax}{a^2 + y^2}, \frac{z}{r}\right). \tag{10.26}$$

For $a = 0$, we get the Schwarzschild solution as

$$\lambda = \frac{2m}{r}; \; l_i = \left(1, \frac{x}{r}, \frac{y}{r}, \frac{z}{r}\right). \tag{10.27}$$

Apart from these forms of the Kerr metric, there should be many other forms that will be discovered in the near future and those forms may help us know more properties of the Kerr solution.

10.8   Some Elementary Properties of the Kerr Solution

The utmost valuable configuration of Kerr metric is Boyer–Lindquist form for the study of the elementary properties of Kerr solution. One can rewrite the Boyer–Lindquist form as

$$ds^2 = A dt^2 - B(d\phi - \omega dt)^2 - C dr^2 - D d\theta^2,$$

where

$$A = \frac{\triangle R^2}{\Sigma^2}, \; B = \frac{\Sigma^2}{R^2} \sin^2\theta, \; C = \frac{R^2}{\triangle}, \; D = R^2, \; \omega = -\frac{g_{t\phi}}{g_{\phi\phi}} = \frac{2mra}{\Sigma^2},$$

with

$$\triangle = r^2 - 2mr + a^2, \; R^2 = r^2 + a^2 \cos^2\theta,$$

$$\Sigma^2 = (r^2 + a^2)^2 - a^2 \triangle \sin^2\theta = \frac{R^4 \triangle - 4a^2 m^2 r^2 \sin^2\theta}{\triangle - a^2 \sin^2\theta}.$$

Here, the fundamental tensor takes the form

$$g_{\mu\nu} = \begin{pmatrix} A - \omega^2 B & 0 & 0 & \omega B \\ 0 & -C & 0 & 0 \\ 0 & 0 & -D & 0 \\ \omega B & 0 & 0 & -B \end{pmatrix}.$$

Here,

$$g_{tt} = A - \omega^2 B = 1 - \frac{2mr}{R^2}, g_{t\phi} = \omega B, g_{rr} = -C, \; g_{\theta\theta} = -D, g_{\phi\phi} = -B. \tag{10.28}$$

The contravariant form of the fundamental tensor is

$$g^{\mu\nu} = \begin{pmatrix} \frac{1}{A} & 0 & 0 & \frac{\omega}{A} \\ 0 & -\frac{1}{C} & 0 & 0 \\ 0 & 0 & -\frac{1}{D} & 0 \\ \frac{\omega}{A} & 0 & 0 & \frac{\omega^2 B - A}{AB} \end{pmatrix}.$$

Note that the metric coefficients do not depend on t and ϕ. This implies that the solution is both stationary and axially symmetric (a solution being axially symmetric means that the solution is invariant under rotation about a fixed axis). As the Kerr metric is stationary and axially symmetric, it possesses the following Killing vectors $t^\alpha = \frac{\partial x^\alpha}{\partial t} = (1, 0, 0, 0)$ and $\phi^\alpha = \frac{\partial x^\alpha}{\partial \phi} = (0, 0, 0, 1)$.

This indicates that the orbits of the Killing vector field $\frac{\partial x^\alpha}{\partial \phi}$ admit the continuous symmetries for the curves $t = $ constant, $r = $ constant, $\theta = $ constant, which are circles. This approves that a spinning source provides the Kerr field. Thus, Kerr solution represents a vacuum field exterior to a spinning source. As $a \to 0$, we reproduce the Schwarzschild line element. Also for $a \to 0$ and $r \to \infty$, we have $g_{ab} \to \eta_{ab}$ so that Kerr solution is asymptotically flat.

The determinant of the metric yields m independent form as

$$\det(g_{\mu\nu}) = -\sin^2\theta (r^2 + a^2\cos^2\theta)^2. \tag{10.29}$$

10.9 Singularities and Horizons

It is obvious that the fundamental tensor of Kerr solution and its contravariant form have singularities at $\triangle = 0$ and $R^2 = 0$. Now one needs to calculate Kretschmann $R^{abcd}R_{abcd}$ of the Kerr spacetime to distinguish between coordinate and curvature singularities. Here,

$$R^{abcd}R_{abcd} = \frac{48m^2(r^2 - a^2\cos^2\theta)(R^4 - 16a^2r^2\cos^2\theta)}{R^{12}}. \tag{10.30}$$

This indicates that the metric has just a coordinate singularity at $\triangle = 0$ (by setting $g^2_{\phi t} - g_{tt}g_{\phi\phi} = 0$). However, the Kerr spacetime has a real physical singularity at $R^2 = 0$. $R^2 = 0$ implies

$$R^2 = r^2 + a^2\cos^2\theta = 0.$$

It follows that

$$r = 0 \ and \ \cos\theta = 0 \ or \ \theta = \frac{\pi}{2}.$$

Now, Eqs. (10.20) and (10.21) yield

$$x^2 + y^2 = a^2, \ z = 0. \tag{10.31}$$

This is a ring-type singularity having radius a that lies in the $z = 0$ equatorial plane.

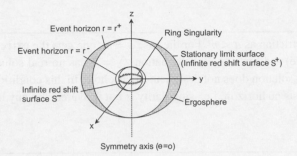

Figure 93 The figure indicates the position of the horizons, ergosurfaces, and curvature singularity in the Kerr black hole spacetime.

The event horizon is a surface at which the radial coordinate r reverses its signature. Thus, the Kerr spacetime has horizons where the four velocity of a viewer tends to zero, or the surface $r =$ constant becomes null. The event horizon is a solution of $\triangle = 0$. This yields

$$r^2 - 2mr + a^2 = (r - r_+)(r - r_-) = 0,$$

i.e.,

$$r = r_\pm = m \pm \sqrt{m^2 - a^2}. \tag{10.32}$$

Note that $R^{abcd}R_{abcd}$ remains finite at $r_\pm$. Thus, the points $r = r_\pm$ are **coordinate singularity** rather than a curvature singularity. Also when $a \to 0$ (nonspinning limit), we obtain $r = 2m$ or 0. $r = 2m$ is the position of the horizon in the Schwarzschild geometry. This indicates that we can recognize $r_\pm$ as the positions of the **inner and outer horizons** ($r = r_+$ is the outer horizon and $r = r_-$ is the inner horizon). One can refer the region $r < r_+$ as the **interior** of the Kerr black hole (see Fig. 93).

Note 10.1

If we assume the surfaces $T = r - constant = 0$, then its normal is

$$n_\mu = T_{,\mu} = (0, 1, 0, 0).$$

Now,

$$n_\mu n_\nu g^{\mu\nu} = g^{rr} = -\frac{\triangle}{R^2}.$$

Hence,

$$n_\mu n^\mu = 0,$$

on the surfaces $r = r_+$ and $r = r_-$, which are **null hypersurfaces, i.e., horizons**.

Note 10.2

We have taken the restriction as $a < m$. For the case $a = m$, one gets the **extremal Kerr black hole** (two horizons coincide) and for $a > m$ the equation $\triangle = 0$ has no real solution. So horizon does not exist, and the Kerr solution does not designate a black hole. In this condition, we obtain a **naked singularity** (there exists no horizon, i.e., singularity is not "covered" by any horizon).

Note 10.3

Kerr black hole solution is regular in the following three regions:

Region I: $r_+ < r < \infty$:
Here the $r = $ constant hypersurfaces are time-like. The metric turns out to be flat for $r \to \infty$, and this region can be considered as the exterior of the black hole.

Region II: $r_- < r < r_+$:
Here the $r = $ constant hypersurfaces are space-like. If a particle enters inside the outer horizon, then it starts falling continuously for decreasing values of r and after reaching the inner horizon it enters region III.

Region III: $0 < r < r_-$:
Here, the $r = $ constant hypersurfaces are again time-like. This region holds the physical singularity. For extremal Kerr black holes, the two horizons coincide and region II dissolves.

10.10 Static Limit and Ergosphere

For constant (r, θ, ϕ), i.e., for a static world-line, the Kerr metric assumes the form

$$ds^2_{|r,\theta,\phi=constant} = g_{tt}dt^2 = \left[1 - \frac{2mr}{R^2}\right] dt^2.$$

The **surfaces of infinite redshift** in the Kerr spacetime can be found by removing the coefficient g_{tt}.

Thus, from Kerr solution, we find

$$g_{tt} = 1 - \frac{2mr}{R^2} = \frac{(r^2 - 2mr + a^2 \cos^2 \theta)}{R^2} = (r - r_{s+})(r - r_{s-}) = 0.$$

This yields the surfaces of infinite redshift at

$$r = r_{s\pm} = m \pm (m^2 - a^2 \cos^2 \theta)^{\frac{1}{2}}. \tag{10.33}$$

Note 10.4

It is termed as infinite redshift surfaces due to the following reason. Let, us put a source at $r = r_{emit}$, which is very close to the black hole that emits a light signal with wavelength λ_{emit}. This signal will be observed at $r = r_{ob}$ with wavelength as

$$\lambda_{ob} = \sqrt{\frac{g_{tt}(r_{ob})}{g_{tt}(r_{emit})}}\, \lambda_{emit}.$$

At $r = r_{s_{\pm}} = r_{emit}$, $g_{tt}(r_{emit}) = 0$ and hence $\lambda_{ob} = \infty$.

One can notice that in the region for $r > r_{s_+}$, the tangent to the static world-line $k = \partial_t$ is time-like, however, the region for $r < r_{s_+}$ the world-line is space-like. The surface $r = r_{s_+}$ is known as **stationary limit surface** or **ergosurface**.

We, now, consider an observer sitting in a rocket ship attempting to come to a halt at a stationary point in the Kerr black hole spacetime. This rocket moves with a four velocity, U^α that is proportional to the Killing vector, $t^\alpha [\equiv \frac{\partial x^\alpha}{\partial t} = (1, 0, 0, 0)]$. Thus,

$$U^\alpha = \gamma t^\alpha.$$

Since the rocket motion is not along geodesic, so to normalize the four velocity, we multiply the factor $\gamma \ [\equiv (g_{\alpha\beta} t^\alpha t^\beta)^{-\frac{1}{2}} = (g_{\alpha\beta}(\partial_\alpha t)(\partial_\beta t))^{-\frac{1}{2}} = (g_{\alpha\beta}(\delta_t^\alpha)(\delta_t^\beta))^{-\frac{1}{2}} = (g_{tt})^{-\frac{1}{2}}]$.

Here, the velocity curves are only time-like outside $r = r_{s_+}$, or inside $r = r_{s_-}$ but becomes null when $\gamma^{-2} = g_{tt} = 0$, i.e., at $r = r_{s_+}$. Thus, the **static limit** is characterized by $g_{tt} = 0$. Hence, the rocket cannot be fixed when $r \le r_{s_+}$, even if the engine provides an arbitrary large force. Rather, the dragging of inertial frame restrains it to rotate with the black hole. Actually, for $r < r_{s_+}$ velocity curves, i.e., the world-line is space-like. One can ignore the other solution $r = r_{s_-}$ since it lies inside the event horizon. Actually, in the region $r_+ < r < r_{s_+}$, the coordinates r and t are both space-like. This finite region between the horizon and the static limit is called the **Ergosphere** of the Kerr spacetime. Note that the static limit does not coincide with the black hole event horizon. In the ergosphere region, the local light cones are tilted in the direction of black hole rotation so that every time-like vector gains a rotational component. Hence, all physical trajectory is compelled to rotate within the ergosphere. Note that outside the ergosurface, the light cones are normal (not titled) (see Fig. 94). For $a \to 0$, i.e., for nonspinning limit, the stationary limit surface coincides with the event horizon and the ergosphere disappears. Also, since the position of ergosphere is at the outside of the outer horizon, it is possible to enter in the inside region of the ergoregion and return. Within the ergoregion, the Killing vector $k^\mu = (1, 0, 0, 0)$ becomes space-like, i.e., $k^\mu k^\nu g_{\mu\nu} = g_{tt} < 0$.

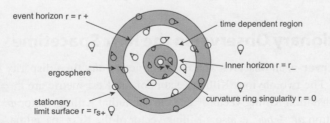

Figure 94 Light cones in Kerr spacetime.

Since $\sqrt{m^2 - a^2 \cos^2 \theta} \geq \sqrt{m^2 - a^2}$, we have

$$r_{s-} \leq r_- \leq r_+ \leq r_{s+}. \tag{10.34}$$

10.11 Zero Angular Momentum Observers in the Kerr Spacetime

Let the observers with zero angular momentum (called **ZAMOs**) have four velocity u^α. Therefore,

$$L = u_\alpha \phi^\alpha = 0.$$

This yields

$$g_{\phi t} \dot{t} + g_{\phi\phi} \dot{\phi} = 0.$$

[dot indicates differentiation with respect to proper time τ].

Also, suppose the observers are moving in the ϕ direction with angular velocity $\Omega \equiv \frac{d\phi}{dt} = \omega$. We note that ZAMOs acquire an angular velocity equal to $\omega = -\frac{g_{t\phi}}{g_{\phi\phi}}$. Hence, one obtain

$$\omega = -\frac{g_{t\phi}}{g_{\phi\phi}} = \frac{2mar}{\Sigma} = \frac{2mar}{[(r^2 + a^2)^2 - a^2 \sin^2 \theta(r^2 - 2mr + a^2)]}$$

$$= \frac{a(r^2 + a^2 - \triangle)}{[(r^2 + a^2)^2 - a^2 \sin^2 \theta \triangle]}.$$

Since

$$(r^2 + a^2)^2 > a^2 \sin^2 \theta(r^2 - 2mr + a^2),$$

we always get $\frac{\omega}{ma} > 0$. Thus, the signs of angular velocity and the angular momentum ma of the black hole are same, i.e., the ZAMOs rotate with the black hole. Once a viewer moves toward the black hole, this angular velocity increases. This is a remarkable characteristic of the Kerr black hole that the motion of the ZAMO in the gravitational field of the Kerr black hole is corotating with the black hole. This feature is called the **dragging of inertial frames**.

For large r, $\omega \simeq \frac{2am}{r^3}$ and the dragging vanishes finally at infinity.

10.12 Stationary Observer in the Kerr Spacetime

A **stationary observer** is an observer who does not notice any time disparity in the black hole's gravitational field. The two obvious Killing vectors of the Kerr metric are $t^\alpha = \frac{\partial x^\alpha}{\partial t}$ and $\phi^\alpha = \frac{\partial x^\alpha}{\partial \phi}$ (known as time translation Killing vector t^α and the azimuthal Killing vector ϕ^α). Then, in general, the linear combination $at^\alpha + b\phi^\alpha$ is also a Killing vector of the Kerr spacetime.

Let, the stationary observer move with a four velocity U^α, which is proportional, in general, to the Killing vector, $t^\alpha + \Omega\phi^\alpha$, i.e.,

$$U^\alpha = \gamma(t^\alpha + \Omega\phi^\alpha),$$

where γ is a new normalization factor given by

$$\gamma^{-2} = g_{\alpha\beta}(t^\alpha + \Omega\phi^\alpha)(t^\beta + \Omega\phi^\beta),$$

$$= g_{tt} + 2\Omega g_{t\phi} + \Omega^2 g_{\phi\phi},$$

$$= g_{\phi\phi}\left(\Omega^2 - 2\omega\Omega + \frac{g_{tt}}{g_{\phi\phi}}\right),$$

where

$$\omega = -\frac{g_{t\phi}}{g_{tt}} \text{ and } \frac{d\phi}{dt} = \Omega = observer's\ angular\ velocity.$$

A stationary observer can exist in the Kerr spacetime provided $\gamma^{-2} > 0$. For $\gamma^{-2} < 0$, this fails to be true.

Now, $\gamma^{-2} > 0$ only when

$$\Omega_- < \Omega < \Omega_+,$$

where

$$\Omega_\pm = \omega \pm \sqrt{\omega^2 - \frac{g_{tt}}{g_{\phi\phi}}}.$$

Using the value of the metric coefficients of Kerr metric, one gets

$$\Omega_\pm = \omega \pm \frac{\sqrt{\triangle}\phi^2}{\sum \sin\theta}.$$

If we have a stationary observer with $\Omega = 0$, then that observer will be the static observer and from the above result we identify that static observers exist only outside the static limit.

It is obvious that Ω_- changes sign at $r = r_{s+}$. Also, one can see that if r decreases more from r_{s+}, Ω_- increases, whereas Ω_+ decreases. Finally, we attain the state $\Omega_- = \Omega_+$, which implies $\Omega = \omega$. At this point, the stationary observer is compelled to rotate with same angular velocity ω of the black hole. This situation will happen when $\triangle = 0$, i.e., when $r^2 - 2mr + a^2 = 0$. The largest root r_+ indicates the event horizon and smallest root r_- specifies the black hole's inner apparent horizon.

The Killing vector $t^\alpha + \Omega \phi^\alpha$ will be null at r_+ and stationary observers never exist inside this surface (black hole event horizon).

The expression

$$\Omega_H = \frac{a}{r_+^2 + a^2},$$

can be treated as the angular velocity of the black hole. It is actually the minimum angular velocity of a particle at the horizon. Stationary observers just outside the horizon are corotating with the black hole possessing the angular velocity Ω_H.

Note 10.5

The Killing vector $t^\alpha + \Omega_H \phi^\alpha$ is null at the event horizon. It is tangent to the horizon's null generators (which is wrapping around the horizon with an angular velocity Ω_H). The event horizon of the Kerr spacetime is actually known as **Killing horizon**.

10.13 Null Geodesics in Kerr Spacetime

We know that freely falling massless particles describe null geodesics. For the Kerr spacetime, we can construct the Lagrangian for the Boyer–Lindquist form (10.13) as

$$L = \frac{\Delta}{(r^2 + a^2 \cos^2 \theta)} (\dot{t} - a \sin^2 \theta \dot{\phi})^2 - \frac{\sin^2 \theta}{(r^2 + a^2 \cos^2 \theta)} [(r^2 + a^2)\dot{\phi} - a\dot{t}]^2$$
$$- \frac{(r^2 + a^2 \cos^2 \theta)\dot{r}^2}{\Delta} - (r^2 + a^2 \cos^2 \theta)\dot{\theta}^2. \tag{10.35}$$

[Dot indicates the derivative with respect to affine parameter p]

The corresponding Euler–Lagrange equations are

$$\frac{d}{dp} \left(\frac{\partial L}{\partial \dot{x}} \right) - \frac{\partial L}{\partial x} = 0, \quad x = t, r, \theta, \phi.$$

We know that the first integral of geodesics equation is

$$g_{\mu\gamma} \frac{dx^\mu}{dp} \frac{dx^\gamma}{dp} = \epsilon, \tag{10.36}$$

where $\epsilon = 1$ for the time-like geodesics and $\epsilon = 0$ for null geodesics.

Note that the Kerr metric is axially symmetric and without any loss of generality we consider the null geodesic in equatorial plane $\theta = constant$, i.e., null geodesics lie in the hypersurface $\theta = constant$, then $\frac{d\theta}{dp} = 0$. Kerr metric coefficients do not contain t and ϕ explicitly which means ϕ and t are cyclic coordinates.

For the above Lagrangian, we get the following equations of motion corresponding to $x = t, \phi$ and θ, respectively,

$$\frac{\Delta}{(r^2 + a^2 \cos^2 \theta)}(\dot{t} - a \sin^2 \theta \dot{\phi}) + \frac{a \sin^2 \theta}{(r^2 + a^2 \cos^2 \theta)}[(r^2 + a^2)\dot{\phi} - a\dot{t}] = l. \tag{10.37}$$

$$\frac{a\Delta \sin^2 \theta}{(r^2 + a^2 \cos^2 \theta)}(\dot{t} - a \sin^2 \theta \dot{\phi}) + \frac{(r^2 + a^2) \sin^2 \theta}{(r^2 + a^2 \cos^2 \theta)}[(r^2 + a^2)\dot{\phi} - a\dot{t}] = n. \tag{10.38}$$

$$\frac{a^2 \Delta}{(r^2 + a^2 \cos^2 \theta)^2}(\dot{t} - a \sin^2 \theta \dot{\phi})^2 - \frac{2a\Delta\dot{\phi}}{(r^2 + a^2 \cos^2 \theta)}(\dot{t} - a \sin \theta \dot{\phi})$$

$$-\frac{(r^2 + a^2)}{(r^2 + a^2 \cos^2 \theta)^2}[(r^2 + a^2)\dot{\phi} - a\dot{t}]^2 + \frac{a^2\dot{r}^2}{\Delta} = 0. \tag{10.39}$$

Here, l and n are integration constants, which are realized as the energy and the component of the angular momentum along the rotation axis, respectively. Also, the first integral (10.36) yields

$$\frac{\Delta}{(r^2 + a^2 \cos^2 \theta)}(\dot{t} - a \sin^2 \theta \dot{\phi})^2 - \frac{\sin^2 \theta}{(r^2 + a^2 \cos^2 \theta)}[(r^2 + a^2)\dot{\phi} - a\dot{t}]^2$$

$$-\frac{(r^2 + a^2 \cos^2 \theta)\dot{r}^2}{\Delta} = 0. \tag{10.40}$$

One can notice that the above four equations contain three unknowns $\dot{r}, \dot{t}$ and $\dot{\phi}$, therefore, there should exist some constraint between n and l. After some algebraic manipulation, we get from (10.37)–(10.40) as

$$n^2 - (al \sin^2 \theta)^2 = 0. \tag{10.41}$$

Taking $n = al \sin^2 \theta$, Eqs. (10.37)–(10.40) yield the following solutions for $\dot{t}$, $\dot{r}$, and $\dot{\phi}$ as

$$\dot{t} = \frac{(r^2 + a^2)l}{\Delta}, \tag{10.42}$$

$$\dot{r} = \pm l, \tag{10.43}$$

$$\dot{\phi} = \frac{al}{\Delta}. \tag{10.44}$$

Eq. (10.43) indicates that the constant l may be immersed into the affine parameter p. Also $\pm$ sign signifies that we can have two null congruences. The negative sign implies that we have ingoing null congruence and positive sign implies that we have outgoing null congruence. For outgoing null congruence, we elect $\dot{r} = +l$ and with the help of Eqs. (10.42) and (10.44) we have

$$\frac{dt}{dr} = \frac{\dot{t}}{\dot{r}} = \frac{(r^2 + a^2)}{\Delta}; \quad \frac{d\phi}{dr} = \frac{\dot{\phi}}{\dot{r}} = \frac{a}{\Delta}. \tag{10.45}$$

For $a^2 < m^2$, one can solve above equations to yield

$$t = r + \left(\frac{mr_+}{\sqrt{m^2 - a^2}}\right) \ln\left|\frac{r}{r_+} - 1\right| - \left(\frac{mr_-}{\sqrt{m^2 - a^2}}\right) \ln\left|\frac{r}{r_-} - 1\right|, \qquad (10.46)$$

$$\phi = \frac{a}{2\sqrt{m^2 - a^2}} \ln\left|\frac{r - r_+}{r - r_-}\right| + constant. \qquad (10.47)$$

Let us consider three regions I $(r > r_+)$, II $(r_- < r < r_+)$, III $(0 < r < r_-)$ in Kerr spacetime. One can note that $\triangle > 0$ in regions I and III and $\triangle < 0$ in region II. Therefore, Eq. (10.45) indicates that $\frac{dr}{dt} > 0$ in region I and so this congruence is known as the **principal congruence of outgoing null geodesics** ($\frac{dr}{dt} > 0 \Rightarrow r$ increases as t increases).

The solution corresponding to $\dot{r} = -l$ provides the congruence, which is known as the **principal congruence of ingoing null geodesics**. Here, one has to replace t by $-t$ and ϕ by $-\phi$ ($\frac{dr}{dt} < 0 \Rightarrow r$ decreases as t increases).

One can note that the null congruences give information about the radial variation of the light cone structure. We notice that the light cone becomes narrower as $r \to r_+$. Here, $r = r_+$ is a coordinate singularity as $r \to r_+$, $t \to \infty$, and $\phi \to \infty$. The above relation (10.46) reduces to Schwarzschild congruences in the nonspinning limit $a \to 0$.

10.14 Kerr Solution in Eddington–Finkelstein Coordinates

Now, we are interested in finding a new coordinate (v, r, θ, ψ) in which ingoing radial null geodesics becomes straight lines,

$$v = t + r^*, \ \psi = \phi + \phi^*, \qquad (10.48)$$

where

$$r^* = \int \frac{r^2 + a^2}{\triangle} dr = r + \left(\frac{mr_+}{\sqrt{m^2 - a^2}}\right) \ln\left|\frac{r}{r_+} - 1\right|$$

$$- \left(\frac{mr_-}{\sqrt{m^2 - a^2}}\right) \ln\left|\frac{r}{r_-} - 1\right|. \qquad (10.49)$$

$$\phi^* = \int \frac{a}{\triangle} dr = \frac{a}{2\sqrt{m^2 - a^2}} \ln\left|\frac{r - r_+}{r - r_-}\right| + constant. \qquad (10.50)$$

Here, we use the transformation $t \to v, \phi \to \psi$, where

$$dv = dt + \frac{(r^2 + a^2)}{\triangle} dr; d\psi = d\phi + \frac{a}{\triangle} dr.$$

For advanced time coordinate (v, r, θ, ψ) the Boyer–Lindquist line element is transformed into

$$ds^2 = \left(1 - \frac{2mr}{R^2}\right) dv^2 - 2dvdr + \frac{4amr\sin^2\theta}{R^2}dvd\psi$$

$$+ 2a\sin^2\theta drd\psi - R^2d\theta^2 - \frac{\Sigma^2\sin^2\theta}{R^2}d\psi^2. \tag{10.51}$$

This form is known as **advanced Eddington–Finkelstein form of the Kerr solution**. These coordinates produce an extension of the Kerr metric across the future horizon. Actually, the coordinates (v, r, θ, ψ) work well on the future horizon, however, not on regular, i.e., singular on the past horizon.

We can describe the outgoing radial null geodesics by using the following coordinate (u, r, θ, χ), where

$$u = t - r^*, \quad \chi = \phi - \phi^*,$$

$$ds^2 = \left(1 - \frac{2mr}{R^2}\right) du^2 + 2dudr + \frac{4mra\sin^2\theta}{R^2}dud\chi - 2a\sin^2\theta drd\chi$$

$$- R^2d\theta^2 - \frac{\Sigma^2\sin^2\theta}{R^2}d\chi^2. \tag{10.52}$$

This form is known as **retarded Eddington–Finkelstein form of the Kerr solution**.

Here, the coordinates (v, r, θ, χ) are regular on the past horizon but singular on the future horizon.

10.15 Maximal Extension of Kerr Spacetime

By using the Eddington–Finkelstein-like coordinates $((v, r, \theta, \psi)$ and $(u, r, \theta, \chi))$, like the Reissner–Nordström black hole, we can extend the Kerr spacetime across the horizons, i.e., it is possible to allow the continuation of the Kerr metric across the event horizon.

The conformal diagram is very similar to Reissner–Nordström black hole except that one could go beyond $r < 0$. One can draw the conformal structure of the Kerr spacetime along the symmetry axis, which is shown in Fig. 95. Here, we have three regions I $(r_+ < r < \infty)$, II $(r_- < r < r_+)$, and III $(-\infty < r < r_-)$ in Kerr spacetime. Region I is flat spacetime (it is the exterior of the black hole), region II includes closed trapped surface (here $r =$ constant, surfaces are space-like), and region III includes the ring singularity as well as closed time-like curves, i.e., it violates the causality condition (it contains actually ring singularity and $r < 0$ space). However, regions I and II obey the causality condition.

The conformal diagram of Kerr spacetime contains infinite copies of patch both with $r > 0$ and with $r < 0$. In this spacetime, a particle entering the inner horizon can either cross the ring or enter the additional copy of the region II, and then another copy of the region I with $r > r_+$ (which is asymptotically flat), etc. These copies of region I are causally connected.

For extremal case $a^2 = m^2$, two horizons r_+ and r_- overlap and therefore, region II does not occur. Thus, the conformal diagram is very similar to extreme Reissner–Nordström black hole case. Conformal structure of the extreme Kerr spacetime along the symmetry axis is shown in Fig. 96.

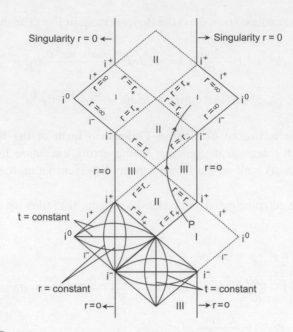

Figure 95 Conformal structure of the Kerr spacetime.

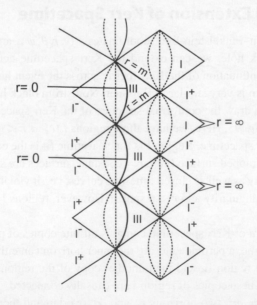

Figure 96 Conformal structure of the extreme Kerr spacetime.

10.16   The Hawking Radiation

Einstein's GR predicts classically that a black hole absorbs everything. However, in 1974, Hawking showed by considering quantum field theory that at the early times despite there being no particles of the quantum field but later on a distant observer will see an outgoing flux of particles having thermal spectrum. This flux of energy emitted by the black hole causes it to loose mass and finally it will evaporate. This phenomena proposed by Hawking is known as **Hawking radiation**.

In Fig. 97, we will give an outline about the process.

According to quantum theory, pairs of virtual particles and antiparticles are continuously created and annihilated in the empty space. When this phenomenon is happening in the vicinity of the horizon of the black hole, then it is affected by the strong gravitational field around the black hole. Among these pairs, some of them would not be able to anhilate, because positive particle in the pair (say an electron with its positive energy and positive mass) may come out in the form of thermal emission radiating from the black hole, whereas the negative particle (say, a positron, with its negative energy and negative mass) may move into the black hole (see Fig. 97). The outgoing particle carries energy when it appears as radiation emitted by the black hole. For a distant observer, this spectacle will seem to have been produced (discharge radiation) by the black hole. So as to recompense for this positive energy, transmitted by the positive particle, the black hole would gradually lose mass. This phenomena is known as **Hawking radiation**.

Note 10.6

A vacuum contains a place with anything but empty! Actually, in vacuum, a phenomena of creation and destruction of pairs of particles/antiparticles happens continuously for very short durations. These particles are called virtual particles as they cannot be directly evaluated by particle detectors during the very short times. Heisenberg's uncertainty principle helps us describe this fact. The total energy of the vacuum is zero. Let the pair of particles and antiparticles having energy ΔE

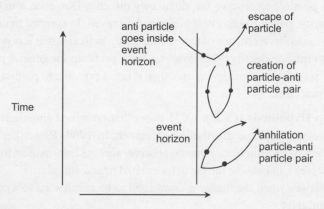

Figure 97   Hawking radiation.

be constantly created for a life time of about $\frac{\hbar}{\Delta T}$. We define it with an uncertainty that

$$\Delta E.\Delta T > \frac{\hbar}{2} \quad (\hbar = \text{reduced Planck constant}).$$

Notice that total energy of the virtual particles and antiparticles remain unchanged as one of the two particles has a positive energy, and the other particle has a negative one. This event is known as **quantum fluctuations** of the vacuum.

Note 10.7

The opposite phenomenon of Hawking radiation is not possible. If the particle with positive energy falls back into the black hole, then the other particle with negative energy (antiparticle) will also have to fall, as a particle with a negative energy cannot exist in our universe.

Note 10.8

As a black hole emits particles, its mass and size gradually decrease. Since the black hole radiates, it will evaporate and vanish.

The No Hair Theorem: The no hair theorem asserts that the spacetime metric for a black hole in matter free space is distinctively expressed by three physical parameters such as the mass M, the angular momentum J, and the electric charge Q (and for magnetic monopole charge if it exists) of the source. These three quantities are conserved. One can determine these quantities by investigating particle motion in empty space.

Naked Singularity: We know black hole has a region around its physical singularity, known as the event horizon and the gravitational force of the singularity is very strong so that even light cannot escape. So, it is not possible to observe the singularity directly. However, a naked singularity is a gravitational singularity, which has no event horizon. As a result, in contrast to the black hole, it can be observed from the outside. Existence of naked singularity indicates that it is possible to detect the collapse of a body to infinite density. This creates a basic problem for general relativity because if there exists a naked singularity in the spacetime, then it is not possible to predict anything about the future evolution of spacetime.

Cosmic Censorship Hypothesis (CCH): CCH states that no naked singularity is formed due to the collapse of physically realistic distribution of matter. In 1969, Roger Penrose proposed this hypothesis. The failure of CCH indicates that an observer receive information from a singularity. In other words, an observer can observe the particles created near a singularity.

A circumstance arises when the matter is compelled to be compacted to a point. This is known as **a space-like singularity**.

Space-like singularities are the feature of nonrotating uncharged black holes.

A situation arises when certain light rays come from a region with infinite curvature. This is known as a **time-like singularity**.

Usually, time-like singularities occur in the exact solutions of charged or rotating black holes .

10.17 Penrose Process

In 1969, Penrose proposed an amazing result that energy can be taken out from a black hole with an ergosphere. It was believed that no material body or even a light ray can come out from a black hole. Now, we will discuss the mechanism proposed by Penrose.

Consider a test particle with rest mass m_0 and four momentum

$$p^\mu = m_0 \frac{dx^\mu}{d\tau} = m_0 U^\mu,$$

is moving towards a Kerr black hole along a geodesic. First, we consider the conserved quantities associated with the Killing vectors

$$K^\mu = \partial_t \text{ and } R^\mu = \partial_\phi.$$

[here the metric coefficients of Kerr metric do not depend on t and ϕ]

We recall Kerr metric in $(-,+,+,+)$ signature

$$ds^2 = -\left(1 - \frac{2mr}{R^2}\right) dt^2 - \frac{4mar\sin^2\theta}{R^2} dt d\phi + \frac{R^2}{\triangle} dr^2 + R^2 d\theta^2 + \frac{\Sigma^2 \sin^2\theta}{R^2} d\phi^2.$$

For a test particle of the above four momentum, we can have two conserved quantities, the energy (E) and angular momentum (L) of the particle as

$$E = -K_\mu p^\mu = m_0 \left(1 - \frac{2mr}{R^2}\right)\frac{dt}{d\tau} + \frac{2m_0 mar}{R^2}\sin^2\theta\frac{d\phi}{d\tau}$$

and

$$L = R_\mu p^\mu = -\frac{2m_0 mar}{R^2}\sin^2\theta\frac{dt}{d\tau} + \frac{m_0(r^2+a^2)^2 - m_0 \triangle a^2 \sin^2\theta}{R^2}\sin^2\theta\frac{d\phi}{d\tau}.$$

As the energy is positive, so we have used minus sign in the definition of E. Since both K^μ and p^μ are time-like, their inner product is negative. However, K^μ is space-like inside the ergosphere region ($r_+ < r < r_{s+}$), therefore a particle with negative energy can exist in the ergosphere region.

$$E = -K_\mu p^\mu < 0.$$

The particles outside the stationary limit surfaces should have positive energies. Particles with negative energy would never be able to escape from this region. To escape from this ergosphere region, it needs to accelerate until its energy is positive. The method in which negative energy particles created in the ergosphere, are used to extract positive energy from a Kerr black hole is called the **Penrose process**.

Suppose a particle with the four momentum $p^{(0)\mu}$ and total positive energy

$$E_0 = -K_\mu p^{(0)\mu}$$

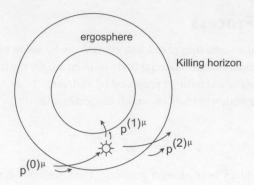

Figure 98   Penrose process.

comes from far away and enters the ergosphere of the Kerr black hole. After entering the ergosphere, region it decays into two new particles with four momenta $p^{(1)\mu}$ and $p^{(2)\mu}$ (see Fig. 98). We know for a freely falling particle total energy, E_0, is constant.

By local conservation of momentum, we have

$$p^{(0)\mu} = p^{(1)\mu} + p^{(2)\mu}.$$

Contracting with Killing vector K_μ yields

$$E^{(0)} = E^{(1)} + E^{(2)}.$$

One of the new particles falls into the horizon with negative energy $E^{(1)}$ since K_μ is space-like within the ergosphere, whereas the other escapes to infinity (i.e., free motion or geodesics) with larger energy, $E^{(2)}$, than that of the original energy, $E^{(0)}$, of the falling object since the energy is conserved. Thus, the particle has returned with more energy ($E^{(0)} + |E^{(1)}|$) than it entered with $E^{(0)}$. Thus, some of the energy of a rotating black hole can be taken out through the above Penrose process and the mass of the black hole inevitability can be diminished to $M - |E^{(1)}|$. In other words, the energy $|E^{(1)}|$ has been taken away from the black hole.

Now, a natural question arises: how much energy can be taken out from a Kerr black hole through the Penrose process?

Penrose energy extraction process is self-preventive. Particles entering the ergosphere region, extract energy from the black hole through the Penrose process, also carry negative angular momentum that is opposite to the black hole. That means, Penrose process extricates energy from the rotating black hole by diminishing its angular momentum $J = ma$ up to zero, whereas the mass of the black hole, m is still finite. Note that ergosphere will disappear when $J = 0$, and then the energy extraction process will stop.

To see this limit on energy extraction, we define a new Killing vector χ^μ as

$$\chi^\mu = K^\mu + \Omega_H R^\mu.$$

The vector χ^μ is future-directed null on the outer horizon and tangent to the horizon (χ^μ is time-like just outside the event horizon). $\Omega_H = \frac{a}{r_+^2 + a^2}$ is the angular velocity on the horizon. If a particle with momentum $p^{(1)\mu}$ crosses the event horizon, then

$$p^{(1)\mu} \chi_\mu < 0,$$

$$\Rightarrow p^{(1)\mu}(K_\mu + \Omega_H R_\mu) < 0,$$

$$\Rightarrow -E^{(1)} + L^{(1)}\Omega_H < 0,$$

$$\Rightarrow L^{(1)} < \frac{E^{(1)}}{\Omega_H}. \tag{10.53}$$

[$L^{(1)} = p^{(1)\mu} R_\mu$ is the conserved angular momentum.]

If the particle's energy ($E^{(1)}$) is negative, then the particle must have a negative angular momentum ($L^{(1)}$) since Ω_H is positive and its motion is opposite to the black hole's rotation. Therefore, the black hole loses its angular momentum during the Penrose process.

Note that after the black hole guzzles a particle of energy ($E^{(1)}$) and angular momentum ($L^{(1)}$), it should settle down to a Kerr solution with parameters adapted by

$$\delta m = E^{(1)}, \ \delta J = L^{(1)}.$$

Thus, Eq. (10.53) yields a restriction on how much a particle can decrease the angular momentum as

$$\delta J < \frac{\delta m}{\Omega_H}. \tag{10.54}$$

Although Penrose process is to extricate energy from the black hole, it does not violate the area theorem, i.e., in this process it is not possible to reduce mass of the black hole below the initial value M_{irr}, where M_{irr} is the irreducible mass. We know the area of the event horizon is nondecreasing. In Penrose process, the mass as well as momentum decrease and its combined effect may cause the area to increase. To confirm this, the area of the outer horizon, which is located at

$$r_+ = m + \sqrt{m^2 - a^2},$$

could be calculated.

On the horizon, the induced metric γ_{ij} is given by

$$\gamma_{ij}dx^i dx^j = ds^2(dt = 0, dr = 0, r = r_+, \ i,j = \theta, \ \phi)$$

$$= (r_+^2 + a^2 \cos^2 \theta)d\theta^2 + \left[\frac{(r_+^2 + a^2)^2 \sin^2 \theta}{r_+^2 + a^2 \cos^2 \theta}\right] d\phi^2.$$

Hence, the horizon area will be the integral of the induced volume element as

$$A = \int \sqrt{|\gamma|} d\theta d\phi = \int \sqrt{g_{\theta\theta} g_{\phi\phi}} d\theta d\phi$$

$$= 4\pi(r_+^2 + a^2).$$

Here, $|\gamma| = (r_+^2 + a^2)^2 \sin^2 \theta.$

The irreducible mass of the black hole is defined by

$$M_{irr}^2 = \frac{A}{16\pi} = \frac{1}{4}(r_+^2 + a^2)$$

$$= \frac{1}{2}(m^2 + \sqrt{m^4 - (am)^2}) = \frac{1}{2}(m^2 + \sqrt{m^2 - J^2}). \tag{10.55}$$

This result yields

$$m^2 = M_{irr}^2 + \frac{J^2}{4M_{irr}^2} \geq M_{irr}^2.$$

Hence, in Penrose process it is impossible to diminish mass of the black hole below the initial value of the irreducible mass M_{irr}.

Now, to find the effect of M_{irr} due to the change in mass or angular momentum, we calculate the variation of M_{irr} as

$$\delta M_{irr} = \frac{1}{4M_{irr}} \frac{a}{\sqrt{m^2 - a^2}} (\Omega_H^{-1} \delta m - \delta J). \tag{10.56}$$

The restriction (10.54) indicates that

$$\delta M_{irr} > 0.$$

The irreducible mass always increases, i.e., it can never be reduced.

Now we can find the maximum amount of extraction of energy from a Kerr black hole before the black hole comes to stop its rotation as

$$m - M_{irr} = m - \frac{1}{\sqrt{2}}(m^2 + \sqrt{m^4 - J^2})^{\frac{1}{2}}.$$

The irreducibility M_{irr} implies that the area $A(= 16\pi M_{irr}^2)$ can never decrease. Eq. (10.56) yields

$$\delta A = 8\pi \frac{a}{\Omega_H \sqrt{m^2 - a^2}} (\delta m - \Omega_H \delta J),$$

$$\Rightarrow \delta m = \frac{\kappa}{8\pi} \delta A + \Omega_H \delta J, \tag{10.57}$$

where

$$\kappa = \frac{\sqrt{m^2 - a^2}}{2m(m + \sqrt{m^2 - a^2})}, \tag{10.58}$$

is known as **surface gravity of Kerr solution**.

Eq. (10.57) provides a clue to think about a correspondence between black hole and thermodynamics. The first law of thermodynamics can be stated as

$$dE = TdS - pdV,$$

where T is the temperature, S is the entropy, p is the pressure, and V is the volume, so that pdV represents work done by the system. It is obvious that the term $\Omega_H \delta J$ in (10.57) indicates the work done by the black hole when a particle is entering it. Hence, people started thinking of establishing black hole thermodynamics by identifying the thermodynamical quantities and black hole's characteristics as

energy $\leftrightarrow$ *mass*; *entropy* $\leftrightarrow$ *area*; *temperature* $\leftrightarrow$ *surface gravity*.

Note 10. 9

Hawking discovered the relation between the entropy of the black hole and the area of the event horizon as

$$S = \left(\frac{kc^3}{G\hbar} \right) \frac{A}{4}.$$

10.18 The Laws of Black Hole Thermodynamics

It is argued that four laws of black hole mechanics are the physical properties of the black hole. Some renowned scientists, B. Carter, S. Hawking, J. Bardeen, and J. Bekenstein discovered that these laws are very similar to the laws of thermodynamics. These laws of black hole mechanics are demonstrated through the following names:

ZEROTH LAW

For a stationary black hole, horizon has constant surface gravity.

[An object experiences gravitational acceleration at its surface, which is known as **surface gravity**.]

THE FIRST LAW

The following expression is known as the first law of black hole thermodynamics:

$$dm = \frac{\kappa}{8\pi} dA + \Omega dJ + \Phi dQ,$$

where m is the mass, A is the horizon area, Ω is the angular velocity, Φ is the electrostatic potential, κ is the surface gravity, and Q is the electric charge.

THE SECOND LAW

The horizon area is nondecreasing function of time, $\frac{dA}{dt} \geq 0$.

THE THIRD LAW

It is impossible to construct a black hole with vanishing surface gravity, i.e., we never achieve $\kappa = 0$.

Now we will discuss very briefly the laws of black hole thermodynamics for Kerr solution.
For Kerr black hole, the horizon area A is given by

$$A = 4\pi(r_+^2 + a^2) = 8\pi m \left\{ m + \left(m^2 - \frac{J^2}{m^2} \right)^{\frac{1}{2}} \right\}.$$

Plugging G and c in the above equation, we get

$$A = \left(\frac{8\pi Gm}{c^2} \right) \left\{ \frac{Gm}{c^2} + \left(\frac{G^2 m^2}{c^4} - \frac{J^2}{m^2 c^2} \right)^{\frac{1}{2}} \right\}. \tag{10.59}$$

This indicates that area of a black hole horizon never decreases, which obeys the second law of black hole thermodynamics.

Now, differential of (10.59) yields

$$d(mc^2) = \left(\frac{\alpha c^2}{8\pi G} \right) dA + \Omega dJ = \left(\frac{\alpha \hbar}{2\pi ck} \right) dS + \Omega dJ, \tag{10.60}$$

where

$$\alpha = \frac{\left(G^2 m^2 - \frac{J^2 c^2}{m^2} \right)}{\left[\left(\frac{2Gm}{c^2} \right) \left\{ \frac{Gm}{c^2} + \left(\frac{G^2 m^2}{c^4} - \frac{J^2}{m^2 c^2} \right)^{\frac{1}{2}} \right\} \right]},$$

$$\Omega = \frac{\frac{J}{m}}{\left[\left(\frac{2Gm}{c^2} \right) \left\{ \frac{Gm}{c^2} + \left(\frac{G^2 m^2}{c^4} - \frac{J^2}{m^2 c^2} \right)^{\frac{1}{2}} \right\} \right]},$$

$$S = \left(\frac{kc^3}{G\hbar} \right) \frac{A}{4}.$$

Now, comparing (10.60) with first law of thermodynamics

$$dU = TdS - pdV,$$

we find the expression for the temperature T as

$$T = \frac{\alpha \hbar}{2\pi kc}. \tag{10.61}$$

Note that α remains the same everywhere on its horizon, therefore, temperature is uniform. In other words, horizon has constant surface gravity, and this is the zeroth law of black hole thermodynamics.

Here, the temperature, i.e., surface gravity will be zero when $m = a$. This is not possible as one needs infinite number of operations to reach the extreme black hole. In other words, it is impossible to construct a black hole with vanishing surface gravity and this is the zeroth law of black hole thermodynamics.

Note 10.10

An outline of the mathematical calculations of generating rotating black hole solutions from a general spherically symmetric black hole solution.

Step 1: Consider a general spherically symmetric metric

$$ds^2 = f(r)dt^2 - \frac{dr^2}{g(r)} - h(r)(d\theta^2 + \sin^2\theta d\phi^2).$$

Step 2: Write down the above metric in the advance null (Eddington–Finkelstein) coordinates (u, r, θ, ϕ) using the transformation

$$du = dt - \frac{dr}{\sqrt{fg}}.$$

The above metric in the null coordinates converts

$$ds^2 = f(r)du^2 + 2\sqrt{\frac{f}{g}}dudr - h(r)(d\theta^2 + \sin^2\theta d\phi^2).$$

Step 3: Now, express the inverse metric $g^{\mu\nu}$ using a null tetrad $Z^{\mu}_{\alpha} = (l^{\mu}, n^{\mu}, m^{\mu}, \overline{m}^{\mu})$ in the form

$$g^{\mu\nu} = l^{\mu}n^{\nu} + l^{\nu}n^{\mu} - m^{\mu}\overline{m}^{\nu} - m^{\nu}\overline{m}^{\mu},$$

where $\bar{m}^{\mu}$ is the complex conjugate of m^{μ}, and the tetrad vectors satisfy the relations

$$l_{\mu}l^{\mu} = n_{\mu}n^{\mu} = m_{\mu}m^{\mu} = l_{\mu}m^{\mu} = n_{\mu}m^{\mu} = 0, \ l_{\mu}n^{\mu} = m_{\mu}\bar{m}^{\mu} = -1.$$

Here, the tetrad vectors satisfying the above relations are given by

$$l^{\mu} = \delta^{\mu}_r, n^{\mu} = \sqrt{\frac{g}{f}}\delta^{\mu}_u - \frac{g}{2}\delta^{\mu}_r, m^{\mu} = \frac{1}{\sqrt{2h}}\left(\delta^{\mu}_{\theta} + \frac{i}{\sin\theta}\delta^{\mu}_{\phi}\right).$$

Step 4: Let us consider a complex transformation in the $r - u$ plane given by

$$r \to r' = r + ia\cos\theta, u \to u' = u - ia\cos\theta,$$

along with the complexification of the metric functions $f(r)$, $g(r)$, and $h(r)$. Subsequently, in the complex transformation, the new tetrad vectors convert as

$$l'^\mu = \delta_r^\mu, \ n'^\mu = \sqrt{\frac{G(r,\theta)}{F(r,\theta)}}\delta_u^\mu - \frac{G(r,\theta)}{2}\delta_r^\mu, \ m'^\mu = \frac{1}{\sqrt{2H(r,\theta)}}\left(ia\sin\theta(\delta_u^\mu - \delta_r^\mu) + \delta_\theta^\mu + \frac{i}{\sin\theta}\delta_\phi^\mu\right),$$

where $F(r,\theta), G(r,\theta)$, and $H(r,\theta)$ are, respectively, the complexified form of $f(r)$, $g(r)$, and $h(r)$.

Step 5: Using the new tetrad, we find the new inverse metric in the advance null coordinates as

$$ds^2 = Fdu^2 + 2\sqrt{\frac{F}{G}}dudr + 2a\sin^2\theta\left(\sqrt{\frac{F}{G}} - F\right)dud\phi - 2a\sqrt{\frac{F}{G}}\sin^2\theta drd\phi$$

$$- Hd\theta^2 - \sin^2\theta\left[H + a^2\sin^2\theta\left(2\sqrt{\frac{F}{G}} - F\right)\right]d\phi^2.$$

Step 6: Now to write down the above metric in Boyer–Lindquist coordinates (where only the off-diagonal term is $g_{t\phi}$), we continue to execute the succeeding global coordinate transformation

$$du = dt' + \chi_1(r)dr, \quad d\phi = d\phi' + \chi_2(r)dr.$$

Implanting the above coordinate transformations in the metric and setting $g_{t'r}$ and $g_{r\phi'}$ to zero, we get

$$\chi_1(r) = -\frac{\sqrt{\frac{G(r,\theta)}{F(r,\theta)}}H(r,\theta) + a^2\sin^2\theta}{G(r,\theta)H(r,\theta) + a^2\sin^2\theta}, \quad \chi_2(r) = -\frac{a}{G(r,\theta)H(r,\theta) + a^2\sin^2\theta}.$$

Remember that the above transformation is possible only when χ_1 and χ_2 depend only on r. Note that if $\chi_1(r)$ and $\chi_2(r)$ depend on θ also, then we cannot perform the above global coordinate transformation. Note that all complexification of the functions need not provide χ_1 and χ_2 that are independent on θ. One needs to guess suitably. In fact, there are several methods to complexify. One possible complexification is as follows:

$$\frac{1}{r} \to \frac{1}{2}\left(\frac{1}{r'} + \frac{1}{\bar{r}'}\right) = \frac{r}{\rho^2}, r^2 \to r'\bar{r}' = \rho^2,$$

where $\rho^2 = r^2 + a^2\cos^2\theta$. Finally, the above global coordinate transformation yields the metric in the Boyer–Lindquist coordinate as

$$ds^2 = Fdt^2 + 2a\sin^2\theta\left(\sqrt{\frac{F}{G}} - F\right)dtd\phi - \frac{H}{GH + a^2\sin^2\theta}dr^2 - Hd\theta^2$$

$$- \sin^2\theta\left[H + a^2\sin^2\theta\left(2\sqrt{\frac{F}{G}} - F\right)\right]d\phi^2,$$

where we have dropped the prime sign from t' and ϕ'.

Elementary Cosmology

11.1 Introduction

Cosmology is the study of the universe as a whole. That means cosmology deals with the structure and evolution of the universe on the largest scales of space and time. Gravity plays the central role for the structure of the universe on these scales and regulates its evolution. Thus, general relativity is the main part of cosmological study and cosmology is one of the most important applications of general relativity. The universe consists of stars, gases, collections of matter (with zero rest mass or massive particles) with known and unknown characters and vacuum energy.

Note 11.1

Zero rest mass particles (photons, gravitons, etc.) are referred to as **radiation** and nonzero rest mass particles (protons, neutrons, electrons, etc.) are called **matter**. It is argued that protons and neutrons are built through more fundamental particles dubbed as **quarks**. The general terms for particles made up of these quarks are **baryons**. The neutrinos are extremely weak interacting particles, which are created, e.g., in radioactive decay. Neutrinos with very small masses behave approximately like radiation in some circumstances and matter in others.

The visible matter in the universe is mostly contained in galaxies—gravitationally bound collections of stars, gas, and dust. A typical galaxy has about 10^{11} stars and a total mass of 10^{12} times of solar mass, i.e., $= 10^{12} M_\odot$. In the observable universe, very roughly 10^{11} galaxies are seen. Assuming the visible matter in the galaxies is smoothed and uniformly distributed over the largest scales, then at the present day the density is approximately estimated as

$$\rho_{visible}(t_0) \ \sim \ 10^{-31} \ gm/cm^3,$$

which is equivalent to one proton per cubic meter.

Apart from galaxies, the universe comprises of radiation consisting of zero rest mass particles—photons and also some Newtonian and gravitational waves. The velocity of this radiation is similar to the velocity of light. The radiation is not gravitationally bound, clustered; that moves up the galaxies. Detected electromagnetic radiation that has been left-over from the hot big bang, with the greatest energy density, is the **cosmic background radiation**. To remarkable experimental precision, this is very similar to the black body spectrum with the temperature of $2.725 \pm .001 \ K$. This result (closest to the black body spectrum) provides the evidence for a big bang. As the peak of this spectrum lies in the microwave band, so this radiation is termed as the **cosmic microwave background radiation (CMBR)**. Similar to black body radiation, the density of the CMBR with a temperature of 2.735 K is

$$\rho_r(t_0) \sim 10^{-34} \; gm/cm^3.$$

In recent time, it is argued that the energy density in this background radiation is much less than the average density of the matter in the galaxies.

Note 11.2

Penzias and Wilson discovered CMBR in 1965. This supports the big bang.

In recent times, some experimental evidence indicates that most of the mass in the universe is neither luminous matter in the galaxies nor radiation, detected so far. This mass can be identified by its gravitational effect, even if it cannot be observed directly. This mass is dubbed **dark matter**. The simplest indication of this dark matter comes from weighing spiral galaxies.

[A **spiral galaxy** is a disc of stars and dust rotating about a central nucleus]

To measure the amount of mass in the universe, we compute the motion of the galaxies and apply Newton's law of gravity. A detailed plot of the orbital speed of the galaxy, i.e., by measuring the Doppler shifts in 21 cm line of neutral hydrogen, the velocities of clouds of this gas in the disc can be mapped as a function of distance r from the center of the galaxy, reveals the distribution of mass within the galaxy. The velocity $V(r)$ at radius r is related to the mass $M(r)$ up to radius r is

$$\frac{GM(r)}{r^2} = \frac{V^2(r)}{r}.$$

Outside of a radius that contains most of the mass, it is expected that $V(r)$ should vary as $r^{-\frac{1}{2}}$. But this is not seen. Rather, in almost all cases, $V(r)$ remains more or less a constant. This indicates that, even in the outer region of the galaxy, $M(r)$ is increasing with r, in fact, $M(r) \propto r$. In fact, almost every galaxy contains a halo of dark, unseen matter, which is 10 times more than the mass seen in visible light.

There are many indications (observational evidences) that the visible matter and detectable radiation contain only one third of the mass of the universe. To know the exact nature of missing mass is an active area of research in cosmology. The remaining two third of the density of the universe must be in some smooth unclustered form, dubbed **dark energy**. In 1998, two groups (Riess et al., 1998, Perlmuter et al., 1999) by observing supernovae reported direct evidence for dark energy. The evidence is based on the difference between the luminosity distance in a universe dominated by dark matter and one dominated by dark energy.

Note 11.3

Riess and Perlmuter got noble prize in 2010 for discovering accelerating phase of the universe.

Note 11.4

The standard distance unit in cosmology is **parsec** (pc).

$$1\ pc = 3.09 \times 10^{18}\ cm = 3.26\ light\ year.$$

However, the range of scales is such that it is useful to deal with the **astronomical unit** (AU), the distance from the sun to the earth, which is 4.85×10^{-6} pc, **microparsec** (μpc) $= 10^{-6}$ pc , **milliparsec** $= 10^{-3}$ pc, **kiloparsec** (kpc) $= 10^{3}$ pc, **megaparsec** (Mpc) $= 10^{6}$ pc, and **gigaparsec** (Gpc) $= 10^{9}$ pc.

Distance to the nearest star $= 1$ pc (Proxima Centauri).

Distance to the Andromeda from the earth is 2,538,000 light years $= 778.5$ kpc (Andromeda galaxy is the nearest to the Milky Way, i.e., our galaxy).

Distance to the nearest large cluster (the Virgo cluster of several thousand galaxies) $= 20$ Mpc

Distance to the edge of the visible universe $= 14$ Gpc.

11.2 Homogeneity and Isotropy

Optical and radio surveys of the sky confirm that the distribution of galaxies is the same in all directions. Also, the precise consistency of the CMBR confirms that, indeed, the universe is isotropic about our position on a large scale.

Based on the observations cosmologists introduced the following two postulates:

Weyl postulate: Let the galaxies be treated as particles (or points). The Weyl postulate states that the world lines of galaxies are recognized with a specific class of observers, known as fundamental observers, have a common point of intersection, form a bundle (or congruence) of geodesics. One can also describe a common time coordinate, which quantifies the proper time of each such observer.

Alternative: The Weyl postulate states that if one designates the world lines of galaxies as fundamental observers, then these world lines form a three bundle of nonintersecting geodesics orthogonal to a series of space-like hypersurfaces.

Now, we are interested in describing the Weyl postulate in terms of coordinate and metric of spacetime. Let us assume that galaxies, i.e., the particles are moving along nonintersecting world lines a, b, c, ... which have no irregularities (see Fig. 99). We can also use x^{μ} ($\mu = 1, 2, 3$) to label a typical world line in the three bundles of galaxy world lines. Here, the three coordinates x^{μ} are space-like. Further, let the coordinate x^0 as time coordinate measures the proper time along each curve, $x^{\mu} = constant$. Hence it is obvious that $x^0 = constant$ is a usual spacelike hypersurface, which is orthogonal to the usual world line characterized by $x^{\mu} = constant$. It is always possible to find a continuum from a discrete set of points (here galaxies) through smooth fluid approximation. Here we use four coordinate x^i, $i = 0, 1, 2, 3$ to describe space and time and in terms of these coordinates, we define the metric tensor as g_{ik}, where the line element is

$$ds^2 = g_{ik}dx^i dx^k.$$

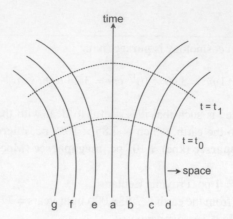

Figure 99 Particles are moving along nonintersecting geodesics.

According to Weyl postulate, the orthogonality condition yields

$$g_{0\mu} = 0.$$

Also, the line x^μ = constant is a geodesic.

We know the geodesic equations as

$$\frac{d^2 x^i}{ds^2} + \Gamma^i_{kl} \frac{dx^k}{ds} \frac{dx^l}{ds} = 0.$$

Putting the condition, $x^i = constant$, $i = 1, 2, 3$ in this geodesic equations, we get

$$\Gamma^i_{00} = 0, \quad i = 1, 2, 3.$$

The above equation implies

$$\frac{dg_{00}}{dx^i} = 0, \quad i = 1, 2, 3.$$

Hence g_{00} depends only on x^0. Rescaling x^0, we can make g_{00} = constant and without any loss of generality, we can take it as unity, i.e.,

$$g_{00} = 1.$$

Hence, the consequence of Weyl postulate transforms the above line element as

$$ds^2 = dx^{0^2} - g_{\mu\nu} dx^\mu dx^\nu = c^2 dt^2 - g_{\mu\nu} dx^\mu dx^\nu, \qquad (11.1)$$

where $ct = x^0$. This time coordinate is the cosmic time and t is the galaxy's proper time.

Cosmological Principle

This principle states that at any given cosmic time, the universe looks homogeneous and isotropic, i.e., there is no ideal position or direction in the universe.

Homogeneity means that the appearance of the universe is the same at each point, while isotropy implies the universe looks identical in all directions.

Note 11.5

The above two do not necessarily imply one another. As for example, the universe occupied with a uniform magnetic field, is homogeneous, because all points are similar, however, it is not isotropic, as directions alongside the field lines can be dissimilar from those perpendicular to them. Also, if one sees a spherically symmetric distribution from its central point, he will observe it isotropic in nature but not essentially homogeneous.

11.3   Robertson–Walker Metric

On the large scale (more than 100 million light years or so), the universe looks homogeneous and isotropic around us. At this scale, the density of galaxies is approximately the same and all directions from us appear to be equivalent. Using Weyl's postulate one can write the general metric as

$$ds^2 = c^2 dt^2 - h_{ij} dx^i dx^j, \quad (i, j = 1, 2, 3) \tag{11.2}$$

where the h_{ij} are functions of (t, x^1, x^2, x^3).

Now, we try to find the metric when the spacetime of the universe is homogeneous and isotropic.

Let us consider two neighboring galaxies whose coordinates are (x^1, x^2, x^3) and $(x^1 + \Delta x^1, x^2 + \Delta x^2, x^3 + \Delta x^3)$, respectively. The distance between two neighboring galaxies on the same hypersurface $t = $ constant is given by

$$d\sigma^2 = h_{ij} \Delta x^i \Delta x^j. \tag{11.3}$$

Consider three widely separated galaxies A, B, and C shown in Fig. 100 at some time t_i. At a later time, t another triangle $(A'B'C')$ is formed by the same galaxies.

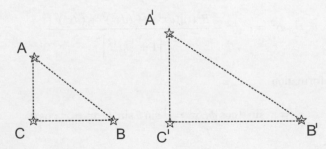

Figure 100   The sides of a triangle are expanded by the same factor.

The postulate of homogeneity and isotropy at all points and directions on a particular hypersurface implies that two triangles must be similar and, also, the increase in length must be independent of the position and direction of the triangle. Three galaxies at A, B, C at time t_i expand away from each other to A^1, B^1, C^1 at time t and the sides of the triangle are expanded by the same factor. Thus, the increase in distance $AB \rightarrow A^1B^1$ be the same as $AC \rightarrow A^1C^1$ (look at the expansion from the point of view of the observer in A). It follows that the expansion is controlled by a single function of time, i.e., the functions h_{ij} must involve the time coordinate t through a common factor $R^2(t)$. Hence the metric (11.2) takes the form

$$ds^2 = c^2 dt^2 - R^2(t) \gamma_{ij} dx^i dx^j, \tag{11.4}$$

where γ_{ij} are the functions of (x^1, x^2, x^3) only.

Now we give our attention to the homogeneous and isotropic three space given by

$$d\sigma^{1^2} = \gamma_{ij} dx^i dx^j. \tag{11.5}$$

This space must be a space of constant curvature (standard theorem of differential geometry).

In such a space, the three-dimensional fourth rank Riemann tensor can be constructed from the three-dimensional metric tensor (11.5)

$$^3R_{ijkl} = k(\gamma_{ik}\gamma_{jl} - \gamma_{il}\gamma_{jk}), \tag{11.6}$$

where k is a constant.

It can be found that the three-dimensional Riemann tensor of the space given by (11.6) has the form (11.5) if γ_{ij} assumes the following form

$$d\sigma^2 = \left(1 + \frac{1}{4}kr^{1^2}\right)^{-2} [(dx^1)^2 + (dx^2)^2 + (dx^3)^2], \tag{11.7}$$

where

$$r^{1^2} = (x^1)^2 + (x^2)^2 + (x^3)^2.$$

Hence the metric (11.4) takes the following form

$$ds^2 = c^2 dt^2 - \frac{R^2(t)[(dx^1)^2 + (dx^2)^2 + (dx^3)^2]}{\left[1 + \frac{1}{4}kr^{1^2}\right]^2}. \tag{11.8}$$

The following transformation

$$x^1 = r^1 \sin\theta \cos\phi, \quad x^2 = r^1 \sin\theta \sin\phi, \quad x^3 = r^1 \cos\theta,$$

yields the following form of Eq. (11.8)

$$ds^2 = c^2 dt^2 - R^2(t) \left[\frac{dr'^2 + r'^2(d\theta^2 + \sin^2\theta d\phi^2)}{\left(1 + \frac{1}{4}kr'^2\right)^2} \right].$$ (11.9)

Finally, the transformation

$$r = \frac{r'}{\left(1 + \frac{1}{4}kr'^2\right)},$$

gives

$$ds^2 = c^2 dt^2 - R^2(t) \left[\frac{dr^2}{1 - kr^2} + r^2(d\theta^2 + \sin^2\theta d\phi^2) \right].$$ (11.10)

This is known as **Robertson–Walker (R–W) metric**.

This is the general metric of the spacetime of the universe if the universe looks isotropic and homogeneous. The metric equation (11.10) discovered independently by Robertson (1936) and Walker (1936) for isotropic and homogeneous spacetime. Here $R(t)$ is called the **scale factor**. It is also called the **curvature radius** of the universe. k is called **curvature parameter**.

$k = -1$ corresponds to the universe with negative spatial curvature, i.e., **open universe**. Here the geometry is hyperbolic.

$k = +1$ corresponds to the universe with positive spatial curvature, i.e., **closed universe**. Here the geometry is spherical.

$k = 0$ corresponds to the **flat universe** with zero curvature. Here, the geometry is Euclidean.

The area of the sphere of coordinate radius r with center $r = 0$ is $4\pi r^2 R^2(t)$.

The proper radius of the sphere from $r = 0$ to $r = r_1$ radius is

$$l = R(t) \int_0^{r_1} \frac{dr}{\sqrt{1 - kr^2}}$$

$$= R(t) \sin^{-1} r_1 > R(t)r_1, \quad for\ k = 1,$$

$$= R(t)r_1, \quad for\ k = 0,$$

$$= R(t) \sinh^{-1} r_1 < R(t)r_1, \quad for\ k = -1.$$

Note 11.6

We can find three-dimensional spatial metric form of Eq. (11.10) as follows:
We know for a space with constant curvature, the Riemannian tensor can be expressed as

$$^3R_{\alpha\beta\sigma\delta} = k(\gamma_{\alpha\sigma}\gamma_{\beta\delta} - \gamma_{\alpha\delta}\gamma_{\beta\sigma}).$$

Here, the constant k is the curvature.

Now, contracting with $\gamma^{\alpha\sigma}$, we get

$$^3R_{\beta\delta} = \gamma^{\alpha\sigma} \, ^3R_{\alpha\beta\sigma\delta} = k\gamma^{\alpha\sigma}(\gamma_{\alpha\sigma}\gamma_{\beta\delta} - \gamma_{\alpha\delta}\gamma_{\beta\sigma}) = k(3\gamma_{\beta\delta} - \gamma_{\beta\delta}) = 2k\gamma_{\beta\delta}. \qquad (i)$$

As we have already considered the three-space is isotropic, therefore, this space should be spherically symmetric about all points. We can write the line element of this three spherically symmetric spacetime as

$$d\sigma^2 = \gamma_{\alpha\beta}dx^\alpha dx^\beta = e^\lambda(r)dr^2 + r^2(d\theta^2 + sin^2\theta d\phi^2).$$

From the above metric, one can easily calculate the nonvanishing components of the Ricci tensor as

$$R_{11} = \lambda'/r, \; R_{22} = cosec^2\theta \, R_{33} = 1 + \frac{1}{2}re^{-\lambda}\lambda' - e^{-\lambda}. \qquad (ii)$$

Equating (i) and (ii), we obtain the following two equations as

$$\lambda'/r = 2ke^\lambda, \; 1 + \frac{1}{2}re^{-\lambda}\lambda' - e^{-\lambda} = 2kr^2.$$

The solution of these equation is

$$e^{-\lambda} = 1 - kr^2.$$

Hence the metric for the three-space of constant curvature can be expressed as

$$d\sigma^2 = \frac{dr^2}{1 - kr^2} + r^2(d\theta^2 + sin^2\theta d\phi^2).$$

11.4   Hubble's Law

In 1929, Edwin Hubble found an interesting result that the wavelength of light from distant galaxies systematically increases. The fractional increase of this wavelength is proportional to the distance D of the galaxy from us. Let λ be the wavelength of light sent out by the galaxy and wavelength of the light received by us be $\lambda + \Delta\lambda = \lambda_0$, then

$$z = \frac{\Delta\lambda}{\lambda} \propto D. \qquad (11.11)$$

The quantity z is called **spectral shift** ($z > 0$ means a redshift, $z < 0$ means a blueshift). Hubble inferred this as a Doppler effect and attributed a velocity of recession $v = cz$ to the source galaxy ($v > 0$ means recession from us, $v < 0$ means approach to us).

From the observation (11.11), Hubble announced the famous result that the speed of the recession of the galaxy is proportional to its distance from us, i.e.,

$$v = HD, \qquad (11.12)$$

where H is the **Hubble's constant**.

The formula for v is **Hubble's law.** The value of Hubble's constant has altered considerably from the beginning due to increases in the accuracy of observations. The currently accepted value of Hubble's constant is

$$H = (50.3 \pm 4.3) \; km \; sec^{-1} \; Mpc^{-1}.$$

Remember: $H^{-1} = 1.8 \times 10^{10}$ years.

11.5 Dynamical Equation of Cosmology

We derive the Einstein equations for the R–W metric, in which the matter exists in the form of the perfect fluid. The energy-momentum tensor for the perfect fluid distribution of matter is taken as

$$T_{ij} = (p + \rho)U_i U_j - p g_{ij}, \tag{11.13}$$

where mass energy density and pressure are ρ and p, respectively, and U_i is the four velocity of the matter given by

$$U_i = \frac{dx^i}{ds},$$

with $x^i(s)$ describing the world line of matter.

We have chosen that we are moving in comoving coordinate and in the comoving frame, the fluid is at rest. Hence,

$$U_i = 0; \; U_0 = 1, \; i = 1, 2, 3.$$

[**Comoving coordianates** are the coordinates that are carried along with the expansion]

From the R–W metric,

$$ds^2 = dt^2 - a^2(t)\left[\frac{dr^2}{1 - kr^2} + r^2 d\theta^2 + r^2 \sin^2\theta d\phi^2\right], \tag{11.14}$$

we have,

$$g_{00} = 1 = g^{00}; \quad g_{11} = -\frac{a^2}{1 - kr^2} = (g^{11})^{-1}$$

$$g_{22} = -a^2 r^2 = (g^{22})^{-1}; \quad g_{33} = \sin^2\theta g_{22} = (g^{33})^{-1}$$

Now, we compute the nonzero Christoffel symbols as

$$\Gamma^0_{ij} = -\frac{1}{2}\frac{\partial}{\partial t}g_{ij}; \quad \Gamma^i_{0j} = \frac{\dot{a}}{a}\delta^i_j$$

$$\Gamma^1_{11} = \frac{kr}{1 - kr^2}; \quad \Gamma^1_{22} = -r(1 - kr^2); \quad \Gamma^1_{33} = \sin^2\theta\Gamma^1_{22}$$

$$\Gamma^2_{12} = \Gamma^3_{13} = \frac{1}{r}; \quad \Gamma^2_{33} = -\frac{1}{2}\sin 2\theta; \quad \Gamma^3_{23} = \cot\theta$$

[over dot implies differentiation with respect to the time t and $i, j = 1, 2, 3$]

Now the components of Ricci tensor are

$$R_0^0 = -\frac{3\ddot{a}}{a}; \; R_j^i = -\frac{1}{a^2}(a\ddot{a} + 2\dot{a}^2 + 2k)\delta_j^i.$$

From this, we calculate,

$$R = g^{ab}R_{ab} = -\frac{6(a\ddot{a} + \dot{a}^2 + k)}{a^2}.$$

The nonzero component of T_{ij} are

$$T_{00} = \rho; \; T_{11} = \frac{pa^2}{1 - kr^2}; \; T_{22} = pr^2a^2; \; T_{33} = pr^2\sin^2\theta a^2.$$

Hence the 00 and 11 components of Einstein Field equation

$$(G_i^j = R_i^j - \frac{1}{2}g_j^iR = 8\pi GT_j^i)$$

are

$$3(\dot{a}^2 + k) = 8\pi G\rho a^2, \tag{11.15}$$

$$2a\ddot{a} + \dot{a}^2 + k = -8\pi Gpa^2. \tag{11.16}$$

[The 22 and 33 components yield equations, which are equivalent to (11.16)]

Note that in the above equations, pressure and density are independent of spatial coordinate. This confirms that the cosmological fluid possesses isotropy and homogeneity.

The conservation equations of energy-momentum tensor $T_{\lambda;\gamma}^\gamma = 0$ yield,

$$\frac{\partial p}{\partial x^\lambda} - \frac{\partial}{\partial x^\gamma}[(p + \rho)U^\gamma U_\lambda] - (p + \rho)\Gamma_{\sigma\gamma}^\gamma U^\sigma U_\lambda + (p + \rho)\Gamma_{\gamma\lambda}^\sigma U^\gamma U_\sigma = 0.$$

Simplifying the above equation, we obtain,

$$\frac{dp}{dx^\lambda} - \frac{d}{dt}[(p + \rho)U_\lambda] - 3(p + \rho)\frac{\dot{a}}{a}U_\lambda = 0.$$

The components $\lambda = 1, 2, 3$ give $0 = 0_j$ and the nontrivial part, corresponding to $\lambda = 0$ is

$$\dot{\rho} + 3(p + \rho)\frac{\dot{a}}{a} = 0. \tag{11.17}$$

Exercise 11.1

Show that Eq. (11.17) is the consequence of the Eqs. (11.15) and (11.16).
Hint: Taking a derivative of the Eq. (11.15) and using Eq. (11.16), one can get Eq. (11.17).

Show that Bianchi identity,

$$G^{\gamma}_{\mu;\gamma} = 0 \Rightarrow T^{\gamma}_{\mu;\gamma} = 0.$$

Hint: Take a derivative of the Eq. (11.15) and use Eq. (11.16).

11.6   Newtonian Cosmology

Let us consider the universe to be an immensely large sphere of gas (that means larger than we can imagine but not infinite). Treating the gas particles as galaxies, i.e., the universe is a huge sphere filled with the gas of galaxies and its volume is very large. Further, we consider that the gaseous sphere is isotropic and homogeneous. An observer or a point which is carried along with the expansion is said to be comoving. As the sphere is isotropic and homogeneous, the expansion is regulated by a single function of time and as a result, we can write the distance between any two comoving points at a time t as

$$r(t) = R(t)r_0, \tag{11.18}$$

where r_0 is a constant for the pair and $R(t)$, called the scale factor, is the universal expansion factor. Differentiating (11.18) with respect to time, we get,

$$v(t) = \dot{r}(t) = H(t)r(t), \tag{11.19}$$

where

$$H(t) = \frac{\dot{R}(t)}{R(t)}. \tag{11.20}$$

$H(t)$ is called the Hubble's parameter. Equation (11.19) is called Hubble's law. Note that H is a function of time.

It is customary to denote its present value by H_0, i.e.,

$$H_0 = \frac{\dot{R}(t_0)}{R(t_0)}, \tag{11.21}$$

where t_0 is the present moment.

Hubble's law is consistent with the observation that all other galaxies are moving away from us. This indicates that the distance between two galaxies is increasing with the time that means the velocity of separation v is a function of time. Let at the present time the separation distance be r, then there must have been a time τ in the past when the distance between them was very small. Thus, according to Eq. (11.19), we have

$$\tau = \frac{r}{v} = \frac{1}{H_0}. \tag{11.22}$$

The quantity $\frac{1}{H_0}$ is known as Hubble time, which is a rough estimate of the time when the expansion of the universe began. In other words, $\frac{1}{H_0}$ can be considered as the scale for the **age of the universe**.

Now, suppose that a particle is separated from us at time t by a distance r. Its equation of motion according to Newton's second law is

$$\ddot{R}(t) = -\frac{4\pi}{3}G\rho(t)R(t),\tag{11.23}$$

where $\rho(t)$ is the density of the matter at time t.

$$\left[-\frac{GMm}{r^2} = P = mf = m\ddot{r}\right.$$

$$\Rightarrow \ddot{r} = -\frac{G}{r^2}\frac{4}{3}\pi r^3 \rho,$$

$$\Rightarrow \ddot{r} = -\frac{4}{3}\pi r G\rho,$$

$$\left.\Rightarrow r_0\ddot{R}(t) = -\frac{4}{3}\pi r_0 R(t)G\rho\right].$$

Let $\rho(t_0)$ be the density at some typical time t_0, then the conservation of matter implies

$$\rho(t_0)R^3(t_0) = \rho(t)R^3(t).\tag{11.24}$$

Equations (11.23) and (11.24) imply

$$R^2\ddot{R} + \frac{4\pi}{3}G\rho(t_0)R^3(t_0) = 0.\tag{11.25}$$

This equation implies we never get $\ddot{R} = \dot{R} = 0$ unless $\rho = 0$. That means **Newtonian cosmology rules out a static universe.** So, the universe must expand (or contract). This result agrees with the general theory of relativity, which came into existence before Hubble actually discovered the expansion of the universe.

Integrating of Eq. (11.25) yields

$$\dot{R}^2 = \frac{8\pi}{3}G\rho(t)R^2 - k,\tag{11.26}$$

where k is the constant of integration (the choice of minus sign is conditional and is a measure of total energy of the particle).

Also, we note that the general theory of relativity (i.e., Einstein equation for R–W metric) leads to the same Eq. (11.15) for the scale factor.

11.7   Cosmological Redshift

The observed wavelengths of the spectral lines from a star are not the same as the original wavelengths of the spectral lines of the star. The lines are shifted to the red or blue due to the relative velocity between the earth and the star. If the star is approaching the earth then we get blue-shift and if the star is receding then one gets redshift. We will discuss how the shifted spectral lines are related to the scale factor.

Consider a distant galaxy situated at a point whose coordinates are (r_1, θ_1, ϕ_1). It emits a light ray that propagates and reaches us $(r = 0)$. Light ray travels along a null geodesic. Without any loss of generality, we consider that the path of the light lies on the plane $(\theta = \theta_1, \phi = \phi_1)$. Suppose the present epoch is denoted by $t = t_0$ and let a light ray leave the source at $t = t_1$. For null geodesic, we have $ds = 0$. Now using $d\theta = 0$, $d\phi = 0$, the R–W metric yields the following condition for the ray to arrive at $r = 0$ at $t = t_0$

$$\int_{t_1}^{t_0} \frac{c\,dt}{a(t)} = \int_0^{r_1} \frac{dr}{(1 - kr^2)^{\frac{1}{2}}}. \tag{11.27}$$

[In the null geodesics (*with* $ds = 0, d\theta = d\phi = 0$),

$$\frac{c\,dt}{a(t)} = \pm \frac{dr}{(1 - kr^2)^{\frac{1}{2}}},$$

we should take minus sign in this relation as r decreases as t increases along this null geodesic]

Light wave starts at $r = r_1$ and reaches us at $r = 0$. Let two successive crests of the wave leave at t_1 and $t_1 + \Delta t_1$ and arrive at t_0 and $t_0 + \Delta t_0$, respectively. Equation (11.27) yields

$$\int_{t_1+\Delta t_1}^{t_0+\Delta t_0} \frac{c\,dt}{a(t)} = \int_0^{r_1} \frac{dr}{\sqrt{1 - kr^2}} = \int_{t_1}^{t_0} \frac{c\,dt}{a(t)}. \tag{11.28}$$

If one assumes $a(t)$ as a slowly changing function, then it necessarily remains unaffected over the small intervals Δt_0 and Δt_1. Thus, we get from (11.28)

$$\frac{c\Delta t_0}{a(t_0)} - \frac{c\Delta t_1}{a(t_1)} = 0. \tag{11.29}$$

$$\left[\text{let } \frac{1}{a(t)} = \frac{dZ(t)}{dt}, \text{ then } \int_{t_1+\Delta t_1}^{t_0+\Delta t_0} \frac{c\,dt}{a(t)} = c[Z(t)]_{t_1+\Delta t_1}^{t_0+\Delta t_0} = cZ(t_0 + \Delta t_0) - cZ(t_1 + \Delta t_1) \right.$$

$$\left. = cZ(t_0) + c\Delta t_0 \frac{dZ}{dt}\Big|_{(t_0)} - cZ(t_1) - c\Delta t_1 \frac{dZ}{dt}\Big|_{(t_1)} = cZ(t_0) + c\frac{\Delta t_0}{a(t_0)} - cZ(t_1) - c\frac{\Delta t_1}{a(t_1)} \right]$$

Here t coordinate is the proper time of the fundamental observer, therefore, we have $\lambda_0 = c\Delta t_0 =$ wavelength measured by the observer and $\lambda_1 = c\Delta t_1 =$ wavelength at the source. One can define the redshift as,

$$z = \frac{\lambda_0 - \lambda_1}{\lambda_1},$$

therefore, we get,

$$1 + z = \frac{\lambda_0}{\lambda_1} = \frac{\nu_1}{\nu_0} = \frac{\Delta t_0}{\Delta t_1} = \frac{a(t_0)}{a(t_1)}. \tag{11.30}$$

Thus, the wavelength of the light wave will be increased by a fraction z in the transmission from galaxy to us if $a(t_0) > a(t_1)$. Hence, Hubble's interpretations of redshift are clarified if one assumes the scale factor $a(t)$ to be an increasing function of time. One refers to the present redshift as a **cosmological redshift**. Thus, the redshift z of a distant galaxy provides information about the amount by which the scale factor has expanded since the light was emitted. Eq. (11.30) indicates that in general, we can write,

$$1 + z = \frac{a(t_0)}{a(t)}, \tag{11.31}$$

i.e., let a light ray leaving the source at any arbitrary time t and received by us in the present epoch be denoted by $t = t_0$. Now, differentiating above equation with respect to t, we get

$$\frac{dz}{dt} = -\frac{a_0}{a^2}\frac{da}{dt} = -\frac{a_0 H}{a},$$

or

$$\frac{dt}{a(t)} = -\frac{dz}{a_0 H(z)}. \tag{11.32}$$

Note 11.7

One will have blue-shift for a contracting universe (i.e., in that case $a(t_0) < a(t_1)$).

11.8 Derivation of Hubble's Law

Consider a light ray propagation from a distant galaxy at (r_1, θ_1, ϕ_1) towards $r = 0$. The equations of a null geodesic imply that this light ray moves along the path $\theta = \theta_1, \phi = \phi_1$. Suppose the present epoch is denoted by $t = t_0$ and let a right ray leave the source at $t = t_1$. Then the condition for the ray to reach at $r = 0$, at $t = t_0$

$$\int_{t_1}^{t_0} \frac{c\,dt}{a(t)} = \int_0^{r_1} \frac{dr}{\sqrt{1 - kr^2}} = f(r_1). \tag{11.33}$$

Assuming that r_1 is small for nearby objects, we then get approximately,

$$f(r_1) \cong r_1 \cong \frac{c(t_0 - t_1)}{a(t_0)}. \tag{11.34}$$

Now Taylor's expansion near t_0,

$$a(t_1) \cong a(t_0) + (t_1 - t_0)\dot{a}(t_0) = a(t_0)\left[1 - (t_0 - t_1)\frac{\dot{a}(t_0)}{a(t_0)}\right],$$

$$= a(t_0)[1 - (t_0 - t_1)H_0] \ where \ H_0 = \frac{\dot{a}(t_0)}{a(t_0)}.$$

Also, we know

$$(1 + z)^{-1} = \frac{a(t_1)}{a(t_0)} \cong 1 - (t_0 - t_1)H_0.$$

For, small redshift z, we have

$$1 - z \cong 1 - (t_0 - t_1)H_0.$$

This implies

$$cz \cong r_1 a(t_0)H_0 = D_1 H_0, \tag{11.35}$$

$$where \ D_1 = r_1 a(t_0), \tag{11.36}$$

may be defined as a proper distance at the epoch t_0.

From a Doppler shift point of view, cz may be identified with the velocity of recession of a galaxy in proportion to its distance from us. This is Hubble's law and H_0 is the Hubble's constant given by

$$H_0 = \frac{\dot{a}(t_0)}{a(t_0)}.$$

11.9 Angular Size

The angular measurement describing how a large sphere or circle looks from a given point of view is the angular diameter or angular size. In Euclidean geometry, the diameter d is related to the observed angle $\Delta\theta$ as

$$\Delta\theta = \frac{d}{r}, \tag{11.37}$$

where r is its distance. However, for curved spacetime, one needs to use R–W spacetime.

Let us consider a galaxy G_1 having linear extend $d(\overline{AB})$ and the angle subtended by this galaxy G_1 at the observer O is $\Delta\theta_1$ (see Fig. 101). Consider two neighboring null geodesics (representing light rays) from the two points A, B at the two extremities of G_1. Without any loss of generality, we can select the coordinates of A and B as (θ_1, ϕ_1) and $(\theta_1 + \Delta\theta_1, \phi_1)$, respectively. For the curved space, we use R–W line element to find the proper distance between A and B. Now, plugging $t = t_1 =$ constant,

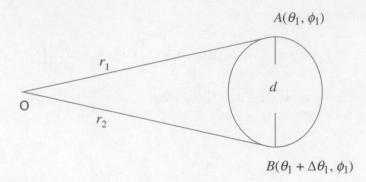

Figure 101 Galaxy has linear extend $d(\overline{AB})$.

$r = r_1 =$ constant, $\phi = \phi_1 =$ constant, and $d\theta = \Delta\theta_1$ in the R–W line element, we get

$$ds^2 = -r_1^2 a^2(t_1)(\Delta\theta_1)^2 = -d^2.$$

[in G_1, the space-like separation $AB = d$, which is a rest frame]

Thus,

$$\Delta\theta_1 = \frac{d}{r_1 a(t_1)} = \frac{d(1+z)}{r_1 a(t_0)}, \quad (using \; Eq. \; (11.30))$$

$$= \frac{d(1+z)}{D_1}. \quad [using \; Eq. \; (11.36)] \tag{11.38}$$

Note that in Euclidean space, $\Delta\theta_1$ is decreasing with an increase in the distance of the galaxy from us. However, in curved space, the result is different. For expanding universe, the scale factor $a(t)$ is increasing with time and consequently, z will be increasing (by Eq. (11.31)). Therefore, it is not always true that $\Delta\theta_1$ decreases with time. Light from a galaxy takes much time to reach us for expanding universe.

11.10 Number Count

The number of discrete luminous sources, i.e., galaxies in a certain region in the sky is governed by the redshift of the galaxies. When one observes galaxies then he sees them as they were at time t when they emitted light.

Let $n(t)$ denote the proper density of the galaxies in the sky. We try to find the number of galaxies observed in a given survey with redshift less than z. From Eq. (11.31), we have

$$1 + z = \frac{a(t_0)}{a(t)},$$

and the corresponding value of r can be obtained as

$$\int_t^{t_0} \frac{cdt}{a(t)} = \int_0^r \frac{dr}{\sqrt{1 - kr^2}}.$$

Consider a shell of radii r and $r + dr$. Therefore, galaxies occupy within the solid angle $d\omega$ with this spherical shell has the following proper volume at time t as

$$dV = \frac{a^3(t)r^2}{\sqrt{1 - kr^2}} d\omega dr,$$

where $d\omega = \sin\theta d\theta d\phi$. Thus, the proper three-volume of this shell at t is given by

$$dV = 4\pi r^2 \frac{dr}{\sqrt{1 - kr^2}} a^3(t).$$

Since $n(t)$ is the proper density of the galaxies, therefore, the number of galaxies in this volume is obtained by multiplying this with $n(t)$

$$dN = dVn(t) = \frac{4\pi r^2 dr}{\sqrt{1 - kr^2}} n(t)a^3(t). \tag{11.39}$$

Hence, the total number of galaxies with redshifts up to z is given by

$$N(z) = \int_0^r \frac{4\pi r^2}{\sqrt{1 - kr^2}} a^3(t)n(t)dr. \tag{11.40}$$

Note that t is related to r and r is related to z through the above equations. Also notice that if there is no new creation of galaxies up to r distances, then one can consider $n(t) = $ constant. Then one can easily find the value of $N(z)$, which depends on the parameter k. For flat universe with $k = 0$, $N(z)\alpha r^3$. For closed universe with $k = 1$, $N(z)$ increases faster than the flat universe, however, for an open universe with $k = -1$, $N(z)$ decreases faster than flat universe.

From, Eq. (11.39), it is obvious that

$$\frac{dN}{dz} = n\frac{dV}{dz}.$$

Let the number of galaxies be conserved, then

$$na^3(t) = n_0 a^3(t_0).$$

This implies

$$n = n_0(1 + z)^3. \quad (by \ (11.31)) \tag{11.41}$$

[n_0 is the number of galaxies at present time]

Hence the number of galaxies ΔN within the redshift range Δz at redshift z can be obtained as

$$\Delta N = \left(\frac{dN}{dz}\right)\Delta z = n_0(1+z)^3\left(\frac{dV}{dz}\right)\Delta z. \tag{11.42}$$

11.11 Luminosity Distance

To know the characteristics of distant stars, one can measure the redshift of light emitted due to the expansion of the universe and this redshift z is associated to the scale factor $a(t)$ as

$$\frac{\lambda_0}{\lambda} = 1 + z = \frac{a(t_0)}{a(t)}.$$

Here, subscript 0 means the value at present.

The luminosity distance is defined as the distance at which a star would lie based on its observed luminosity. It is actually related to the amount of light received from the star. The star's actual luminosity can be determined by using the inverse-square law. However, the luminosity distance is different from the actual distance because the real universe is not flat; as it is expanding the inverse square law fails to give the exact distance measurement. Let L be the luminosity of the star (i.e., the energy emitted per unit time) and F be the measured flux density (i.e., energy received by the observer, per unit time per unit area, from the star) then, the luminosity distance D is defined by the relation

$$4\pi D^2 = \frac{L}{F}. \tag{11.43}$$

If the space is static, then the luminosity distance $D = r_1 a(t_0)$ may be defined as a proper distance at the present epoch t_0 (see Eq. 11.36). Then Eq. (11.43) yields the simple relation

$$F = \frac{L}{4\pi a_0^2 r_1^2}.$$

Since the universe is expanding, it affects the observations in two ways namely, each individual photon loses energy due to cosmological redshift and secondly due to change of distance, the photons reach the observer fewer than expected. These two affect the flux received, which is proportional to (1+z) in each case and hence allowing these two factors for the flux received, we estimate

$$F = \frac{L}{4\pi a_0^2 r_1^2 (1+z)^2}. \tag{11.44}$$

Equations (11.43) and (11.44) yield

$$D = a(t_0)r_1(1+z). \tag{11.45}$$

The luminosity distance depends on what kind of Friedmann–Walker universe we live in. In the following, we will show that luminosity distance actually depends on redshift and the present values of Hubble parameter, $H_0 = \frac{\dot{a}(t_0)}{a(t_0)}$ and deceleration parameter, $q_0 = \frac{-a(t_0)\ddot{a}(t_0)}{\dot{a}^2(t_0)}$.

From Eqs. (11.31), (11.33), and (11.34), we have

$$\int_{t_1}^{t_0} \frac{dt}{a(t)} = \int_0^{r_1} \frac{dr}{\sqrt{1 - kr^2}} = f(r_1) \cong r_1, \quad 1 + z = \frac{a(t_0)}{a(t_1)}.$$

[assumed that r_1 is small for nearby objects and $c = 1$]

Let, $\frac{Z(t)}{dt} = \dot{Z} = \frac{1}{a(t)}$, then the left-hand side of the above integral yields

$$\int_{t_1}^{t_0} \frac{dt}{a(t)} = Z(t_0) - Z(t_1) = Z(t_0) - Z(t_1 - t_0 + t_0),$$

$$= Z(t_0) - Z(t_0) - (t_1 - t_0)\dot{Z}(t_0) - \frac{1}{2}(t_1 - t_0)^2\ddot{Z}(t_0) - \dots$$

$$= (t_0 - t_1)\frac{1}{a(t_0)} + (t_0 - t_1)^2\frac{\dot{a}(t_0)}{2a^2(t_0)} + \dots$$

Thus,

$$r_1 a(t_0) = (t_0 - t_1) + \frac{1}{2}(t_0 - t_1)^2 H(t_0) + \dots \tag{11.46}$$

Now,

$$z = \frac{a(t_0)}{a(t_1)} - 1 = a(t_0)\left[\frac{1}{a(t_1 - t_0 + t_0)}\right] - 1.$$

Taylor's expansion near t_0 yields

$$z = a(t_0)\left[\frac{1}{a(t_0)} + (t_1 - t_0)\left(\frac{1}{a}\right)_{t_0}^{\cdot} + \frac{(t_1 - t_0)^2}{2}\left(\frac{1}{a}\right)_{t_0}^{\cdot\cdot} + \dots\right] - 1,$$

$$= a(t_0)\left[\frac{1}{a(t_0)} + (t_1 - t_0)\left(\frac{-\dot{a}(t_0)}{a^2(t_0)}\right) + \frac{(t_1 - t_0)^2}{2}\left(\frac{-\ddot{a}(t_0)}{a^2(t_0)} + \frac{2\dot{a}^2(t_0)}{a^3(t_0)}\right) + \dots\right] - 1$$

$$or, \ z = (t_0 - t_1)H_0 + (t_0 - t_1)^2 H_0^2\left(\frac{q_0}{2} + 1\right) + \dots$$

Now expressing $(t_0 - t_1)H_0$ (which is small) in terms of z, we get

$$(t_0 - t_1)H_0 = z - \left(\frac{q_0}{2} + 1\right)z^2 + \dots \tag{11.47}$$

Now, substituting $(t_0 - t_1)$ in Eq. (11.46), we obtain

$$r_1 a(t_0) = \frac{1}{H_0}\left[z - \frac{z^2}{2}(q_0 + 1) + \ldots\right]. \tag{11.48}$$

Hence from Eq. (11.45), one can get the luminosity

$$D = (1 + z)r_1 a(t_0) = \frac{(1 + z)}{H_0}\left[z - \frac{z^2}{2}(q_0 + 1) + \ldots\right],$$

or

$$D = \frac{1}{H_0}\left[z + \frac{z^2}{2}(1 - q_0) + \ldots\right]. \tag{11.49}$$

Note that this result is model independent, i.e., the result is true for all models.

11.12 Olbers' Paradox

German astronomer Heinrich Wilhelm Olbers (1758–1840) raised an interesting query that can help us understand a fundamental property of the universe.

Why is the sky dark at night?

The question is usually called **Olbers' paradox.**

According to his calculations where he assumed that the universe is static, infinite, eternal, and uniformly filled with stars, the sky should be infinitely bright.

Let us assume that the universe is unchanging and infinite in size and the stars fill the universe uniformly and all the stars are alike and have a luminosity L and the number density of stars per unit volume is n, which is constant through the whole space. Then the number of stars in the portion of a spherical shell of radius r, thickness dr, and solid angle dw will be

$$r^2 nL\,dr\,dw.$$

According to the inverse square law, these stars contribute the flux as

$$\frac{r^2 nL\,dr\,dw}{4\pi r^2}.$$

Then the total intensity of the light from the stars will be

$$I = \int_0^\infty \left(\frac{4\pi nL}{4\pi}\right) dr = \infty.$$

This indicates that the whole sky will be a great flash of light in all directions. However, the sky looks dark at night. This contradiction is known as **Olbers' paradox.**

Olbers tried to express a possible resolution by postulating the existence of an interstellar medium that absorbs the light from stars, which is responsible for the large value of the intensity of the light from the stars. This resolution is not justified. There were also a few other possible explanations that were proposed before Hubble discovered that the universe is expanding, namely,

   (i) Existence of clouds of dust that create an obstacle to seeing the distant stars.

   (ii) There are only a finite number of stars that are present in our universe.

   (iii) Stars are irregularly distributed so that many stars lie behind one another so that only a finite angular area is subtended by them.

   (iv) The universe is young as a result of light from distant stars yet to come.

These resolutions are also not properly justified.

Resolution of Olbers' Paradox: In 1929, Hubble discovered that the universe is expanding and this expanding universe hypothesis provides a beautiful resolution of the Olbers' paradox.

Let $n(t_1)$ be the proper density of the galaxies, therefore, the number of galaxies in the shell $(r_1 \leq r \leq r + dr_1)$ is (by (11.39))

$$dN = dVn(t_1) = \frac{4\pi r_1^2 dr_1}{\sqrt{1 - kr_1^2}} n(t_1)a_1^3(t).$$

Since the total radiation received per unit area per unit time is

$$\frac{L}{4\pi D_1^2(1 + z_1)^2} = \frac{L}{4\pi a^2(t_0)r_1^2(1 + z_1)^2},$$

therefore, the stars within the shell provide the flux, which received at $r = 0$ is

$$\frac{L}{4\pi a^2(t_0)r_1^2(1 + z_1)^2} \cdot \frac{4\pi r_1^2 a^3(t_1)n(t_1)dr_1}{\sqrt{1 - kr_1^2}},$$

$$= \frac{L}{a^2(t_0)} \cdot \frac{a^3(t_1)n(t_1)}{(1 + z_1)^2} \cdot \frac{dr_1}{\sqrt{1 - kr_1^2}}.$$

Then the total intensity of the light from the stars will be

$$I = \int_0^\infty \frac{L}{a^2(t_0)} \cdot \frac{a^3(t_1)n(t_1)}{(1 + z_1)^2} \cdot \frac{dr_1}{\sqrt{1 - kr_1^2}}. \qquad (11.50)$$

Consider the models of the universe where no new creation of stars take place at any stage, i.e.,

$$a^3(t_1)n(t_1) = constant.$$

Stars lying at a very large distance cause the factor $(1 + z_1)^2$ to be sufficiently large so that the above integral (11.50) converges. In other words, the expanding cosmological models make the integral (11.50) convergent and this leads to the finite brightness of the sky. Truly, the convergence is very quick so that the expected brightness of the sky is very small. This confirms the observed darkness of the night sky.

11.13 Friedmann Cosmological Models

The universe comprises of stars, galaxies, cluster of galaxies, and superclusters and the distribution of these objects is inhomogeneous compared on a galactic scale. However, for sufficiently large scale ($\sim$10 Mpc or more) it appears to be homogeneous. For modeling of the universe, we assume that the universe is homogeneous and isotropic and, also, for an ideal situation, we consider the universe to be filled with the perfect fluid distribution of matter. As in Eq. (11.13), we can write the energy-momentum tensor for perfect fluid distribution of matter as

$$T^{ik} = (p + \rho)U^i U^k - pg^{ik},$$

with $U^i = (1, 0, 0, 0)$ being the fluid velocity vector [Weyl postulate indicates that all galaxies are expected to have this velocity vector].

Here the components of the energy-momentum tensor are given by

$$T^0_0 = \rho, T^1_1 = T^2_2 = T^3_3 = -p.$$

Now, Einstein field equations for R–W metric are

$$2\frac{\ddot{a}}{a} + \frac{\dot{a}^2 + k}{a^2} = -8\pi Gp, \tag{11.51}$$

$$\frac{3(\dot{a}^2 + k)}{a^2} = 8\pi G\rho. \tag{11.52}$$

The conservation law $T^{ik}_{;k} = 0$ implies

$$\dot{\rho} + 3(p + \rho)\frac{\dot{a}}{a} = 0. \tag{11.53}$$

Equations (11.51) and (11.52) yield

$$\frac{\ddot{a}}{a} = -\frac{4\pi G}{3}(\rho + 3p). \tag{11.54}$$

This is known as the **acceleration equation**. Note that for normal matter both p and ρ are positive and as a result, $\rho + 3p > 0$ (i.e., matter distribution satisfies the strong energy condition). This implies that $\ddot{a} < 0$. Since at present ($t = t_0$) we get redshift, therefore, $a(t_0) > 0$ as well as $\frac{\dot{a}(t_0)}{a(t_0)} > 0$. These yield the curve $a(t)$ with respect to time, which is concave downward in nature. This nature of the curve indicates that $a(t) = 0$ at finite time in past and one can refer this time as $t = 0$, i.e., $a(0) = 0$. Usually, this time is referred to the beginning of the universe, which is known as the **big bang**.

To get a physically viable model of the universe, one will have to solve the Einstein field equations (11.51) and (11.52). We have, actually in hand, two independent equations as conservation Eq. (11.53) is not an independent equation (Bianchi identity leads to the conservation equation). Note that these two independent equations contain three unknowns, namely a, ρ, p. So to get exact solutions, one needs to introduce a relation between the unknowns. Most commonly, the equation of state $p = p(\rho)$ is used to get exact solutions.

For further study, we choose the equation of state as

$$p = m\rho. \tag{11.55}$$

For the above equation of state, Eq. (11.53) implies

$$\rho \propto a^{-3(1+m)}. \tag{11.56}$$

This indicates that the energy density in matter dominant universe falls off as $\rho \propto a^{-3(1+m)}$.

For different values of the equation of state parameter m, we get different cosmological models of the universe.

Case I: $m = \frac{1}{3}$

When the cosmological fluid is **dominated by radiation**, then the equation of state can be taken as

$$p = \frac{1}{3}\rho. \tag{11.57}$$

Here most of the energy density of the universe is due to radiation and the energy density in radiation falls off as

$$\rho \propto a^{-4}. \tag{11.58}$$

Case II: $m = 0$

When the cosmological fluid is **dust**, then the equation of state parameter will be zero, i.e., $p = 0$. Here the fluid pressure is negligible as compared to energy density ρ. The energy density due to dust falls off as

$$\rho \propto a^{-3}. \tag{11.59}$$

Note that the energy density of radiation falls off faster than the dust-filled universe. The entire era for which $p = 0$ and $\rho \neq 0$ is known as **matter dominant era**.

Case III: $m = -1$

For **vacuum energy**, $m = -1$, i.e.,

$$p = -\rho. \tag{11.60}$$

Equation (11.53) yields

$$\rho = constant. \tag{11.61}$$

Thus, the vacuum energy is constant, i.e., it remains same at all times. Note that dust and radiation energy densities are decreasing with the expansion of the universe but vacuum energy remains constant and, therefore, will dominate the universe.

Observationally, it is found that the present universe is dominated by nonrelativistic matter, i.e., $p << \rho$. In other words, the later stage of the evolution, the dominant constituent of the universe is pressureless dust. Therefore, in the following, we will consider the models of dust-filled universe.

11.14 Dust Model

For dust-filled universe, $p = 0$. Eq. (11.53) yields

$$\frac{8\pi G\rho}{3} = \frac{A}{a^3}. \tag{11.62}$$

[A = integration constant]
Equation (11.51) implies

$$\dot{a}^2 = -k + \frac{A}{a}. \tag{11.63}$$

Now we consider three different cases for different values of curvature parameter k.

(i) Einstein–de Sitter model: $(k = 0)$

$k = 0$ corresponds to the flat universe with zero curvature. The model with $k = 0$ is often called the Einstein–de Sitter flat model.
Equation (11.63) yields

$$\dot{a}^2 = \frac{A}{a} \implies a = a_0 \left(\frac{3H_0 t}{2} \right)^{2/3}. \tag{11.64}$$

Here $A = a_0^3 H_0^2$ and a_0 and H_0 are the present values of scale factor and Hubble constant, respectively,

$$a_0 = a(t_0), \quad \left(\frac{\dot{a}}{a} \right)_{t=t_0} = H_0.$$

Now, we can find the present epoch, which is given by

$$t = t_0 = \frac{2}{3} H_0^{-1}. \tag{11.65}$$

This present epoch t_0 is the **age of the universe**.

One can define the Hubble parameter at any epoch t by $H(t)$ as

$$H(t) = \frac{\dot{a}}{a} = \frac{2}{3t}. \tag{11.66}$$

We also deduce the energy density of the universe

$$\rho = \frac{3H^2(t)}{8\pi G} = \frac{1}{6\pi G t^2}. \tag{11.67}$$

Thus, for Einstein–de Sitter model

$$a(t) \propto t^{2/3}, \quad \rho(t) \propto \frac{1}{t^2}.$$

At the epoch $t = 0$, the scale factor vanishes, i.e., at this epoch $t = 0$, the spacetime becomes singular [here, $R_{ijkm}R^{ijkm}$, $R_{ik}R^{ik}$, etc., diverge as $t \longrightarrow 0$].

It is assumed that the universe came into existence at, $t = 0$, i.e., the epoch $t = 0$ is the beginning of the universe. According to the present estimate of $H_0 \approx 50$ (in the units used earlier), the age of the universe (according to Eq. (11.65)) is approximately 13.3 billion years.

The deceleration parameter q is found as

$$q = -\frac{\ddot{a}/a}{(\dot{a}/a)^2} = \frac{1}{2}. \tag{11.68}$$

Thus, deceleration parameter $q(t)$ remains fixed at the value $q = \frac{1}{2}$.

(ii) Closed elliptic model ($k = 1$)

$k = +1$ corresponds to the universe with positive spatial curvature.

Now, Eq. (11.63) yields

$$\dot{a}^2 = -1 + \frac{A}{a}. \tag{11.69}$$

This equation implies that $\dot{a} = 0$ at $a = A$. This indicates that the universe expands up to $a = A$ and then starts contracting to $a = 0$. Now, we can write the Eqs. (11.51) and (11.52) in terms of H and q as

$$(1 - 2q)H^2 + \frac{1}{a^2} = 0, \tag{11.70}$$

$$H^2 + \frac{1}{a^2} = \frac{8\pi G\rho}{3}. \tag{11.71}$$

Equation (11.70) indicates that we always get deceleration parameter, which is greater than $\frac{1}{2}$, i.e.,

$$q > \frac{1}{2}.$$

The Eqs. (11.70) and (11.71) yield

$$\rho = \frac{3H^2}{4\pi G}q. \tag{11.72}$$

Equation (11.69) $\Longrightarrow A = (1 + \dot{a}_0{}^2)a_0$ at present epoch t_0, i.e.,

$$A = \frac{2q_0 H_0^{-1}}{(2q_0 - 1)^{3/2}}. \tag{11.73}$$

$$[A = a_0(1 + \dot{a}_0{}^2) = a_0(a_0^2 H_0^2 + 1) = a_0 + a_0^3 H_0^2$$

$$= \frac{1}{\sqrt{2q_0 - 1}H_0} + \frac{H_0^2}{(2q_o - 1)^{3/2}H_0^3} = \frac{2q_0}{H_0(2q_0 - 1)^{3/2}} = \frac{2q_0 a_0}{2q_0 - 1}]$$

$$[a_0^{-2} = (2q_0 - 1)H_0^2 \Rightarrow a_0 = \frac{1}{\sqrt{(2q_0 - 1)}H_0}]$$

Equation (11.69) can be written as

$$\frac{\dot{a}^2}{a_0^2} = H_0^2 \left[1 - 2q_0 + \frac{2q_0 a_0}{a} \right]. \tag{11.74}$$

$$[(11.69) \Longrightarrow \frac{\dot{a}^2}{a_0^2} = -\frac{1}{a_0^2} + \frac{A}{a_0^2 a} = (1 - 2q_0)H_0^2 + \frac{(\frac{1}{a_0^2} + H_0^2)a_0}{a}$$

$$= (1 - 2q_0)H_0^2 + \frac{[(2q_0 - 1)H_0^2 + H_0^2]a_0}{a} = (1 - 2q_0)H_0^2 + \frac{2q_0 a_0 H_0^2}{a}.]$$

From Eq. (11.74), we obtain

$$t_0 = H_0^{-1} \int_0^{a_0} \left[1 - 2q_0 + \frac{2q_0 a_0}{a} \right]^{-\frac{1}{2}} \frac{da}{a_0}. \tag{11.75}$$

Here we choose the limits 0 and a_0 ($t = 0$ is the initial value of t for which $a(t) = 0$). The present age of the universe is taken as $t = t_0$ and value of $a(t)$ at $t = t_0$, i.e., a_0 is taken as the upper limit of the above integral.

The solution of $a(t)$ from the above indefinite form of the integral can be written as

$$a(t) = \frac{q_0 a_0}{2q_0 - 1}(1 - \cos\theta), \tag{11.76}$$

and

$$tH_0 = q_0(2q_0 - 1)^{-\frac{3}{2}}(\theta - \sin\theta). \tag{11.77}$$

Now, using the integration limits, we finally get,

$$t_0 = H_0^{-1} \left[\frac{q_0}{(2q_0 - 1)^{3/2}} \left\{ \sin^{-1} \left(\frac{q_0 - 1}{q_0} \right) + \frac{\pi}{2} \right\} - \frac{1}{(2q_0 - 1)} \right]. \qquad (11.78)$$

Note that for different values of q_0, one can get the different present age of the universe, e.g.,

$$q_0 = 1, t_0 = \left(\frac{\pi}{2} - 1 \right) H_0^{-1} \cong 0.57 H_0^{-1},$$

$$q_0 = \frac{2}{3}, t_0 \cong 0.615 H_0^{-1}.$$

Equation (11.76) indicates the function $a(t)$ evolves from $a = 0$ to $a = \frac{2q_0 a_0}{2q_0 - 1} = A$ and then turns back to $a = 0$ after the elapse of a time interval

$$T = \frac{2\pi q_0}{H_0} (2q_0 - 1)^{-\frac{3}{2}}.$$

Here $a \longrightarrow 0, \rho \longrightarrow \infty$ at the singularity. Such a state occurs twice in this model at $\theta = 0$ and $\theta = 2\pi$. This indicates that it is a closed model of the universe.

We can also find the deceleration parameter

$$q(t) = (1 + \cos \theta)^{-1}.$$

$$[(11.62) \Rightarrow 8\pi G\rho = \frac{3A}{a^3}; \ (11.70) \Rightarrow H^2 = \frac{1}{a^2(2q-1)}; \ (11.71) \Rightarrow \frac{8\pi G\rho}{3} = 2qH^2$$

$$\textit{eliminating } \rho, \textit{ we get, } \frac{1}{2q} = 1 - \frac{a}{A}; \textit{ use } a(t) = \frac{A}{2}(1 - \cos \theta) \textit{ by } (11.76)]$$

For the period of one round θ varies from 0 to 2π, $q(t)$ increases from $\frac{1}{2}$ to ∞ and then falls to $\frac{1}{2}$ again.

(iii) The open hyperbolic model ($k = -1$):

$k = -1$ corresponds to the universe with negative spatial curvature, i.e., **open universe**. Here the geometry is hyperbolic.

Equation (11.63) yields

$$\dot{a}^2 = 1 + \frac{A}{a}. \qquad (11.79)$$

Now, we can write the Eqs. (11.51) and (11.52) in terms of H and q as

$$(1 - 2q)H^2 - \frac{1}{a^2} = 0, \qquad (11.80)$$

$$H^2 - \frac{1}{a^2} = \frac{8\pi G\rho}{3}. \qquad (11.81)$$

Equation (11.80) indicates $q < \frac{1}{2}$. As before,

$$A = \frac{2q_0 H_0^{-1}}{(1 - 2q_0)^{3/2}}.$$

Equation (11.79) can be written as

$$\frac{\dot{a}^2}{a_0^2} = H_0^2 \left(1 - 2q_0 + \frac{2q_0 a_0}{a} \right), \tag{11.82}$$

or

$$t_0 = H_0^{-1} \int_0^{a_0} \left[1 - 2q_0 + \frac{2q_0 a_0}{a} \right]^{-\frac{1}{2}} \frac{da}{a_0}. \tag{11.83}$$

The solution of $a(t)$ from the above indefinite form of the integral can be written as

$$a(t) = \frac{q_0 a_0}{1 - 2q_0} (\cosh U - 1) \tag{11.84}$$

and

$$H_0 t = q_0 (1 - 2q_0)^{-3/2} (\sinh U - U). \tag{11.85}$$

Now, using the integration limits, we finally get,

$$t_0 = H_0^{-1} \left[\frac{1}{1 - 2q_0} - \frac{q_0}{(1 - 2q_0)^{3/2}} \ln \left(\frac{1 - q_0}{q_0} + \frac{(1 - 2q_0)^{1/2}}{q_0} \right) \right]. \tag{11.86}$$

When $U = 0$, we have $t = 0$ and the scalar factor $a(t) = 0$. However, for $U \longrightarrow \infty$, one can see $t \longrightarrow \infty$ and the scale factor is extremely large, i.e., $a(t) \longrightarrow \infty$. Here, no turning point occurs during the expansion as $a(t)$ has no extreme point.

For $q_0 = 0$, $t_0 = H_0^{-1}$. Note that the age of the universe, t_0 increases with the decrease of q_0. It has a maximum value H_0^{-1}.

We can also find the deceleration parameter as before

$$q(t) = (1 + \cosh U)^{-1}.$$

During the evolution, U varies from 0 to ∞, $q(t)$ decreases from $\frac{1}{2}$ to 0 .

Note 11.8

The volume of universe can be calculated using R–W metric as

$$V = a^3 \int_{r=0}^{\infty} \int_{\theta=0}^{\pi} \int_{\phi=0}^{2\pi} \frac{r^2 dr \sin\theta \, d\theta \, d\phi}{(1 - kr^2)^{1/2}}.$$

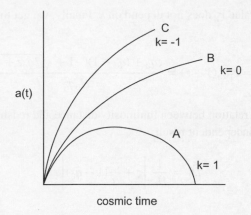

Figure 102 Universe with $k = 0, -1, 1$.

For closed universe with $k = 1$, r must lie between 0 and 1. However, for a flat and open universe with $k = 0, -1$ r lies between 0 and ∞. For a closed universe, we use $r = \sin\psi$, which yields

$$V = 4\pi a^3 \int_0^\pi \sin^2\psi\, d\psi = 2\pi^2 a^3(t).$$

Obviously, it is finite. But for $k = 0, -1$, the integral diverges. Thus for $k = 1$, the universe is said to be closed and for $k = 0, -1$, it is open (see Fig. 102).

Note 11.9

We can also find the luminosity distance to arbitrary large redshifts. We know, a light ray reaches us at time t_0, which was emitted at time t_1 possesses z redshift, then, we can relate this z with scale factor $a(t)$ as

$$a_1 = \frac{a_0}{1+z}.$$

Now, using Eq. (11.27), we have

$$\int_0^{r_1} \frac{dr}{(1-kr^2)^{\frac{1}{2}}} = \int_{t_1}^{t_0} \frac{dt}{a(t)} = \int_{a_1}^{a_0} \frac{da}{a\dot{a}}$$

$$= \frac{1}{a_0 H_0} \int_{(1+z)^{-1}}^{1} \left(1 - 2q_0 + \frac{2q_0}{x}\right)^{-\frac{1}{2}} \frac{dx}{x} \quad [\textit{using } (11.74) \textit{ or } (11.82) \textit{ and } x = \frac{a}{a_0}]$$

Solving this, we get,

$$r_1 = \frac{zq_0 + (q_0 - 1)(-1 + \sqrt{2q_0 z + 1})}{a_0 H_0 q_0^2(1+z)}.$$

One can observe that the value r_1 does not depend on k. Finally, we get the luminosity distance, D_L as

$$D_L = r_1 a_0(1+z) = \frac{zq_0 + (q_0 - 1)(-1 + \sqrt{2q_0 z + 1})}{H_0 q_0^2}.$$

For fixed q_0, we can find a relation between luminosity distance and redshift. Also, we can compare this result with the model independent result

$$D_L = \frac{1}{H_0}\left[z + \frac{z^2}{2}(1 - q_0)\right].$$

11.15 Cosmology with Λ

In 1915, Einstein formulated the general theory of relativity. At that time he, as well as people, believed that the universe was static. Since at that time expansion of the universe had not been discovered, so Einstein wanted to get static solutions of the universe from his field equation. Einstein field equations for R–W metric indicate that the scale factor $a(t)$ can only be constant if

$$\rho = -3p = \frac{3k}{8\pi G a^2}. \tag{11.87}$$

$$[\textit{field equations} \Rightarrow \ddot{a} = \frac{-4\pi G}{3}(\rho + 3p)a \ ; \ 2\frac{\ddot{a}}{a} + \frac{\dot{a}^2 + k}{a^2} = -8\pi Gp;$$

$$\textit{and put } a(t) = a = \textit{ constant}]$$

It is well known that energy density $\rho > 0$, so Eq. (11.87) indicates that the pressure p must be negative. Definitely this is not a realistic solution. Also if one considers $p = 0$, then, $\rho = 0$. This is also not possible. In order to avoid this unrealistic situation, in 1917, Einstein introduced a constant term Λ in his field equation. In the modified field equation, the adding term is known as the **cosmological constant.** The modified field equation is written as

$$R_{ab} - \frac{1}{2}g_{ab}R - \Lambda g_{ab} = 8\pi G T_{ab}. \tag{11.88}$$

Then the field equations for R–W metric can be written explicitly as

$$3(\dot{a}^2 + k) = 8\pi G \rho a^2 + \Lambda a^2, \tag{11.89}$$

$$2a\ddot{a} + \dot{a}^2 + k = -8\pi G p a^2 + \Lambda a^2. \tag{11.90}$$

[Note that $[\Lambda] = L^{-2}$]

11.16   Einstein Static Universe

In 1917, immediately after the discovery of general relativity, Einstein tried to find a static model of the universe from his general theory of relativity but he failed. Then he proposed an outer pressure in the field equation in the form of the cosmological constant. In this case, he had put $a = a_0 = $ constant and neglected pressure. It is obvious that $\dot{a} = 0$, $\ddot{a} = 0$.

Then from Eqs. (11.88) and (11.89), one can easily get

$$\Lambda = \frac{k}{a_0^2}; \quad \frac{3k}{a_0^2} - \Lambda = 8\pi G\rho.$$

Thus to get a feasible solution, one should consider $k = 1$ and $\Lambda > 0$, i.e., the universe with positive spatial curvature. Hence, we get,

$$\Lambda = 4\pi G\rho = \frac{1}{a_0^2}. \tag{11.91}$$

This is Einstein static universe, which is closed, finite in volume. Note that this universe is not bounded. For $t = $ constant, the three volume of the universe is given by

$$V = 2\pi^2 a_0^3 = 2\pi^2 \Lambda^{-3/2}. \tag{11.92}$$

Remember that observations do not support the Einstein static universe. Actually, in Einstein's static model there is matter but there is no motion.

11.17   The de Sitter Universe

Since the Einstein static universe was not feasible from the observational point of view, people were interested in finding a viable solution and soon after the Einstein's proposals, de Sitter found a new solution to the Einstein field equations. He assumed the following conditions

$$k = 0, \rho = 0, p = 0 \ and \ \Lambda > 0.$$

The field Eq. (11.89) yields

$$\dot{a}^2 = \frac{\Lambda}{3}a^2.$$

This implies

$$a \propto \exp\left[\left(\frac{\Lambda}{3}\right)^{\frac{1}{2}} t\right]. \tag{11.93}$$

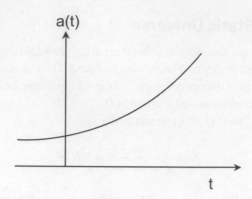

Figure 103 Behavior of $a(t)$ in de Sitter model.

Hence the R–W line element takes the form

$$ds^2 = dt^2 - e^{2Ht}[dr^2 + r^2(d\theta^2 + \sin^2\theta d\phi^2)], \tag{11.94}$$

where $H = \sqrt{(\frac{\Lambda}{3})}$.

Note an interesting property of the metric given in (11.94) is that this is a singularity free metric. That means that the scale factor, $a(t)$ does not vanish for any finite value of t (either past or future). The de Sitter universe given by (11.94) represents an expanding empty universe (see Fig. 103). Actually, in de Sitter model there is motion without matter.

Taking suitable transformations,

$$T = t - \frac{\ln(1 - H^2R^2)}{2H}, \ r = e^{-Ht}R,$$

the line element (11.94) is rewritten in Schwarzschild coordinate R and T as

$$ds^2 = (1 - H^2R^2)dT^2 - \frac{dR^2}{1 - H^2R^2} - R^2(d\theta^2 + \sin^2\theta d\phi^2). \tag{11.95}$$

The event horizon is given by the surface $R = \frac{1}{H}$.

Note 11.10

In general FRW line element with flat spacetime, one can write in Schwarzschild coordinate.

$$ds^2 = dt^2 - e^{2f(t)}[dr^2 + r^2 d\Omega_2^2.]$$

Let us use the transformation

$$R = e^{f(t)}r \ or, \ r = e^{-f(t)}R,$$

therefore,

$$dr = e^{-f(t)}dR - e^{-f(t)}f'(t)Rdt.$$

Hence we get

$$ds^2 = dt^2 - e^{2f(t)}[e^{-f(t)}dR - e^{-f(t)}f'(t)Rdt]^2 - R^2d\Omega_2^2,$$

or

$$ds^2 = [1 - R^2f'^2(t)]dt^2 - dR^2 + 2Rf'(t)dRdt - R^2d\Omega_2^2.$$

Writing,

$$\Phi = 1 - R^2f'^2(t),$$

we get

$$ds^2 = \Phi dt^2 - dR^2 + 2Rf'(t)dRdt - R^2d\Omega_2^2,$$

or

$$ds^2 = \Phi\left[dt + \frac{Rf'(t)dR}{\Phi}\right]^2 - dR^2\left[1 + \frac{R^2f'^2(t)}{\Phi}\right] - R^2d\Omega_2^2.$$

This implies

$$ds^2 = \Phi dT^2 - \left(1 + \frac{1 - \Phi}{\Phi}\right)dR^2 - R^2d\Omega_2^2,$$

where

$$dT = dt + \frac{Rf'(t)dR}{\Phi}.$$

Hence finally we get

$$ds^2 = \Phi dT^2 - \frac{dR^2}{\Phi} - R^2d\Omega_2^2.$$

11.18 Perfect Cosmological Principle

The perfect cosmological principle states that the universe appears to be the same not only at all points and in all directions but in all epochs. This hypothesis leads to a steady state model of the universe. According to the perfect cosmological principle, the observable parameter, Hubble constant, $\frac{\dot{a}}{a}$ must be independent of the present time t_0, i.e., its value remains constant throughout the evolution of the universe. Let us represent the fixed value, H of the Hubble constant. Then we have,

$$\frac{\dot{a}}{a} = H, \ \forall \ t. \tag{11.96}$$

This implies

$$a(t) = a(t_0) \exp[Ht]. \tag{11.97}$$

In this model the deceleration parameter assumes the fixed value as

$$q = -\frac{a\ddot{a}}{\dot{a}^2} = -1. \tag{11.98}$$

Thus, the perfect cosmological principle provides the model of the universe with the following line element

$$ds^2 = dt^2 - a^2(t_0)e^{2Ht}[dr^2 + r^2(d\theta^2 + \sin^2\theta d\phi^2)]. \tag{11.99}$$

11.19 Particle and Event Horizon

Consider a galaxy G_1 at (r_1, θ_1, ϕ_1) emitting light waves towards us at time t_1. Let the light wave arrive at $r = 0$ at time $t = t_0$. Then null geodesic equation $(ds = 0)$ for R–W metric

$$\int_{t_1}^{t_0} \frac{dt}{a(t)} = \int_0^{r_1} \frac{dr}{(1 - kr^2)^{1/2}}.$$

Then redshift as well as luminosity distance are given by

$$z = \frac{a(t_0)}{a(t_1)} - 1, \quad D_1 = r_1 a(t_0)(1 + z).$$

Also one can write the **proper distance** from $r = 0$ to $r = r_1$ as

$$d_p(t) = \int_0^{r_1} \sqrt{g_{rr}} dr = a(t) \int_0^{r_1} \frac{dr}{\sqrt{1 - kr^2}} = a(t) \int_{t_1}^{t_0} \frac{dt}{a(t)}. \tag{11.100}$$

The limit on the proper distance up to which we can observe is called the **particle horizon**. Note that if the t-integral converges, then our vision is restricted by the particle horizon. Let the limiting value of r_1 as $z \to \infty$, be r_l. Hence the limiting proper distance is

$$R_l = a(t) \int_0^{r_l} \frac{dr}{(1 - kr^2)^{1/2}}. \tag{11.101}$$

If one gets a finite value of R_l, then we say that the universe has a particle horizon. It is not possible to see the particles at present with $r_1 > r_l$. Note that the particle horizon creates a barricade to communication from the past.

As above let a galaxy at $r = r_1, t = t_0$ send light signal to an observer at $r = 0$. Suppose t_1 is the time of arrival. Then, we have

$$\int_{t_0}^{t_1} \frac{dt}{a(t)} = \int_0^{r_1} \frac{dr}{(1 - kr^2)^{1/2}}.$$

Suppose the left-hand side of the above integral converges to a finite value as $t_1 \to \infty$. This finite value is achieved by the right-hand side integral for $r_1 = r_H$ (say). Hence it is obvious that for $r_1 > r_H$ the above relation does not hold good. As a result, no signal from $r_1 > r_H$ will come to the observer at r_0. Hence no light from a distant galaxy beyond a proper distance

$$R_H = a(t) \int_{t_0}^{\infty} \frac{dt}{a(t)}, \tag{11.102}$$

will reach the observer at r_0. This limit is known as the **event horizon**. Friedmann models do not possess an event horizon. However, the de Sitter model contains an event horizon at $\frac{1}{H_0^2}$. Note that the event horizon creates a barricade to communication from the future.

11.20 Radiation Model

When the universe is dominated by radiation, then the equation of state can be taken as

$$p = m\rho = \frac{1}{3}\rho,$$

where the equation of state parameter $m = \frac{1}{3}$.

Equations (11.54) and (11.56) yield

$$\frac{\ddot{a}}{a} = -\frac{1 + 3m}{6} A a^{-3(1+m)}.$$

[A = integration constant]

Multiplying both sides by $2\dot{a}$, we get

$$2\dot{a}\ddot{a} = -\frac{1 + 3m}{3} A \dot{a} a^{(-2-3m)}.$$

After integrating, we obtain

$$\dot{a}^2 = \frac{1}{3} A a^{-(1+3m)} + D,$$

where,

$$D = a_0^2 H_0^2 - \frac{1}{3} A a_0^{-(1+3m)}.$$

Hence, putting $m = \frac{1}{3}$, we have

$$\dot{a}^2 = \frac{A}{3a^2} + D.$$

Solving this we get

$$a^2(t) = \frac{1}{D} \left[D^2 t^2 + \sqrt{\frac{4AD^2}{3}} t \right]. \qquad (11.103)$$

This indicates that the universe follows big-bang singularity and is expanding in nature.

This solution is not physically interesting as the present universe is far from radiation dominated, however, the behavior near $t = 0$ is interesting.

Behavior near $t = 0$:
Equation (11.58)$\Longrightarrow$

$$\rho \propto a^{-4}.$$

Thus for small $a(t)$, we are expecting that the $\frac{k}{a^2}$ term is not important compared to $\frac{1}{a^4}$ term. Thus from Eq. (11.52), we can write

$$\frac{\dot{a}^2}{a^2} \cong \frac{8\pi G\rho}{3} = \frac{A}{a^4}. \qquad (11.104)$$

Solution of this equation yields

$$a(t) = (4A)^{\frac{1}{4}} t^{\frac{1}{2}}. \qquad (11.105)$$

The radiation density can be found as

$$\rho = \frac{3}{32\pi G} \frac{1}{t^2}. \qquad (11.106)$$

One can also calculate the radiation temperature T_{rad} as

$$\rho = \sigma_0 T_{rad}^4,$$

where σ_0 is the radiation density constant. Replacing the values of the various physical constants in cgs units, one can get

$$T_{rad} = \left(\frac{3}{32\pi G\sigma_0}\right)^{1/4} t^{-1/2} \cong \frac{1.52 \times 10^{10}}{(t_{sec})^{1/2}} \,\, {}^0K.$$

For example, when the age of the universe was 1 sec, the temperature of the universe was as high as $1.52 \times 10^{10} \, K$.

$$\sigma_0 = \frac{8\pi^5 k^4}{15 c^3 h^3} = 7.5641 X 10^{-15} \, erg \, m^{-3} \, K^{-4}.$$

Boltzmann constant, $k = 1.380662 \times 10^{-16} \, erg \, K^{-1}$, $G = 6.56 X 10^{-8} \, dyne \, cm^2 \, g^{-2}$.

Note that in the case $k = 0$, Eq. (11.52) holds exactly.

11.21 Cosmological Inflation

In 1981, Alan Guth proposed an important concept, known as cosmological inflation, which is a theory of the exponential expansion of space in the early universe. The inflationary scenario occurred around 10^{-36} second after the big bang. Following the inflationary period, there was a phase in which the universe continued to expand much faster (exponentially), than the rate given by standard cosmology. It is argued that in the inflationary phase, the vacuum energy density of the scalar field (V_0) dominates energy density (ρ) grows, i.e., we can use $\rho \approx V_0$. From Eq. (11.52), we have

$$\frac{3(\dot{a}^2 + k)}{a^2} = 8\pi G\rho,$$

$$\Rightarrow \dot{a}^2 = \frac{8\pi G a^2 V_0}{3} - k.$$

In the inflationary phase, the square of the scale factor $a^2(t)$ dominates curvature term, hence

$$\dot{a}^2 = \frac{8\pi G V_0}{3} a^2.$$

Solving this we get,

$$a(t) = a_0 \exp(Ht), \tag{11.107}$$

where, $H = \sqrt{\frac{8\pi G V_0}{3}}$ and a_0 is the value of the scale factor when inflation began. Hence the universe expands exponentially due to inflation and it is often called a period of accelerated expansion as the distance between two fixed points in the universe is increasing exponentially. This new model is interesting as it can resolve some fundamental problems in cosmology like horizon and flatness problems. It is known that the universe is continually expanding or closed depending on whether the density parameter $\Omega(t)(= \frac{\rho(t)}{\rho_c}$, where $\rho_c = \frac{3H^2}{8\pi G}$ is the critical density), $\Omega(t) \leq 1$ or $\Omega(t) > 1$.

From Eq. (11.52), we have

$$\frac{\dot{a}^2}{a^2} = \frac{8\pi G\rho}{3} - \frac{k}{a^2},$$

or

$$H^2 = \frac{8\pi G\rho}{3} - \frac{k}{a^2},$$

or

$$1 = \Omega - \Omega_k, \tag{11.108}$$

where, $\Omega_k = \frac{k}{H^2 a^2}$.

Thus, for $k > 0$, $k = 0$, $k < 0$ implies

$$\Omega > 1, \ \Omega = 1, \ \Omega < 1.$$

It is anticipated that after 10^{-43} sec from big bang, i.e., at the Planck time

$$|\Omega - 1| < 10^{-57}.$$

Therefore, one needs to do some fine tuning and this is known as **flatness problem**. We note from Eq. (11.108) that $k = 0$ implies $\Omega = 1$, i.e., the universe is very much flat.

In inflationary scenario, $a(t) \approx e^{\lambda t}$ and $\rho \approx V_0$, therefore, Eq. (11.52) yields

$$\lambda^2 + \frac{k}{a^2} = \frac{8\pi G V_0}{3}.$$

Note that the first term of left- and right-hand side terms are constants and only the second term of the left-hand side is variable. For consistency, we must take $k = 0$. Thus, inflationary scenario resolves flatness problem.

Cosmological principle indicates that the universe is homogeneous and isotropic. Also, it is argued that the region that evolves into the visible universe would have been causally connected. However, in the early universe, it is possible to use only the physical forces capable of violating causality to account for the homogeneity and isotropy. This is known as the **horizon problem**. In the inflationary scenario, one can resolve this horizon problem by choosing the initial size of the universe to be much smaller than the horizon distance, which inflates and evolves into the present-day visible universe.

11.22 Cosmography Parameters

In recent years, particularly after the discovery of the accelerating universe, people have shown interest in cosmography analysis. It is a part of cosmology with minimal dynamical assumptions. In this scheme, it is assumed that the universe is homogeneous and isotropic on the large scale and no specific dynamical theory is assumed *a priori*. One can expand the scale factor in Taylor series about t_0 as,

$$\frac{a(t)}{a(t_0)} = 1 + H_0(t - t_0) + \frac{1}{2!}q_0 H_0^2(t - t_0)^2 + \frac{1}{3!}j_0 H_0^3(t - t_0)^3 + \frac{1}{4!}s_0 H_0^4(t - t_0)^4 + \frac{1}{5!}l_0 H_0^5(t - t_0)^5$$

$$+ \frac{1}{6!}m_0 H_0^6(t - t_0)^6 + 0(|t - t_0|^7),$$

where t_0 is the present time and the suffix 0 indicates the value of the parameters at t_0. Here, the coefficients of distinct powers of t are designated as,

$$H = \frac{\dot{a}}{a}, q = -\frac{\ddot{a}}{aH^2}, j = \frac{\dddot{a}}{aH^3}, s = \frac{1}{aH^4}\frac{d^4 a}{dt^4}, l = \frac{1}{aH^5}\frac{d^5 a}{dt^5}, m = \frac{1}{aH^6}\frac{d^6 a}{dt^6},$$

and are known as Hubble (H), deceleration (q), jerk (j), snap (s), lerk (l), m-parameters, respectively. Using these parameters, one can find the distance-redshift relations. The redshift, z is defined in terms of the scale factor $a(t)$ as

$$\frac{a}{a_0} = \frac{1}{1 + z}.$$

Note that

$$a \to 0 \implies z \to \infty, \quad a \to \infty \implies z \to -1 \text{ and when } a = a_0, \ z = 0.$$

This means that a increases from 0 to ∞ and z decreases from ∞ to -1. Observing the sign of q, one can predict whether the universe is accelerating or decelerating. $q < 0$ indicates acceleration and $q > 0$ indicates deceleration. Also, the change of sign of jerk parameter j in an expanding model indicates that the acceleration starts increasing or decreasing.

One can express the above cosmography parameters in terms of the deceleration parameter as,

$$j = -\frac{dq}{dx} + q + 2q^2,$$

$$s = \frac{dj}{dx} - (2 + 3q)j,$$

$$l = \frac{ds}{dx} - (3 + 4q)s,$$

and

$$m = \frac{dl}{dx} - (4 + 5q)l,$$

where $x = \ln a$.

From the observational point of view, it is important to express the deceleration parameter in terms of the redshift parameter z as,

$$q(z) = q_0 + (-q_0 - 2q_0^2 + j_0)z + \frac{1}{2}(2q_0 + 8q_0^2 + 8q_0^3 l - 7q_0 j_0 - 4j_0 - s_0)z^2 + 0(z^3).$$

Elementary Astrophysics

12.1 Stellar Structure and Evolution of Stars

> "Twinkle Twinkle little star
> How I wonder what you are!!"

Everyone, whether he or she is literate or illiterate, has a query what the stars are!

Our universe consists of billions of galaxies of various shapes: elliptical, spiral, irregular, etc., and the universe contains 10^{11} galaxies and each galaxy consists of 10^{11} stars. We live in a spiral galaxy visible in the night sky, known as Milky Way. It is convex shaped with diameter, 10^5 l.y. and thickness about 5000 l.y.

The apparent brightness and color serve as the characteristics for identifying a star. All stars do not have the same brightness. In second-century BC, the Greeks had grouped naked eye visible stars into six classes, called the **magnitudes** [brightest, the first magnitude; faintest stars being the sixth magnitude].

All measurements of the brightness and other characteristics of a star are relative, i.e., one has to compare either two stars with each other or a star with an artificial standard source of light. For easy access, people use the sun as standard one, mass: $M_\odot = 1.98892 \times 10^{30}$ kg, diameter: 1,391,000 km, radius: $R_\odot = 695,500$ km, the surface gravity of the sun: 27.94 g, volume of the sun: 1.412×10^{18} km^3, the density of the sun: 1.622×10^5 kg/m^3, luminosity of sun $L_\odot = 3.846 \times 10^{26}$ W.

Before 1840, people used the eye; in 1840-1940, people used the photographic method. From 1940 onwards, people use the photoelectric method (Fig. 104) to measure the brightness (light) of a star. For the latter case, starlight is allowed to fall on the cathode of a photocell and the resulting photocurrent is amplified and recorded on a pen recorder. The deflection is then directly proportional to the intensity.

Two of the basic observable parameters in stars are:

(i) The luminosity L (or absolute magnitude M)

(ii) The effective temperature T_E (or spectrum)

In the years 1911 and 1913, two scientists Hertzsprung and Russell individually observed that the luminosities and surface temperatures of stars are interrelated. This interrelation is generally demonstrated in a two-dimensional diagram, known as **Hertzsprung–Russell or H-R diagram** in which the vertical axis represents the luminosity and the horizontal axis represents the surface temperature (Fig. 105). The stars can be identified by a point with coordinate (T_E, L) on this diagram.

Most stars (about 75 percent) occupy the region on a diagonal running from top left to bottom right. These stars are known as **main sequence stars**. Actually, these stars are hydrogen-burning

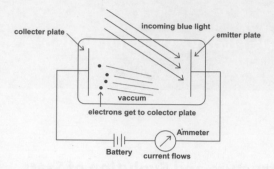

Figure 104 Photoelectric method.

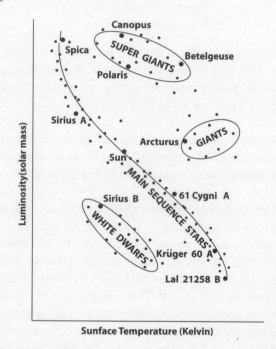

Figure 105 H-R diagram.

stars like the sun. About 1.5 percent large and luminous stars occupy the region in the upper right corner region, known as **giant stars**. It is argued that the hydrogen burning in the core of the stars has been stopped but continues in a thin shell, which moves outward. During the evolution, the core of the stars contracts and heats up, but the outer layers inflate and take a new form of high luminosity and low surface temperature. These stars are known as **red giant stars**. Betelgeuse is the best known example of these stars. The remaining 23.5 percent small and faint stars occupy a region in the lowest region towards the left and are known as **white dwarfs**. These stars are sustained by degenerate electrons. Actually, white dwarfs are the end stage of stellar evolution whose mass is comparable to the sun. The famous white dwarf is Sirius B.

Note 12.1

A usual red giant star with luminosity $L = 1000L_\odot$ and $T_E = 4000K$ has a radius $70R_\odot$, whereas a white dwarf with luminosity $L = L_\odot/100$ and surface temperature $T_E = 16000K$ has the radius $R_\odot/70$ (these are calculated by using $L = 4\pi R^2 \sigma T_E^4$). Actually, the H-R diagram has played an important part in relating the observations with theoretical calculations of stellar evolution, i.e., these diagrams help in the theoretical construction of stellar models. The evolution of a more massive star causes to form a neutron star or a black hole.

In the following, we study the main sequence stars, i.e., hydrogen-burning stars like the sun.

The parameters mass (M), radius (R), luminosity (L), effective temperature (T_E), and chemical compositions vary from star to star but they show certain well-known correlations among them as

$$\left(\frac{L}{L_\odot}\right) = \left(\frac{M}{M_\odot}\right)^{3.5} ; \left(\frac{R}{R_\odot}\right) = \left(\frac{M}{M_\odot}\right)^{0.75} ; \left(\frac{T_E}{T_{E\odot}}\right) = \left(\frac{M}{M_\odot}\right)^{0.5} ; \left(\frac{L}{L_\odot}\right) = \left(\frac{T_E}{T_{E\odot}}\right)^{6.9} ,$$

where $\odot$ represents the corresponding quantities for sun.

Now, to explain the internal structure of stars, we assume the following:

(1) Stars are assumed to be **spherically symmetric and static** (i.e., no rotation, no pulsation, no oscillation, etc.). This indicates that all the physical quantities like density, pressure, temperature, and luminosity will be functions of only one variable r, the distance from the center of the star.

(2) Internal temperatures of the stars are very high so that they must be wholly gaseous. As a result, in steady state, they must be in **hydrostatic equilibrium**.

(3) We assume that interior of the stars have the same **chemical compositions** as 70.4% hydrogen, 28% helium, and 1.6% other heavy elements like oxygen, carbon, iron, etc. These can be written as $X + Y + Z = 1$, where $X = .704$, $Y = .28$, and $Z = .016$ (Fig. 106).

(4) We assume the mass of the stars is constant during a large portion of their life time. That means, we assume the **constancy of mass** during the evolution of the stars.

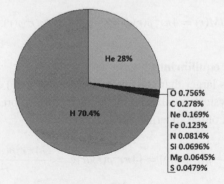

Figure 106 Chemical compositions of stars.

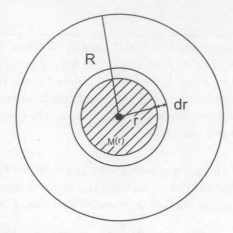

Figure 107 Internal structure of the star.

12.2 Equation of Stellar Structure

Let us consider the star as a sphere of radius R. We consider a spherical shell at a distance r from the center with thickness dr (see Fig. 107). Let $M(r)$ be the mass inside the radius r of the star. The mass of the shell is

$$dM(r) = 4\pi r^2 \rho dr,$$

where ρ is the density at radius r. Let $p(r)$ and $p(r + dr)$, respectively, be the pressures at radii r and $r + dr$. The luminosity $L(r)$ is the total output of energy from the spherical surface of radius r, i.e., $L(r) = 4\pi r^2 F(r)$, where $F(r)$ is the flux per unit area.

Now we accumulate the set of equations that represent the basic principles governing stellar structure:

(i) Equation of continuity or mass equation

The mass of the spherical shell at a distance r from the center is given by

$$dM(r) = 4\pi r^2 \rho(r)dr \Rightarrow \frac{dM(r)}{dr} = 4\pi r^2 \rho(r). \tag{12.1}$$

(ii) Equation of hydrostatic equilibrium

The inner pressure $p(r)$ is larger than the outer pressure $p(r + dr)$ of the shell by the weight of the spherical shell, i.e., the system is balanced by the difference of pressures between its inner and outer surface with the weight of the spherical shell. Therefore,

$$4\pi r^2 [p(r) - p(r + dr)] = [4\pi r^2 \rho(r)dr]g = \frac{G4\pi r^2 \rho(r)dr M(r)}{r^2},$$

$$\Rightarrow \frac{dp(r)}{dr} = -\frac{GM(r)\rho(r)}{r^2}. \tag{12.2}$$

(iii) Equation of thermal equilibrium

It is argued that the energy generated by nuclear reactions in the core of the star flows outward. Let $L(r)$ be the total outward energy flux per unit time across a spherical surface of radius r and it is the luminosity of the star. The additional energy flux, i.e., the increase in the luminosity made by the spherical shell as one goes through the shell from inside to the outward direction is equal to the energy produced within the shell. If $\epsilon(r)$ is the energy generation per unit mass per unit time, we have

$$dL(r) = 4\pi r^2 \rho(r)\epsilon(r)dr,$$

$$\Rightarrow \frac{dL(r)}{dr} = 4\pi r^2 \rho(r)\epsilon(r). \tag{12.3}$$

(iv) Equation of energy transfer

Energy flux is determined by the temperature gradient inside the star. We consider two modes of heat transfer namely radiation and convection.

[a] Let us consider an interior region of a star where heat is flowing outward only due to radiative transfer. The energy flux $F(r)$ across the spherical surface of radius r is given by

$$L(r) = 4\pi r^2 F(r) = -4\pi r^2 \frac{c}{\rho\kappa} \frac{d}{dr}\left(\frac{aT^4}{3}\right).$$

This equation yields the temperature variation due to radiative flux as

$$\frac{dT}{dr} = -\frac{3\kappa(r)\rho(r)L(r)}{16\pi acr^2 T^3(r)}. \tag{12.4}$$

[κ = opacity of the material, $a = \frac{4\sigma}{c}$, σ = Stefan's constant, equation of state, $p_{rad}(r) = \frac{1}{3}aT^4(r)$.]

[b] If the stars are in convective equilibrium, i.e., it involves only up and down motions of the gas, then

$$\frac{d\log T}{d\log p} = \frac{\gamma - 1}{\gamma} \Rightarrow \frac{dT}{dp} = \frac{\gamma - 1}{\gamma}\frac{T}{p}.$$

Using Eq. (12.2), we get

$$\frac{dT}{dr} = -\frac{\gamma - 1}{\gamma}\frac{GT(r)M(r)\rho(r)}{p(r)r^2}. \tag{12.5}$$

[γ is the ratio of specific heats at constant pressure and at constant volume ($\gamma = \frac{c_p}{c_v}$).]

Eqs. (12.1)-(12.5) are the fundamental equations to describe a model of a star, i.e., to get a stellar structure. To determine the stellar parameters by solving the above differential equations, we need some **boundary conditions**

$$M(r) = 0 \text{ at } r = 0, \ M(r) = M \text{ at } r = R, \ L(r) = 0 \text{ at } r = 0, \ p(r) = 0 \text{ at } r = R. \tag{12.6}$$

12.3   Simple Stellar Model

To get some of the physical features of the stellar structure, one can use Eqs. (12.1)–(12.5). By eliminating $M(r)$ from (12.1) and (12.2), one gets

$$-\frac{1}{G}\frac{d}{dr}\left[\frac{r^2}{\rho}\frac{dp}{dr}\right] = 4\pi r^2 \rho. \qquad (12.7)$$

This equation shows how p and ρ vary with radius r in a star.

Note 12.2

We know the Tolman–Oppenheimer–Volkoff (TOV) equation is the stellar structure equation and can be written as

$$\frac{dp}{dr} = -\frac{(p+\rho)(4\pi G p r^3 + M(r))}{r(r - 2M(r))},$$

where the mass is defined as

$$M(r) = \int_0^r 4\pi r^2 \rho(r) dr.$$

In Newtonian limit, energy density is dominating and as a result,

$$\rho(r) >> p(r) \quad and \quad M(r) >> 4\pi r^3 p(r).$$

Also, gravitational potential is small everywhere and hence

$$\frac{2M(r)}{r} << 1.$$

These conditions reduce the TOV equation as

$$-r^2 \frac{dp}{dr} = GM(r)\rho(r).$$

Taking derivative on both sides, we get

$$-\frac{1}{G}\frac{d}{dr}\left[\frac{r^2}{\rho}\frac{dp}{dr}\right] = \frac{dM}{dr} = 4\pi r^2 \rho.$$

This equation coincides with Eq. (12.7) of hydrostatic equilibrium for Newtonian stars.

This equation involves two unknowns, $\rho(r)$ and $p(r)$. Therefore, we need to specify a relation between $\rho(r)$ and $p(r)$, i.e., equation of state. Since it was believed that stars are in convective equilibrium, one can adopt an adiabatic relation between $\rho(r)$ and $p(r)$ as

$$p \propto \rho^{\gamma} = \rho^{1+\frac{1}{n}}, \tag{12.8}$$

where $\gamma = 1 + \frac{1}{n} = \frac{c_p}{c_v}$.

The above imposed model (Eq. (12.8)) is known as **polytropic model** for stellar structure. Here, $n = \frac{1}{\gamma-1}$ is called **polytropic index**.

$$Case - i: \ n = 0 \Rightarrow \rho \propto p^0 = constant,$$

which represents the star with **uniform density**.

$$Case - ii: \ n = 1.5 \Rightarrow p \propto \rho^{\frac{5}{3}},$$

which links to a star in **convective equilibrium.**

$$Case - iii: n = \infty \Rightarrow p \propto \rho,$$

which signifies that the star is in **isothermal configuration**.

Now if we substitute the value of $p = k\rho^{1+\frac{1}{n}}$, k is a constant, then Eq. (12.7) transforms to

$$\frac{1}{r^2} \frac{d}{dr} \left[\frac{r^2}{\rho} \frac{d}{dr} (k\rho^{1+\frac{1}{n}}) \right] = -4\pi\rho G. \tag{12.9}$$

To solve Eq. (12.9), we consider the following transformations:

$$\rho = \rho_c y^n \ and \ p = p_c y^{n+1}, \tag{12.10}$$

where ρ_c and p_c are central density and pressure, respectively, and y is a common parameter. Now Eq. (12.9) becomes

$$\frac{1}{x^2} \frac{d}{dx} \left(x^2 \frac{dy}{dx} \right) + y^n = 0, \tag{12.11}$$

where $r = \alpha x$ and $\alpha = \sqrt{\frac{p_c(n+1)}{4\pi G\rho_c^2}}$.

This equation is known as **Emden's equation**. Solutions y's are known as **Lane–Emden functions**.

To get the solution of Lane–Emden's equation, we assume two further conditions as $\rho = \rho_c$ and $\frac{d\rho}{dr} = 0$ at $r = 0$. These boundary conditions imply the following boundary conditions on y as

$$y = 1 \ and \ \frac{dy}{dx} = 0 \ at \ x = 0.$$

For $n = 0, 1$ and 5, one get exact solution for y.

$$n = 0; \quad y = 1 - \frac{1}{6}x^2,$$

$$n = 1; \quad y = \frac{\sin x}{x},$$

$$n = 5; \quad y = \frac{1}{\left(1 + \frac{1}{3}x^2\right)^{\frac{1}{2}}}.$$

Hence, one gets all the values of the parameters like $\rho, p,$ and M.

Note 12.3

For other values of n, one has to use numerical method to get the Lane–Emden functions. One can rewrite Eq. (12.11) as

$$y'' + \frac{2y'}{x} + y^n = 0,$$

where prime denotes $\frac{d}{dx}$.

This equation has power series solution

$$y = 1 - \frac{1}{6}x^2 + \frac{n}{120}x^4 - \frac{n(8n-5)}{15120}x^6 + \dots\dots$$

The Lane–Emden function y that vanishes (i.e., $p = 0$) at a finite value of x (say x_1) would represent the surface of the stellar configuration. Therefore, the stellar radius is given by $R = \alpha x_1$. The polytropic index $n = 0$ indicates that the star has a uniform density (finite radius), whereas, for $n = 5$, the star has an infinite radius, i.e., x_1 increases with n starting from $x_1 = \sqrt{6}$ for $n = 0$ to infinity for $n = 5$. Therefore, polytropes with $n < 5$ are bound system and polytropes with $n \geq 5$ represents unbounded or unrealistic configurations.

Radius of the star is

$$R = \alpha x_1. \tag{12.12}$$

The total mass of the star is

$$M = 4\pi \int_0^R r^2 \rho(r) dr = 4\pi \alpha^3 \int_0^{x_1} \rho_c y^n x^2 dx.$$

Now using L-E equation $\left(y^n x^2 = -\frac{d}{dx}[x^2 \frac{dy}{dx}]\right)$, we obtain

$$M = -4\pi \alpha^3 \rho_c \int_0^{x_1} \frac{d}{dx}\left(x^2 \frac{dy}{dx}\right) dx,$$

i.e.,

$$M = 4\pi\alpha^3 \rho_c \left[-x^2\frac{dy}{dx}\right]_{x=x_1=\frac{R}{\alpha}}. \tag{12.13}$$

The mean density of the star can be obtained as

$$\bar{\rho} = \frac{M}{\frac{4}{3}\pi R^3} = \frac{3\rho_c}{x_1^3}\left[-x^2\frac{dy}{dx}\right]_{x=x_1}. \tag{12.14}$$

Also, rearranging the above equation, we obtain the **central density** as

$$\rho_c = \frac{Mx_1}{4\pi R^3 \left[-\frac{dy}{dx}\right]_{x=x_1}}. \tag{12.15}$$

For numerical calculation of central density (ρ_c), the above equation assumes the form

$$\rho_c = \frac{4.7\times10^2\left(\frac{M}{M_\odot}\right)x_1}{\left(\frac{R}{R_\odot}\right)^3\left[-\frac{dy}{dx}\right]_{x=x_1}}kg\,m^{-3}. \tag{12.16}$$

The **central pressure** ($p_c = k\rho_c^{1+\frac{1}{n}}$) can be expressed as (eliminating k from (12.12–12.13))

$$p_c = \frac{GM^2}{R^4}\left\{4\pi(n+1)\left(\left[\frac{dy}{dx}\right]_{x=x_1}\right)^2\right\}^{-1}. \tag{12.17}$$

For numerical calculation of central pressure (p_c), the above equation assumes the form

$$p_c = 8.96\times10^{13}\frac{\left(\frac{M}{M_\odot}\right)^2}{\left(\frac{R}{R_\odot}\right)^4}\left\{(n+1)\left(\left[\frac{dy}{dx}\right]_{x=x_1}\right)^2\right\}^{-1}Pa. \tag{12.18}$$

The physical parameters are expressed solely in terms of x_1 and $\left[-x^2\frac{dy}{dx}\right]_{x=x_1}$. Therefore, it is essential to know the values of x_1 and $\left[-x^2\frac{dy}{dx}\right]_{x=x_1}$ for different values of n. Performing numerical integration, one can get the following values for Lane–Emden functions (Table II).

We now derive the expression for the energy of the star. The total gravitational potential energy Ω of the star

$$-\Omega = 4\pi\int_0^R\frac{GM(r)\rho(r)}{r}r^2dr = \int_0^R\frac{GM(r)dM(r)}{r},$$

Table II Solution of Lane–Emden equation.

n	x_1	$\left[-x^2\frac{dy}{dx}\right]_{x=x_1}$
0	2.4495	4.8990
0.5	2.7527	3.7887
1	3.1416	3.1416
1.5	3.6538	2.7141
2	4.3529	2.4111
2.5	5.3553	2.1872
3	6.8968	2.0182
3.5	9.5358	1.8906
4	14.972	1.7972
4.5	31.836	1.7378
5	∞	1.7321

$[dM = 4\pi r^2 \rho dr]$

$$or, \quad -\Omega = \frac{GM^2}{2R} + \frac{1}{2}\int_0^R \frac{GM^2(r)dr}{r^2}.$$

For a polytrope ($\frac{p}{\rho} = k\rho^{\frac{1}{n}}$), we have

$$\frac{d}{dr}\left(\frac{p}{\rho}\right) = \frac{1}{n+1}\frac{1}{\rho}\frac{dp}{dr},$$

$$\Rightarrow \frac{d}{dr}\left(\frac{p}{\rho}\right) = -\frac{1}{n+1}\frac{GM(r)}{r^2} \quad \left(as, \frac{1}{\rho}\frac{dp}{dr} = -\frac{GM}{r^2}\right).$$

$$Therefore, \quad -\Omega = \frac{GM^2}{2R} - \frac{n+1}{2}\int_0^R M(r)\frac{d}{dr}\left(\frac{p}{\rho}\right)dr,$$

$$= \frac{GM^2}{2R} - \frac{n+1}{2}\left[\frac{Mp}{\rho}\right]_0^R + \frac{n+1}{2}4\pi\int_0^R pr^2 dr,$$

$$= \frac{GM^2}{2R} + \frac{n+1}{2}4\pi\int_0^R pr^2 dr.$$

$$Now, \quad \int_0^R pr^2 dr = \left[\frac{pr^3}{3}\right]_0^R - \int_0^R \frac{r^3}{3}\frac{dp}{dr}dr = \frac{1}{3}\int_0^R r\rho GM(r)dr = -\frac{\Omega}{12\pi}.$$

[Note that $M(0) = 0$ and $p(R) = 0$.]

Hence, we have

$$-\Omega = \frac{3}{5-n}\frac{GM^2}{R} \equiv \frac{3(\gamma-1)}{5\gamma-6}\frac{GM^2}{R}. \tag{12.19}$$

The first law of thermodynamics states that

$$dQ = dU + pdV,$$

where U, p, V, and Q are internal energy, pressure, volume, and quantity of heat, respectively. For adiabatic process, $dQ = 0$. Thus, we have

$$dU + pdV = 0.$$

Also, for adiabatic process

$$pV^\gamma = constant = k(say).$$

Hence, we obtain the internal energy

$$U = \frac{V}{\gamma - 1}p.$$

Total internal energy of the star

$$T = \frac{1}{\gamma - 1}4\pi \int_0^R pr^2 dr \Rightarrow T = -\frac{\Omega}{3(\gamma - 1)}. \tag{12.20}$$

Therefore, total energy E or the binding energy of the polytropic star is

$$E = T + \Omega = \frac{3\gamma - 4}{3(\gamma - 1)}\Omega = -\frac{3\gamma - 4}{5\gamma - 6}\frac{GM^2}{R} = -\frac{3 - n}{5 - n}\frac{GM^2}{R}. \tag{12.21}$$

This leads to the fundamental result concerning stability. Note that for $n > 5$, the energy $E < 0$ and hence the stellar configuration is unbounded. Thus, only the stellar structures with $n < 5$ are the stable configurations.

Ritter's Theorem

Equilibrium configurations with $\gamma > \frac{4}{3}$ for which $E > 0$ are stable, whereas those with $\gamma < \frac{4}{3}$ for which $E < 0$ are unstable.

Note 12.4

Virial Theorem

Consider a collection of self-gravitating, spherical distribution of equal mass objects (stars, galaxies, etc.), $m_i, i = 1, 2, \ldots n$. Then, the Virial theorem asserts that for a stable configuration, there should exist a relation between the kinetic and potential energies of the system . More precisely, the total kinetic energy of the objects is equal to minus $\frac{1}{2}$ times of the total gravitational potential energy.

Proof: The equations of motion for the ith particle among the configuration of n particles under mutual gravitational attraction and some external forces (F_i) are

$$m_i \ddot{\mathbf{r}}_i = \mathbf{F}_i.$$

The scalar moment of inertia I of the system about the origin is defined by the equation

$$I = \sum_i m_i r_i^2.$$

Consider that the masses are constant, then the time derivative with respect to time of this moment of inertia yields

$$\frac{1}{2}\frac{dI}{dt} = \sum_i m_i \dot{r}_i \cdot r_i.$$

Taking derivative with respect to time once more, we get

$$\frac{1}{2}\frac{d^2 I}{dt^2} = \sum_i m_i |\dot{\mathbf{r}}_i|^2 + \sum_i m_i \mathbf{r}_i \cdot \mathbf{r}_i.$$

Here, one can note that $\frac{1}{2}\sum_i m_i |\dot{\mathbf{r}}_i|^2 = \frac{1}{2}\sum_i m_i v_i^2 = K$, where K is the kinetic energy.

Thus, we obtain

$$\frac{1}{2}\frac{d^2 I}{dt^2} = 2K + \sum_i \mathbf{r_i} \cdot \mathbf{F_i}.$$

Note that

$$\mathbf{F}_i = \sum_j Gm_i m_j \frac{\mathbf{r}_j - \mathbf{r}_i}{r_{ij}^3}, \quad where \ r_{ij} = |r_j - r_i|.$$

The entire force F_i acting on the ith particle is the sum of all the forces from the other j^{th} particles in the system.

$$\mathbf{F}_i = \sum_{j=1}^{n} \mathbf{F}_{ji},$$

where F_{ji} is the force applied by particle j on particle i. As no particle acts on itself (i.e., $\mathbf{F}_{jj} = 0$ for $1 = j = n$), and using Newton s third law of motion ($\mathbf{F}_{ij} = -\mathbf{F}_{ji}$, equal and opposite reaction), we get

$$\sum_{j=1}^{n} \mathbf{F}_i \cdot \mathbf{r}_i = \sum_{i=2}^{n}\sum_{j=1}^{i-1} F_{ji} \cdot \mathbf{r}_i + \sum_{i=1}^{n-1}\sum_{j=i+1}^{n} \mathbf{F}_{ji} \cdot \mathbf{r}_i = \sum_{i=2}^{n}\sum_{j=1}^{i-1} \mathbf{F}_{ji} \cdot (\mathbf{r}_i - \mathbf{r}_j).$$

Hence,

$$\sum_i \mathbf{r}_i \cdot \mathbf{F}_i = \sum_i \sum_j G m_i m_j \frac{\mathbf{r}_i \cdot (\mathbf{r}_j - \mathbf{r}_i)}{r_{ij}^3}.$$

Interchanging the dummy labels i and j in the summation, and adding the two terms we obtain

$$\sum_i \mathbf{r}_i \cdot \mathbf{F}_i = -\sum_i \sum_j^{\star} G m_i m_j |\mathbf{r}_j - \mathbf{r}_i|^2 / r_{ij}^3,$$

where the $\star$ means that each pair is taken only once. Therefore, we get

$$\sum_i \mathbf{r}_i \cdot \mathbf{F}_i = -\sum_i \sum_j^{\star} G m_i m_j / r_{ij} = V,$$

where V is the potential energy.

Thus, we have

$$\frac{1}{2} \frac{d^2 I}{dt^2} = 2K + V.$$

This equation was derived by Joseph-Louis Lagrange and extended by Carl Jacobi.

In an equilibrium system, dI/dt is minimum and $d^2I/dt^2 = 0$. Therefore, finally we get

$$2K + V = 0 \quad or \quad K = -\frac{1}{2} V.$$

This is known as the Virial theorem.

There are so many applications of the Virial theorem to astronomical problems. This theory can be applied to the study of the period of a pulsating star and to compute the velocities of dust particles in dark nebulae. The Virial theorem has been applied also to the study of the equilibrium configuration and the stability of interstellar clouds. To develop a relation between gravitational potential energy and thermal kinetic energy (i.e., temperature), virial theorem plays an important role in the cores of stars. In the main sequence stars, hydrogen converts into helium in their cores, and it must contract to sustain adequate pressure to support its own weight. This tightening shrinks its potential energy, and the Virial theorem asserts an increase in its thermal energy.

12.4   Jeans Criterion for Star Formation

The star clusters comprising the hot blue stars indicate that stars are taking birth at the present time also. Thus, star formation in our galaxy (and in also other galaxies) is a continuous process. Observation-based theories suggest that stars are made as a result of large-scale gravitational instability buildup in the central region of massive molecular clouds. Gravitational instability leads to collapse and flouting into portions of the pioneering cloud. Each subunit afterwards propagates further collapse. Finally, the disintegration leads to the birth of a group of protostars within the cloud enclosed by the diffuse cover.

The mathematical treatment of the problem of gravitational instability and collapse was first considered by Sir James Jeans in 1902. We will follow Virial theory to achieve it (Virial theory states that for a stable cloud $2K + \Omega = 0$, where $2K$ and Ω are kinetic energy and potential energy of the gas cloud). The stars are formed by self gravitation collapse of a cloud of gas. Now to find the condition for gravitational collapse, we use the notion of Virial theorem. For a scalar moment of inertia I of the system, if one assumes $\frac{d^2 I}{dt^2}$ is negative, then $\frac{dI}{dt}$ will decrease with time and as a result I will also decrease with time. Here, $I = \sum m_i r_i^2$ (a configuration of n particles under mutual gravitational attraction) and the entire dimensions of the gas cloud too reduce with time, i.e., the cloud will shrink. Therefore, to start the constriction, one should have $2K < -\Omega$. Let us consider the desired cloud to be spherically isothermal and uniformly dense. Now, from Eq. (12.19), the potential energy $(-\Omega = \frac{3}{5-n} \frac{GM^2}{R})$ takes the form for constant density as

$$-\Omega = \frac{3}{5} \frac{GM^2}{R}.$$

[for constant density, the polytropic index n is zero]

For perfect gas, one can calculate the kinetic energy as follows:

From kinetic theory of gases

$$P = \frac{1}{3} \mu m_a n_a c_{rms}^2 = \frac{1}{3} \rho c_{rms}^2 = \frac{1}{3} \frac{M}{V} c_{rms}^2,$$

where, $\mu, m_a, n_a, \rho, M, V$ are, respectively, mean molecular weight, mass of the molecule, number of molecules per volume, density of the gas, total mass, and volume of the gas.

Hence,

$$PV = \frac{1}{3} M c_{rms}^2 = \frac{2}{3} K, \quad K = kinetic\ energy.$$

Therefore,

$$K = \frac{3}{2} PV = \frac{3}{2} n\Re T = \frac{3}{2} \frac{\Re TM}{\mu}.$$

Here $M = \mu n$, where n = number of moles of the gas.

Thus, $2K < -\Omega$ implies

$$\frac{3\Re TM}{\mu} < \frac{3GM^2}{5R},$$

$$\Rightarrow M^3 > \frac{125\Re^3 T^3 R^3}{G^3 \mu^3}.$$

However, $M = \frac{4}{3} \pi R^3 \rho = \frac{4}{3} \pi R^3 \mu n_a m_a$. Hence, the above equation implies

$$M^3 > \frac{125\Re^3 T^3}{G^3 \mu^3} \times \frac{3M}{4\pi m_a n_a \mu},$$

or

$$M > \frac{9.3 \times 10^4}{\mu^2} \sqrt{\frac{\left(\frac{T}{100}\right)^3}{n_a}} M_\odot = M_J.$$

Here, M_J is defined as **Jeans mass**.

Hence, one can say any cloud with mass bigger than Jeans mass M_J, will be gravitationally unstable and will suffer a continuous collapse so long as the isothermal state is maintained in the collapsing cloud. This criterion is known as **Jeans criterion** for self-gravitation of an isothermal gas cloud.

In terms of Boltzmann constant κ, the Jeans mass is written as

$$M_J = \left[\frac{125\kappa^3 T^3}{G^3 \mu^3 m_a}\right]^{\frac{1}{2}} \times \left[\frac{3}{4\pi\rho}\right]^{\frac{1}{2}}.$$

Jeans length is defined as the critical radius of a cloud where thermal energy is counterbalanced by gravity (here the former causes the cloud to expand and later causes the cloud to collapse). The formula for Jeans length is

$$R_J = \left[\frac{15\kappa T}{4\pi G\mu m_a \rho}\right]^{\frac{1}{2}}.$$

When the cloud's radius coincides with the Jeans length, then thermal energy per particle equals gravitational work per particle and at this critical length, the cloud neither expands nor contracts, i.e., equilibrium is reached.

When Jeans mass is converted into a corresponding critical density, then that density is called the **Jeans density**. Jeans density is written as

$$\rho_J = \left[\frac{3}{4\pi M^2}\right] \left[\frac{5\kappa T}{G\mu m_a}\right]^3.$$

When a cloud is flattened to a region where the Jeans density is surpassed, it turns out to be unstable to gravitational collapse.

Note 12.5

If the cloud's mass exceeds this critical value M_J, which produces gravity that overcomes the gas's kinetic energy the star will start collapsing. In other words, if the cloud is compacted so that its density exceeds this critical value ρ_J, then it initiates collapsing. Jeans criteria imply that there is a minimum mass below which the thermal pressure prevents gravitational collapse. One can notice that

$$M_J \propto \rho^{-\frac{1}{2}} T^{\frac{3}{2}},$$

and this implies high-density favors collapse, whereas high temperature favors larger Jeans mass. Remember that for a cloud of gas

$$2K + \Omega = 0, \Rightarrow static\ equilibrium,$$

$$2K < -\Omega. \Rightarrow collapse,$$

$$2K > -\Omega, \Rightarrow expansion.$$

12.5 The Birth of Star

Initially, stars consist of clouds of dust and gas, i.e., they start their life as clouds of dust and gas. Gravity attracts these clouds and accumulates them. A small protostar forms due to this collapsing material. If the mass of the collapsing material is less than 0.08 times the sun's mass, then it cannot achieve the phase of nuclear fusion at their core. In its place, they become **brown dwarfs**, stars that twinkle but not ever. But if the collapsing material has enough mass, the collapsing gas and dust burn hotter, ultimately conquering temperatures enough to fuse hydrogen into helium. The stars are born as main sequence stars, powered by hydrogen fusion. The inward pressure caused by gravity is balanced by the outward pressure produced by fusion (Fig. 108) and stabilizes the star. How long can a main sequence star survive? This depends on how massive it is. A massive star has more mass, however, it burns quickly due to higher core temperatures caused by higher gravitational forces. Examples: sun's lifetime is 10 billion years. A star with 10 times sun's mass has lifetime 20 billion years. A star with half sun's mass, e.g., the red dwarf has lifetime 80-100 billion years. Due to the long lifetime, red dwarfs could be considered as worthy sources for planets hosting life.

A main sequence star like our sun gets its luminosity from nuclear reactions, generally the transformation of hydrogen to helium. A star is always radiating energy and it needs the nuclear reactions to compensate that energy loss in order to remain static. When all hydrogen is exhausted, i.e., converted to helium, then this energy source turns off, and the inner region (core) of the star

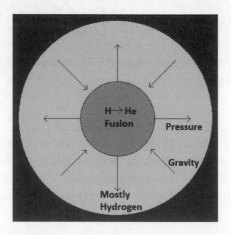

Figure 108 Fusion creates an external pressure that stabilizes the inward pressure caused by gravity, steadying the star.

begins to shrink. This shrinking compresses the core and as a result core's temperature gradually increases. It reachs such a high temperature that the core starts to ignite further reaction that convert helium into carbon and oxygen providing more energy.

This new outward energy expands the outer layers of the star and the star becomes a new "core halo" structure, called a **red giant**.

After the helium is exhausted, the star might then pass through phases of turning carbon into silicon and silicon into iron, which has maximum stability among all nuclei. Therefore, every reaction transforming iron into something else engrosses energy rather than releasing it. When this phenomenon happens, the subsequent evolution of the star depends solely on three things: the **star's mass, angular momentum, and magnetic field.**

A star with a mass around the sun's mass that reaches a state by evolving itself smoothly is known as the **white dwarf**. This is a specific type of star whose pressure originates not from thermal effects but from quantum mechanical ones, which is discussed later.

A massive star will also evolve smoothly through the hydrogen burning phase as in main sequence stars, but it is difficult to comprehend what will happen after that. During the nuclear cycle, either the star goes through a nuclear explosion or the core of the star becomes hydrodynamically unstable and collapses to a compact object such as a **neutron star** or **black hole.** Note that massive star means here 8-10 $M_\odot$.

12.6   White Dwarfs

Highly compact stars with initial masses $M < 4M_\odot$ will end as white dwarfs. White dwarfs are abundant in the universe. In an arbitrary volume of space, about 25% of all stars are white dwarfs. A typical white dwarf with one solar mass has a radius of about 5×10^3 km and its mean density is of the order of 10^6 gm/cm^3. At these densities, the electrons are completely degenerated, i.e., the matter will be completely ionized. The argument for this phenomenon of pressure ionization is as follows: consider the matter within the core of the star being gradually reduced. Due to inward pressure caused by gravity, the atoms get packed more and more tightly and finally a phase is reached when the atomic electrons start coming out from their electronic shells and discharge. Therefore, the matter is entirely ionized. Here, these free electrons provide the bulk of the gas pressure. This gas is considered as absolute zero temperature Fermi gas since the real temperature within the star is very low compared to the Fermi temperature at these high densities. When nuclear fuel is exhausted, these degenerate electrons stop further contraction under a star's own gravitation and stabilize the white dwarfs. Thus, white dwarfs are compact objects supported by degenerate-electron pressure.

Note 12.6

After the nuclear fuel has been exhausted, the star contracts, all the lowermost electron energy levels are occupied and the electrons are forced into higher and higher energy levels, filling the lowest unfilled energy levels. This generates an effective pressure that averts further gravitational collapse.

The mass density will be dominated by nucleons (neutrons and protons are jointly called **nucleons**).

The density is given by

$$\rho = \mu m n, \tag{12.22}$$

where μ (mean molecular weight) is the number of nucleons per electron, m is the nucleon mass, and n the number of electrons per unit volume.

Degenerate electrons follow the **Pauli exclusion principle** (no two electrons can reside in the same state, even under the pressure of a collapsing star of numerous solar masses). Thus, Pauli exclusion principle indicates that at zero temperature, all electronic states fill up to Fermi momentum k_F. Thus,

$$n = \frac{2}{h^3} \int_0^{k_F} 4\pi x^2 dx = \frac{8\pi}{3} \frac{k_F^3}{h^3}. \tag{12.23}$$

Uncertainty principle implies each electron occupies a volume of h^3 in phase space. Since an electron has two spin status for each momentum, the extra factor 2 occurs. The pressure p_e of degenerate electrons is given by

$$p_e = \left(\frac{1}{3}\right) < n\vec{p} \cdot \vec{v} >,$$

which is the momentum imparted to the unit area of a wall per unit time. Here, factor 3 comes from the averaging over three directions of space. Thus, pressure for a zero-temperature Fermi gas is

$$p_e = \frac{8\pi}{3h^3} \int_0^{k_F} \frac{x^2}{\sqrt{(x^2 + m_e^2)}} x^2 dx, \tag{12.24}$$

m_e is the electron mass and we have used the relations

$$\vec{v} = \frac{\vec{p}c^2}{E} \ and \ E^2 = (p^2c^2 + m_e^2c^4) \ with \ c = 1.$$

Now, we consider **two limiting cases**:

Case (i) $k_F << m_e$ **(nonrelativistic case):**

$$(12.24) \Rightarrow p_e = \frac{8\pi k_F^5}{15 m_e h^3}. \tag{12.25}$$

Using (12.22) and (12.23), we get

$$p_e = K\rho^{\frac{5}{3}}, \tag{12.26}$$

where

$$K = \frac{8\pi h^2}{15 m_e} \left(\frac{3}{8\pi m \mu}\right)^{\frac{5}{3}}. \tag{12.27}$$

Note that equation of state (12.26) is polytropic (with index $n = \frac{3}{2}$). Therefore, using Eqs. (12.12) and (12.13), one can find the radius and mass of white dwarfs as

$$R = \left(\frac{5K}{8\pi} \right)^{\frac{1}{2}} x_1 \rho_c^{-\frac{1}{6}},$$ (12.28)

$$M = 4\pi \left(\frac{5K}{8\pi G} \right)^{\frac{1}{2}} \rho_c^{\frac{1}{2}} \left(-x^2 \frac{dy}{dx} \right)_{x=x_1}.$$ (12.29)

From the numerical calculations of the L-E function, Table II provides the values of x_1 and $\left[-x^2 \frac{dy}{dx} \right]_{x=x_1}$ as 3.6538 and 2.7141, respectively for $n = \frac{3}{2}$. Since, here, $\gamma = \frac{5}{3} > \frac{4}{3}$, Ritter's theorem indicates that these white dwarf configurations are stable. Using all values of parameters in units of the solar mass, we find the mass radius relation for nonrelativistic degeneracy,

$$\frac{M}{M_\odot} \mu^2 = \frac{6.5 \times 10^{-5}}{(\mu \frac{R}{R_\odot})^3}.$$ (12.30)

Case (ii) $k_F \gg m_e$ **(relativistic case)**:
In this case,

$$p_e = \frac{8\pi}{3h^3} \frac{k_F^4}{4},$$ (12.31)

or

$$p_e = K\rho^{\frac{4}{3}},$$ (12.32)

where

$$K = \frac{h}{24\pi^3} \left(\frac{3\pi^2}{\mu m} \right)^{\frac{4}{3}}.$$ (12.33)

Note that here, also, the equation of state (12.31) is polytropic (with index $n = 3$). Eqs. (12.12) and (12.13) yield the radius and mass of white dwarfs as

$$R = \left(\frac{K}{\pi G} \right)^{\frac{1}{2}} \rho_c^{-\frac{1}{3}} x_1,$$ (12.34)

$$M = 4\pi \left(\frac{h}{24\pi^4 G} \right)^{\frac{3}{2}} \left(\frac{3\pi^2}{m\mu} \right)^2 \left(-x^2 \frac{dy}{dx} \right)_{x=x_1} \rho_c^{-2}.$$ (12.35)

Table II provides the values of x_1 and $\left[-x^2 \frac{dy}{dx} \right]_{x=x_1}$ as 6.8968 and 2.0182, respectively, for $n = 3$.

Using all values of parameters in units of the solar mass, we find

$$M = \frac{5.84}{\mu^2} M_\odot. \tag{12.36}$$

One can notice that the mass of the white dwarf is independent of the central density and hence the mass is kept fixed at the value given in (12.36) (as ρ_c is increased without limit). Hence, the mass given in (12.36) signifies the maximum mass that a white dwarf can have. This limiting mass is known as the **Chandrasekhar limit**.

Now, one can see that for different values of μ for different elements, one can get different Chandrasekhar limits of white dwarfs. For example, hydrogen, $\mu = 1$, helium, $\mu = 2$, and iron, $\mu = \frac{56}{26}$. Thus, for a star made up of hydrogen, Chandrasekhar limit is 5.84 $M_\odot$. Also the Chandrasekhar limits of stars made with helium and iron are 1.461 $M_\odot$ and 1.24 $M_\odot$, respectively.

Note 12.7

Since in real stars, some electrons are relativistic and some nonrelativistic, so to get a correct picture, one has to solve Eq. (12.24). Eq. (12.24) can be solved analytically to yield

$$p_e = constant. \left[x(2x^2 - 3m_e^2)(\sqrt{x^2 + m_e^2}) + m_e^4 \ln(x + \sqrt{x^2 + m_e^2}) \right]_0^{k_F}.$$

Putting the value of p_e in (12.7), one can get, in principle, the value of ρ and other parameters. After some suitable transformations, the Eq. (12.7) assumes the following form

$$\frac{d}{dz}\left[z^2 \frac{dy}{dz} \right] + z^2(y^2 - 1)^{\frac{3}{2}} = 0,$$

$$where \quad r = \alpha z, \quad \alpha = \frac{1}{\mu m}\left(\frac{3h^3}{32\pi^2 m_e^2} \right)^{\frac{1}{2}}.$$

This is a highly complicated differential equation. However, it can be solved numerically.

12.7 Neutron Stars

There are three types of super dense compact objects that are found when all possible nuclear fuels of the stars have been exhausted. They are white dwarfs, neutron stars, and black holes. White dwarfs are supported by degenerate electrons having mean densities of the order 10^5 to 10^6 gm/cm^3. These white dwarfs have a maximum mass limit (Chandrasekhar limit). If the mass of a white dwarf star is greater than the Chandrasekhar limit, then this white dwarf star supported by the pressure of cold degenerate electrons cannot be in equilibrium. A star whose mass is above the Chandrasekhar limit falls to collapse as its internal pressure fails to counterbalance the gravitational inward force. According to Pauli's exclusion principle, there is a restriction on the possible number density of free electrons. This compels the excess electrons to combine with protons to form neutrons. When the nuclear densities become of the order of 10^{14} gm/cm^3, all electrons combine with protons producing a **neutron star.** At this state, neutrons become completely degenerate. A neutron star is very similar to a white dwarf. The difference is that it is supported by cold degenerate neutrons. Here, all electrons and protons have been converted into neutrons through the following relation:

$$p + e \rightarrow n + \gamma,$$

where the neutrinos γ are escaping from the star. Low-mass neutron stars are very similar to white dwarfs having the same mass, apart from the fact that neutron degeneracy pressure replaces electron degeneracy pressure. The neutrons inside a neutron star behave like an ideal Fermi gas. One can write the density and pressure of this Fermi gas at zero absolute temperature

$$\rho = \frac{8\pi}{h^3} \int_0^{k_F} \sqrt{x^2 + m^2} x^2 dx, \tag{12.37}$$

$$p = \frac{8\pi}{3h^3} \int_0^{k_F} \frac{x^2}{\sqrt{x^2 + m^2}} x^2 dx. \tag{12.38}$$

The above two equations provide an implicit relation of equation of state.

Now, we also consider **two limiting cases**:

Case (i) $k_F \ll m$ **(nonrelativistic case):**

Here, one can find

$$\rho = \frac{8\pi m k_F^3}{3h^3}, \tag{12.39}$$

$$p = \frac{8\pi k_F^5}{15mh^3}. \tag{12.40}$$

Hence, we find the relation between p and ρ as

$$p = K\rho^{\frac{5}{3}}, \tag{12.41}$$

where

$$K = \frac{8\pi h^2}{15m} \left(\frac{3}{8\pi m} \right)^{\frac{5}{3}}. \tag{12.42}$$

In this nonrelativistic case, the density of the neutron star is low and we have a Newtonian polytrope-like a white dwarf with polytropic index $n = \frac{3}{2}$. All the parameters are the same as given in Eqs. (12.28) and (12.29). Thus, a low-mass neutron star is very similar to a white dwarf with the same mass.

Case (ii) $k_F \gg m$ **(relativistic case):**

Here, one can find

$$\rho = \frac{8\pi k_F^4}{4h^3}, \tag{12.43}$$

$$p = \frac{8\pi k_F^4}{12h^3}. \tag{12.44}$$

The relation between p and ρ, i.e., equation of state is given by

$$p = \frac{1}{3}\rho,$$ (12.45)

Now, one can use TOV equation to find the relativistic stellar structure through the equation

$$-r^2 \frac{d\rho}{dr} = 4\rho(r)\left(M(r) + \frac{4\pi}{3}r^2\rho(r)\right)\left(1 - \frac{2M(r)}{r}\right)^{-1}.$$ (12.46)

From this TOV equation, one can find

$$\rho(r) = \frac{3}{56\pi r^2}.$$ (12.47)

This model is not realistic as we cannot get a stellar configuration of finite radius.

12.8   Gravitational Collapse

Gravitational collapse may occur in the star when its nuclear fuel has been exhausted. Degenerate electron pressure can support star masses up to the Chandrasekhar limit. A massive star undergoes continual gravitational collapse when its nuclear fuel comes to an end. One can develop a model in general relativity with a suitable energy-momentum tensor. Since pressure is not likely to be important in gravitational collapse, one may take matter distribution as **dust distribution with uniform density.** To discuss the time evolution, we use the Einstein equations. The interior spacetime is matched to the Schwarzschild geometry to complete the full model of gravitational collapse.

12.9   Oppenheimer–Snyder Nonstatic Dust Model

The pressureless fluid, i.e., dust is characterized by the energy-stress tensor as

$$T^{\mu\nu} = \rho u^\mu u^\nu, \quad u^\mu = four\ velocity\ of\ the\ particle\ and\ \rho = energy\ density.$$

We know the general metric describing the spherical symmetry is (see Eq. (6.7))

$$ds^2 = e^\nu dt^2 - e^\mu dR^2 - R^2(d\theta^2 + \sin^2\theta d\phi^2),$$

where ν, μ, and R are functions of r and t.

Now, conservation law of energy-momentum tensor is

$$T^{\mu\nu}_{;\nu} = 0 \Rightarrow (\rho u^\nu)_{;\nu}u^\mu + \rho u^\nu u^\mu_{;\nu} = 0.$$

Since,

$$u^\mu u_\mu = 1 \Rightarrow (u_\mu u^\mu)_{;\nu} = 0.$$

Hence, we have

$$(\rho u^\nu)_{;\nu} = 0 \Rightarrow (\rho \sqrt{-g} u^\nu)_{,\nu} = 0 \ \textit{by result} \ (8) \ \textit{in Section} \ (3.4).$$

Thus,

$$u^\mu_{;\nu} u^\nu = 0.$$

This means each particle of the dust moves on a geodesic, i.e., coordinates are **comoving** and hence spatial part remains unchanged along the geodesics. Therefore, we can have four velocity as

$$u^\mu = (u^0, 0, 0, 0), \ \textit{where} \ u^0 = \frac{dt}{ds}.$$

The geodesic equation reduces to

$$\frac{du^\mu}{ds} + \Gamma^\mu_{00}(u^0)^2 = 0 \Rightarrow \Gamma^\mu_{00} = 0 \Rightarrow g_{00,\mu} = 0 \Rightarrow g_{00} = \alpha(t).$$

Now, after rescaling the above metric, we have

$$ds^2 = dt^2 - e^\mu dR^2 - R^2(d\theta^2 + \sin^2\theta d\phi^2).$$

Hence, the Einstein field equations for the dust are written as

$$e^{-\mu}(2RR'' + R'^2 - RR'\mu') - (R\dot{R}\dot{\mu} + \dot{R}^2 + 1) = -8\pi G\rho R^2, \tag{i}$$

$$e^\mu(2R\ddot{R} + \dot{R}^2 + 1) - R'^2 = 0, \tag{ii}$$

$$2\dot{R}' - R'\dot{\mu} = 0. \tag{iii}$$

(iii) implies

$$(2\ln R' - \mu)^\cdot = 0,$$

$$\Longrightarrow e^\mu = \frac{R'^2}{1+f}, \tag{iv}$$

f is an arbitrary function of r.

    (ii) implies

$$2R\ddot{R} + \dot{R}^2 - f = 0 \Longrightarrow (R\dot{R}^2 - fR)^\cdot = 0,$$

$$\Longrightarrow \dot{R}^2 - f = \frac{F(r)}{R}, \tag{v}$$

F is an arbitrary function of r.

Using (iv) and (v), (i) yields

$$\frac{R}{R'}(f' - 2\dot{R}\dot{R}') - (\dot{R}^2 - f) = -8\pi G\rho R^2,$$

$$or, \quad \frac{F'}{R'} = 8\pi G\rho R^2. \tag{vi}$$

Now suppose that at $t = 0$, the ball of dust had a uniform density ρ_0 and was at rest. Thus,

$$\dot{R} = \dot{\mu} = 0 \ at \ t = 0.$$

At $t = 0$, we have the surface of a sphere $r =$ constant, having area $4\pi R^2(r, 0)$. We now choose r-coordinate such that $R^2(r, 0) = r^2$. The element is applicable for $r \leq r_b$ ($r_b =$ radius of the star). For $r > r_b$, the spacetime is supported by vacuum and may be described by standard Schwarzschild line element.

$$(vi) \Rightarrow F(r) = \frac{8\pi G\rho_0}{3}r^3 = \alpha r^3,$$

where $\alpha = \frac{8\pi G\rho_0}{3}$.

For $t = 0$, $R^2 = r^2$ and integrating constant is taken to be zero because it corresponds to a point at origin.

Also, $\dot{R} = 0$ at $t = 0$.

(v) gives

$$f(r) = -\frac{1}{r}F(r) = -\alpha r^2.$$

For $t > 0$, write $R = rS$ and substituting in (v), we get

$$\dot{S}^2 = \alpha\frac{(1 - S)}{S}.$$

The solution of this equation can be found as

$$t = \frac{\psi + \sin \psi}{2\sqrt{\alpha}}, \ S = \frac{1}{2}(1 + \cos \psi), \ a \ cycloid$$

S vanishes when $\psi = \pi$. This gives

$$t_0 = \frac{\pi}{2\sqrt{\alpha}}.$$

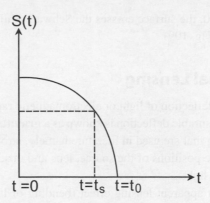

Figure 109 Collapsing star.

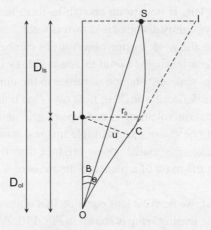

Figure 110 Gravitational lensing diagram.

Thus, we have $S = S(t)$, with $S(0) = 1$ and $S(t_0) = 0$. The function S is plotted given in Fig. 109. It decreases to zero in the time $t_0 = \frac{\pi}{2\sqrt{\alpha}}$. The function e^μ is obtained as

$$e^\mu = \frac{S^2(t)}{1 - \alpha r^2},$$

so that line element becomes

$$ds^2 = dt^2 - S^2(t) \left[\frac{dr^2}{1 - \alpha r^2} + r^2 d\Omega_2^2 \right].$$

From the line element, we see that the ball of dust collapses to a point and volume becomes zero in a finite time. The time t_0 is ~20 minutes for a starting density 1 gm/cm^3. The Schwarzschild radius is reached when $r_b S = 2GM = \alpha r_b^3$, i.e., $S = \alpha r_b^2$ for ball of dust $M = \frac{4\pi}{3}\rho_0 r_b^3$.

The collapse starts at $t = 0$, the surface crosses the Schwarzschild barrier a $t = t_s$ and reaches the singularity state at $t = t_0$ (Fig. 109).

12.10 Gravitational Lensing

Gravitational lensing is the deflection of light or electromagnetic radiation in a gravitational field and an object producing a measurable deflection is known as a **gravitational lens**. Usually, there are three different types of lensing that are used in literature namely, strong lensing, weak lensing, and microlensing depending on the positions of the source, lens and observer, and the mass and shape of the lens (which deflects the light).

Strong lensing: The most apparent lensing effect (bending of light) is when the lens is very massive and the source is very near to it and in this case light can take different paths to the observer and multiple images or arcs and rings of the source will appear.

Weak lensing: When the lens is not strong enough to form multiple images or arcs but can produce an image which is only slightly distorted is known as weak lensing. Since in weak lensing the image is only slightly distorted, the weak lensing observations can be performed through observing a large number of sources. Weak lensing is useful to assess galaxy clusters. It can also be used to probe the inhomogeneous large-scale distribution of matter in the universe. The derivation of weak lensing is same as to derive the deflection angle of light rays (see note 7.1).

Microlensing: Gravitational microlensing is one type of gravitational lens effects in which the image is so small or faint that one doesn't see multiple images, however, the additional light that bends toward the observer makes the source appear brighter than the original one. It is useful to detect objects that range from the mass of a planet to the mass of a star, irrespective of the lights they emit.

Strong Field Limit At first, we derive a lens equation that allows for the large bending of light near an astrophysical body. The lensing setup is shown in Fig. 110. We assume the reference (optic) axis as the line joining the observer O and the lens L. We consider the spacetime such that lens (deflector) causing strong curvature is asymptotically flat and both the observer as well as the source are located in the flat region, i.e., lens (deflector) has no gravitational effect on observer and source.

The source S emits light, which is deflected by lens L. After deflection, this light reaches the observer O making an angle θ to the optical axis OL. If there is no lens, then it makes β angle. α ($\angle SCI$) is the **Einstein deflection angle** (actually, α is the deflection angle due to gravitating body, which may be a black hole or wormhole or star). We symbolize D_{os} as the observer-source distance, D_{ls} as the lens-source distance, and D_{ol} as the observer-lens distance.

By simple trigonometry, one can write the lens equation as

$$tan\beta = tan\theta - \frac{D_{ls}}{D_{os}} [tan\theta + tan(\alpha - \theta)].$$

The lens diagram gives

$$\sin\theta = \frac{u}{D_{ol}},$$

where u is the impact parameter (**impact parameter** u is defined as the perpendicular distance between the deflected path of a light ray and the center of the lens).

The lens deflects a light ray and causes a change in the cross section of a bundle of rays due to its gravitational field.

The ratio of the flux of the image to the flux of the unlensed source is actually the **magnification of an image**. As a consequence of Liouville s theorem, it is argued that the surface brightness is well maintained in gravitational light deflection. As a result, we can define the magnification of an image at the observer, which is the ratio of the solid angles of the image and of the unlensed source. Hence, the magnification of an image (for a circularly symmetric gravitational lensing) is given by

$$\mu = \left(\frac{\sin \beta}{\sin \theta} \frac{d\beta}{d\theta} \right)^{-1}.$$

The **parity of the image** is defined by the sign of the magnification of an image. The **critical curves (CCs)** are the singularities in the magnification lying in the lens plane. The term **caustics** is defined as the corresponding values in the source plane. **Critical images** are recognized as the images of 0-parity. One can define the **tangential (μ_t) and radial (μ_r) magnifications** as

$$\mu_t = \left(\frac{\sin \beta}{\sin \theta} \right)^{-1}, \quad \mu_r = \left(\frac{d\beta}{d\theta} \right)^{-1}.$$

The singularities in the magnification lying in the lens plane are the **tangential critical curves (TCCs)** and **radial critical curves (RCCs)**, respectively, Similarly, we can define the terms **tangential caustic (TC)** and **radial caustics (RCs)**, respectively, as the corresponding values in the source plane. $\beta = 0$ yields the TC and the corresponding values of θ are the TCCs.

12.11 General Spherically Symmetric Spacetime and the Deflection Angle

A general spherically symmetric body with asymptotically flat metric has the line element

$$ds^2 = A(x)dt^2 - B(x)dx^2 - C(x)(d\theta^2 + \sin^2\theta d\phi^2),$$

with

$$A(x) \longrightarrow 1 - \frac{2M}{x}, \ when \ x \longrightarrow \infty,$$

$$B(x) \longrightarrow 1 + \frac{2M}{x}, \ when \ x \longrightarrow \infty,$$

$$C(x) \longrightarrow x^2, \ when \ x \longrightarrow \infty.$$

[Here $x = \frac{r}{r_h}$ is the radial coordinate in units of horizon radius; $A(r) = 0$ gives $r = r_h$, i.e., radius of the event horizon. This is so because every spherically symmetric gravitating body is asymptotically Schwarzschild.]

Now we define a **photon sphere**, which is a spherical region of space where photons are forced to travel in orbits due to gravity. The lower bound for any stable orbit is the radius of the photon sphere. The required condition is that

$$\frac{C'(x)}{C(x)} = \frac{A'(x)}{A(x)},$$

has at least one positive root. The largest root of the above equation is the **radius of the photon sphere** x_m. A photon coming from an asymptotic flat region having impact parameter u will be swerved while impending the gravitating body (lens). The photon or light coming from the source will reach a minimum distance x_0, from the lens. This minimum distance, x_0 is known as **closest approach distance**. By conservation of angular momentum, one can relate the closest approach distance with impact parameter u as

$$u = \sqrt{\frac{C(x_0)}{A(x_0)}} \equiv \sqrt{\frac{C_0}{A_0}}.$$

The geodesic equation [see Eq. (7.54)]

$$\frac{d\phi}{dx} = \frac{\sqrt{B}}{\sqrt{C}\sqrt{\frac{CA_0}{C_0 A} - 1}}.$$

This yields the angular shift of the photon, which is a function of the radial distance from the center and the deflection angle is determined by the function of closest approach as

$$\alpha(x_0) = I(x_0) - \pi,$$

where

$$I(x_0) = \int_{x_0}^{\infty} \frac{2\sqrt{B}\,dx}{\sqrt{C}\sqrt{\frac{CA_0}{C_0 A} - 1}}.$$

In asymptotically limit, i.e., $A, B \longrightarrow 1$ and $C \longrightarrow x^2$, $\alpha(x_0)$ vanishes.

There are two approaches where deflection angle can be approximated.

1. **Weak field limit:** For $x_0 \gg x_m > 1$, a first nonzero order Taylor expansion in $\frac{1}{x_0}$ is made.

    This approach is used in lensing by stars, galaxies, or black hole with a large impact parameter.

    The deflection angle depends on the closest approach x_0. It increases with the decrease of closest approach x_0 and for a specific value of x_0, the value of deflection angle will be 2π, and as a result, the light ray makes a complete circle around the lens. Further, if x_0 decreases, then the light ray does not reach the observer directly, rather it winds many times around the lens prior to reaching the viewer. Finally, when $x_0 = x_m$, the radius of the photon sphere, the deflection angle will get infinite value, i.e., diverge and the light ray will be trapped by the lens (gravitating body).

2. **Strong field limit**: The above integral diverges when $x_0 = x_m$. However, for $0 < x_0 - x_m \ll 1$, it can be approximated by a logarithm

$$\alpha(x_0) = -a_1 \ln(x_0 - x_m) + a_2,$$

where a_1 and a_2 are constants.

This approach is used for black hole.

Note 12.7

An outline of mathematical calculation of lensing in the strong field limit context:

Step 1: Consider a general spherically symmetric metric.

$$ds^2 = A(x)dt^2 - B(x)dx^2 - C(x)(d\theta^2 + \sin^2\theta d\phi^2).$$

Step 2: $A(r) = 0 \Rightarrow r = r_h$ is radius of event horizon.

Write $x = \dfrac{r}{r_h}$, $x_0 = \dfrac{r_0}{r_h}$. [If the change is in terms of Schwarzschild radius, then $r_h = 2M$]

Step 3: Find radius of photon sphere, from the equation $\dfrac{C'(x)}{C(x)} = \dfrac{A'(x)}{A(x)} \Rightarrow x = x_m$.

Step 4: Find impact parameter from $J = \sqrt{\dfrac{C(x_0)}{A(x_0)}} = D_{ol}\sin\theta \cong D_{ol}\theta$.

Step 5: Define two variables:

$$y = A(x), \quad z = \frac{y - y_0}{1 - y_0}, \quad y_0 = A(x_0) = A_0, \quad B(x_0) = B_0, \quad C(x_0) = C_0.$$

Step 6: Now the deflection angle integral becomes

$$I(x_0) = \int_0^1 R(z, x_0)f(z, x_0)dz,$$

where

$$R(z, x_0) = \frac{2\sqrt{A(x)B(x)}}{C(x)A'(x)}(1 - A_0)\sqrt{C_0},$$

and

$$f(z, x_0) = \frac{1}{\sqrt{A_0 - \dfrac{A(x)C_0}{C(x)}}}.$$

i.e.,

$$R(z, x_0) = \frac{2\sqrt{By}}{CA'}(1 - y_0)\sqrt{C_0} \qquad [x \text{ should be replaced by } x = A^{-1}[(1 - y_0)z + y_0]]$$

$$f(z, x_0) = \frac{1}{\sqrt{y_0 - [(1 - y_0)z + y_0]\frac{C_0}{C}}}.$$

Here, the integral diverges for $z \longrightarrow 0$, i.e., one approaches the photon sphere.

Now within the square root portion of $f(z, x_0)$, one can expand Taylor series up to second order in z as

$$\frac{1}{\sqrt{F(z, x_0)}} \equiv f(z, x_0) \cong \frac{1}{\sqrt{p(x_0)z + q(x_0)z^2}} = f_0(z, x_0),$$

$$where, \ F(z, x_0) = y_0 - [(1 - y_0)z + y_0]\frac{C_0}{C},$$

[Taylor's series: $F(z, x_0) = F(0, x_0) + zF'(0, x_0) + \frac{z^2}{2!}F''(0, x_0)$. Here, $F(0, x_0) = 0$]

$$p(x_0) = \frac{1 - A_0}{C_0 A_0'}[C_0' A_0 - C_0 A_0'],$$

$$q(x_0) = \frac{(1 - A_0)^2}{2C_0 A_0'^3}[2C_0 C_0' A_0'^2 + (C_0 C''_0 - 2C_0'^2)A_0 A_0' - C_0 C_0' A_0 A''_0].$$

[Here p and q are known quantities]

When p is zero, the integral, i.e., $\int_0^1 \frac{dz}{z}$ diverges. We see from step 3 that $p = 0$ at $x_0 = x_m$, radius of photon. For $x_0 < x_m$, the light ray is captured by the gravitating body and cannot emerge back (i.e., then loop is formed).

Step 7: To solve the integral in step 6, we split it into two parts

$$I(x_0) = I_D(x_0) + I_R(x_0),$$

where

$$I_D(x_0) = \int_0^1 R(0, x_m)f_0(z, x_0)dz,$$

[this is divergence part]

$$I_R(x_0) = \int_0^1 g(z, x_0)dz,$$

[this is regular integral and can be evaluated numerically], where

$$g(z, x_0) = R(z, x_0)f(z, x_0) - R(0, x_m)f_0(z, x_0).$$

Now we can see that $I_D(x_0)$ can be solved exactly,

$$I_D(x_0) = R(0, x_m)\frac{2}{\sqrt{q}} \ln \frac{\sqrt{q} + \sqrt{p+q}}{\sqrt{p}}.$$

Step 8: Since we are interested in terms up to $0(x_0 - x_m)$,

$$p = \frac{2\beta_m A'_m}{1 - y_m}(x_0 - x_m) + 0(x_0 - x_m)^2,$$

where

$$\beta_m = \beta|_{(x_0 = x_m)} = \frac{C_m(1 - y_m)^2(C''_m y_m - C_m A''_m)}{2y_m^2 C'^2_m}.$$

Substituting this in $I_D(x_0)$, we find

$$I_D(x_0) = -a \ln \left(\frac{x_0}{x_m} - 1 \right) + b_D + 0(x_0 - x_m),$$

where

$$a = \frac{R(0, x_m)}{\sqrt{\beta_m}}, \quad b_D = \frac{R(0, x_m)}{\sqrt{\beta_m}} \ln \frac{2(1 - y_m)}{A'_m x_m}.$$

Step 9: Now for regular part

$$I_R(x_0) = \int_0^1 g(z, x_0)dz,$$

we expand $g(z, x_0)$ in power of $(x_0 - x_m)$ as

$$g(z, x_0) = g(z, x_m) + g'(z, x_m)(x_0 - x_m) + \frac{1}{2!}g''(z, x_m)(x_0 - x_m)^2.$$

Now,

$$I_R(x_0) = \int_0^1 g(z, x_m)dz + 0(x_0 - x_m).$$

Then finally we obtain

$$\alpha(x_0) = I_D(x_0) + I_R(x_0) - \pi.$$

Step 10: Changing $\alpha(x_0) \longrightarrow \alpha(\theta)$ as

$$\alpha(\theta) = -a \ln \left(\frac{\theta D_{ol}}{x_m} - 1 \right) + b_D + I_R(x_m) + 0(x_0 - x_m) - \pi.$$

[when θ is small, $u \cong x_0$]

Note 12.8

Modeling of a star comprising anisotropic fluid distribution

To describe a model of a star, we assume a spherically symmetric interior spacetime line element in curvature coordinates as follows:

$$ds^2 = e^{\nu(r)}dt^2 - e^{\lambda(r)}dr^2 - r^2(d\theta^2 + \sin^2\theta d\phi^2).$$

It is argued that the density of a highly compact star is larger than the nuclear mass density and, therefore, it is assumed that the matter within the star is anisotropic in nature and correspondingly the energy-momentum tensor is described by

$$T_{\mu\nu} = \rho U_\mu U_\nu + p_r \chi_\mu \chi_\nu + p_t (U_\mu U_\nu - \chi_\mu \chi_\nu - g_{\mu\nu}),$$

where the symbols have their usual meanings.

The Einstein field equations for the line element and the above energy-momentum tensor are given by

$$8\pi\rho(r) = \frac{1 - e^{-\lambda}}{r^2} + \frac{\lambda' e^{-\lambda}}{r},$$

$$8\pi p_r(r) = \frac{\nu' e^{-\lambda}}{r} - \frac{1 - e^{-\lambda}}{r^2},$$

$$8\pi p_t(r) = \frac{e^{-\lambda}}{4} \left(2\nu'' + \nu'^2 - \nu'\lambda' + \frac{2\nu'}{r} - \frac{2\lambda'}{r} \right),$$

where primes represent differentiation with respect to the radial coordinate r. We have utilized geometrized units, where G and c are taken to be unity.

The generalized TOV equation is written as

$$\frac{dp_r}{dr} = -(\rho + p_r) \left[\frac{m(r) + 4\pi r^3 p_r}{r\{r - 2m(r)\}} \right] + \frac{2}{r}(p_t - p_r),$$

where

$$m(r) = \int_0^r 4\pi r^2 \rho \, dr.$$

Note that we have three independent equations (as TOV equation comes from above three field equations) with five unknowns. Therefore, to get exact analytical solutions of the field equations, we need two more equations related to the physical parameters.

The exact solution describing a static anisotropic matter distribution is absolutely determined by means of the two generating functions Π and Z given by

$$\Pi = 8\pi(p_r - p_t),$$

and

$$e^{\nu(r)} = e^{\int \left[2Z(r) - \frac{2}{r} \right] dr}.$$

Using some mathematical manipulations, the metric potential takes the following form:

$$e^\lambda = \frac{Z^2(r) e^{\int \left[\frac{4}{r^2 Z(r)} + 2Z(r) \right] dr}}{r^6 \left[-2 \int \frac{Z(r)(1 + \Pi r^2) e^{\int \left\{ \frac{4}{r^2 Z(r)} + 2Z(r) \right\} dr}}{r^8} dr + C \right]},$$

where C is an integration constant.

One can solve the field equations for modeling of a star in two ways: either try to solve the TOV equation or start directly to solve Einstein field equations. Einstein field equations comprise the geometry of the spacetime and matter distribution that produces this spacetime geometry. If we know some part of geometry and some part of matter distribution, then we can find all the physical parameters through the Einstein field equation. This is the beauty of Einstein general theory of relativity.

For well-behaved nature of the solutions for the star comprising an anisotropic fluid, the following conditions should be satisfied:

1. The solutions should be free from physical and geometric singularities. The metric potentials are finite inside the radius of the star, moreover, the fluid sphere should satisfy $e^{\nu(0)} > 0$ and $e^{-\lambda(0)} = 1$. The central pressure and central density also have finite values.

2. The density ρ and pressures p_r and p_t should be positive inside the fluid configuration.

3. The radial pressure gradient $dp_r/dr \leq 0$ and density gradient $d\rho/dr \leq 0$ for $0 \leq r \leq R$. This implies pressure and density should be maximum at the center and monotonically decrease towards the surface.

4. The radial pressure p_r must vanish but the tangential pressure p_t may not necessarily vanish at the boundary $r = R$. However, the radial pressure is equal to the tangential pressure

at the center of the fluid sphere, i.e., pressure anisotropy vanishes at the center, $\Delta(0) = p_r(0) - p_t(0) = 0$.

5. The casualty condition should be obeyed, i.e., the radial and tangential adiabatic speeds of sound should be less than the speed of light. In the unit $c = 1$, the causality conditions take the form $0 < v_{sr}^2 = dp_r/d\rho \leq 1$ and $0 < v_{st}^2 = dp_t/d\rho \leq 1$. In addition to the above, the velocity of sound should be decreasing towards the surface, i.e., $\frac{d}{dr}\frac{dp_r}{d\rho} < 0$ or $\frac{d^2p_r}{d\rho^2} > 0$ and $\frac{d}{dr}\frac{dp_t}{d\rho} < 0$ or $\frac{d^2p_t}{d\rho^2} > 0$ for $0 \leq r \leq R$, i.e., the velocity of sound is increasing with an increase in density and it should be decreasing outward.

6. The adiabatic index, $\Gamma = \frac{\rho+p_r}{p_r}\frac{dp_r}{d\rho}$ for realistic matter should be $\Gamma > 1$.

7. For realistic stars, the compression modulus $\kappa_e = p_r\Gamma$ must be decreasing outward.

8. For physically stable static configuration, the energy conditions like null energy condition (NEC), weak energy condition (WEC), and strong energy condition (SEC) need to be satisified throughout the interior region, i.e.,

$$\rho \geq 0; \ \rho + p_r \geq 0; \ \rho + p_t \geq 0; \ \rho + p_r + 2p_t \geq 0.$$

9. The interior solution should continuously match with the exterior Schwarzschild solution.

Extrinsic Curvature or Second Fundamental Form

A **hypersurface** Σ is a three-dimensional sub-manifold of a four-dimensional spacetime manifold M. The hypersurface may be time-like, space-like or null.

Let us consider that x^μ ($\mu = 1, 2, 3, 4$) are the coordinate of the four-dimensional spacetime, and $y^a (a = 1, 2, 3)$ are the inherent coordinates of the hypersurface.

The parametric equation of the hypersurface is

$$x^\alpha = x^\alpha(y^a). \tag{A1}$$

The three basis vectors $e_{(a)} = \frac{\partial}{\partial y^a}$ provide the tangent vectors to Σ as

$$e^\mu_{(a)} = \frac{\partial x^\mu}{\partial y^a}.$$

These basis vectors provide the **induced metric or first fundamental form** of the hypersurface by the following scalar product

$$h_{ab} = e_{(a)} \cdot e_{(b)} = g_{\mu\nu} e^\mu_{(a)} e^\nu_{(b)}.$$

[Actually, the elementary distance "ds" between two neighboring points in Σ (which are, therefore, also in M) is given by

$$
\begin{aligned}
ds^2 = g_{\alpha\beta} dx^\alpha dx^\beta &= g_{\alpha\beta} \left(\frac{\partial x^\alpha}{\partial y^a} dy^a \right) \left(\frac{\partial x^\beta}{\partial y^b} dy^b \right) \\
&= \left(g_{\alpha\beta} \frac{\partial x^\alpha}{\partial y^a} \frac{\partial x^\beta}{\partial y^b} \right) dy^a dy^b \\
&= h_{ab} dy^a dy^b,
\end{aligned}
$$

where $h_{ab} = g_{\alpha\beta} \frac{\partial x^\alpha}{\partial y^a} \frac{\partial x^\beta}{\partial y^b}$ is the induced metric tensor in the hypersurface Σ.]

A unit normal n_α can be defined if the hypersurface is not null.

$$n^\alpha n_\alpha = \epsilon = -1, \; \textit{if } \Sigma \textit{ is space-like,}$$

$$= 1, \; \textit{if } \Sigma \textit{ is time-like.}$$

Let us consider the surface Σ, defined by the Eq. $(A1)$, which can be written in the form

$$f(x^{\mu}(y^{a})) = 0.$$

Hence, the unit normal vector n^{μ} is defined as

$$n_{\mu} = \epsilon \left| g^{\alpha\beta} \frac{\partial f}{\partial x^{\mu}} \frac{\partial f}{\partial x^{\mu}} \right|^{-\frac{1}{2}} \frac{\partial f}{\partial x^{\mu}} \tag{A2}$$

Here, $n_{\alpha}n^{\alpha} = \epsilon$ and $n_{\mu}e^{\mu}_{(a)} = 0$.

Note that unit normal is defined when Σ is non-null as for null surface $g^{\mu\nu}f_{,\mu}f_{,\nu} = 0$.

The **extrinsic curvature or second fundamental form** of the hypersurface Σ is defined by

$$k_{ab} = n_{\mu\,;\,\nu}\, e^{\mu}_{(a)} e^{\nu}_{(b)}$$

or

$$k_{ab} = -\epsilon n_{\mu} \left[\frac{\partial^{2} x^{\mu}}{\partial y^{a} \partial y^{b}} + \Gamma^{\mu}_{\alpha\beta} \frac{\partial x^{\alpha}}{\partial y^{a}} \frac{\partial x^{\beta}}{\partial y^{b}} \right].$$

Extrinsic curvature is symmetric tensor, i.e., $k_{ab} = k_{ba}$.

Another form

$$k_{ab} = \frac{1}{2} \left(L_{n} g_{\alpha\beta} \right) e^{\alpha}_{(a)} e^{\beta}_{(b)}$$

Here, L_{n} stands for Lie Derivative.

$k \equiv h^{ab} k_{ab} = n^{\alpha}_{;\alpha} \equiv$ trace of the extrinsic curvature.

Result

(i) If $k > 0$, then the hypersurface is convex

(ii) If $k < 0$, then the hypersurface is concave

h_{ab} is purely the inherent property of a hypersurface geometry, whereas k_{ab} is concerned with extrinsic aspects.

Remember

$$g^{\alpha\beta} = \epsilon n^{\alpha} n^{\beta} + h^{ab} e^{\alpha}_{(a)} e^{\beta}_{(b)},$$

where h^{ab} is the inverse of the induced metric.

It is possible that any arbitrary tensor $T^{\alpha\beta}$ can be projected down to the hypersurface with nonzero tangential components. The quantity that effects the projection is

$$h^{\alpha\beta} \equiv h^{ab} e^{\alpha}_{(a)} e^{\beta}_{(b)} = g^{\alpha\beta} - \epsilon n^{\alpha} n^{\beta}.$$

Note A1

One can make a 3 + 1 split of the spacetime M in a slightly different manner as follows (see Appendix C for more details):

Let us consider an arbitrary scalar field $t(x^\alpha)$ such that $t = $ constant describes a family of nonintersecting space-like hypersurfaces Σ_t. Thus,

$$M = R \times \Sigma_t.$$

If one varies the time, then the induced metric h_{ab} on Σ_t will also vary. For each Σ_t, we can define a unit normal vector n^α (which is time-like as Σ_t is space-like). Let t^α be a time-like vector, then one can decompose this into

$$t^\alpha = Nn^\alpha + \mathbf{N}^\alpha,$$

where, $\mathbf{N}^\alpha$ is the tangent to Σ_t. The scalar function N is known as **lapse** and the vector function $\mathbf{N}^\alpha$ is called **shift** vector. If one takes t without any restriction, then the shift and lapse can be arbitrary vector and scalar function, respectively. This is known as **gauge freedom**. This is used frequently in the Lagrangian formulation of general relativity as reflecting the general covariance of the theory. In fact, the set of variables $h_{ab}, N, \mathbf{N}^\alpha$ can be used instead of using the metric component $g_{\mu\nu}$. Notice that

$$\sqrt{-g} = N\sqrt{h}, \quad where \quad h = det(h_{ab}).$$

Note A2

We call the degree of freedom of a physical system as the number of independent parameters that define its configuration. The phase space is the set of all dimensions of a system and degrees of freedom are occasionally mentioned as its dimensions.

Einstein equations in vacuum, $G_{\mu\nu} = 0$, yields 10 equations for the metric tensor field $g_{\mu\nu}$ and $g_{\mu\nu}$ have 10 degrees of freedom. Bianchi identity $\nabla^\mu G_{\mu\nu} = 0$ represents four constraint equations, which decrease four degrees of freedom of $g_{\mu\nu}$. If a metric is a solution of Einstein equations in one coordinate system, then it should be a solution in any other coordinate system. This reveals that the freedom of coordinate transformations (gauge freedom) kill further four degrees of freedom in $g_{\mu\nu}$. Hence ultimately, we have two degrees of freedom in $g_{\mu\nu}$.

Note A3

The initial data problem of the gravitational field equations is known as the Cauchy problem. Suppose, we are given the values of $g_{\mu\nu}$ and its derivatives on a space-like hypersurface Σ in a 3 + 1 decomposition of the spacetime M. From these given initial data, one can find the second and higher derivatives of $g_{\mu\nu}$ through Einstein field equation. Thus, one can determine the gravitational field in the neighborhood of Σ.

B | Lagrangian Formulation of General Relativity

Practically, every fundamental equation in physics can be found with the support of a variational principle, taking appropriate Lagrangian or action in different cases.

Hamilton's variational principle asserts that $\delta S = \delta \int_{t_1}^{t_2} L\,dt = 0 \Rightarrow$ Lagrange's equation of motion. $\delta S = 0 \Rightarrow$ extremization of $S \Leftrightarrow$ equations of motion. Here, boundary conditions are provided externally. This can be generalized from Newtonian mechanics to classical field theory as in Maxwell's electrodynamics or Einstein's general relativity.

1. Newtonian Mechanics

Here, the action functional is given by

$$S[q, \dot{q}] = \int_{t_1}^{t_2} L(q, \dot{q})\,dt.$$

Here, the integration is over a specific path of the generalized coordinates $q(t)$. For a variation $\delta q(t)$ of this path, $\delta q(t_1) = \delta q(t_2) = 0$.

Extremization of the action function $\Rightarrow \delta S = 0$, i.e.,

$$
0 = \delta S = \int_{t_1}^{t_2} \delta L\,dt = \int_{t_1}^{t_2} \left(\frac{\partial L}{\partial q}\delta q + \frac{\partial L}{\partial \dot{q}}\delta \dot{q} \right) dt
$$

$$
= \left[\frac{\partial L}{\partial \dot{q}}\delta q \right]_{t_1}^{t_2} + \int_{t_1}^{t_2} \left(\frac{\partial L}{\partial q} - \frac{d}{dt}\frac{\partial L}{\partial \dot{q}} \right) \delta q\,\delta t,
$$

$$
\Rightarrow \frac{d}{dt}\frac{\partial L}{\partial \dot{q}} - \frac{\partial L}{\partial q} = 0,
$$

which is Euler–Lagrangian equation for a one-dimensional mechanical system.

2. Field Theory

Here, we are interested in the dynamics of a field $q(x^\alpha)$ in curved spacetime.

Let us consider an arbitrary region W of the spacetime manifold, bounded by a closed hypersurface ∂W. The Lagrangian $L(q, q_{,\alpha})$ depends on a scalar function of the field and its first derivative.

Thus, action function is given by

$$S[q] = \int_W L(q, q_{,\alpha}) \sqrt{-g} \, d^4 x.$$

The variation of q is arbitrary within W but vanishes on ∂W, $[\delta q]_{\partial W} = 0$.

Now,

$$\delta S = \int_W \left[\frac{\partial L}{\partial q} \delta q + \frac{\partial L}{\partial q_{,\alpha}} \delta q_{,\alpha} \right] \sqrt{-g} \, d^4 x$$

$$= \int_W \left[L' \delta q + (L^\alpha \delta q)_{,\alpha} - L^\alpha_{;\alpha} \delta q \right] \sqrt{-g} \, d^4 x$$

$$\left(L' \equiv \frac{\partial L}{\partial q}; L^\alpha \equiv \frac{\partial L}{\partial q_{,\alpha}} \right)$$

$$= \int_W (L' - L^\alpha_{;\alpha}) \delta q \sqrt{-g} \, d^4 x + \oint_{\partial W} L^\alpha \delta q \, d\Sigma_\alpha.$$

(by Gauss divergence theorem)

Thus, using $[\delta q]_{\partial W} = 0$ we get

$$\delta S = 0 \Rightarrow \nabla_\alpha \frac{\partial L}{\partial q_{,\alpha}} - \frac{\partial L}{\partial q} = 0,$$

which is Euler–Lagrange equation for a single scalar field q.

Examples of Lagrangian for some fields

(a) A scalar field ψ

This can represent, e.g., the π^0 meson. The Lagrangian

$$L = \frac{1}{2} \left(g^{\mu\nu} \psi_{,\nu} \psi_{,\mu} + \frac{m^2}{\hbar^2} \psi^2 \right)$$

Euler–Lagrange equations are

$$g^{\alpha\beta} \psi_{\,;\,\alpha\beta} = \frac{m^2}{\hbar^2} \psi$$

$$\left(L^\alpha = -g^{\alpha\beta} \psi_{,\beta} \;;\; L^\alpha_{\,;\,\alpha} = -g^{\alpha\beta} \psi_{\,;\,\alpha\beta} \;;\; L' = -\frac{m^2}{\hbar^2} \psi \right)$$

This is Klein–Gordon equation in curved space.

(b) A charged scalar field Ψ

Here $\Psi = \Psi_1 + i\Psi_2$, which could represent, e.g., π^+ and π^- meson.

The total Lagrangian of the scalar field and electromagnetic field is

$$L = \frac{1}{2} \left(\Psi_{,a} + ieA_a\Psi \right) g^{ab} \left(\overline{\Psi}_{,b} - ieA_b\overline{\Psi} \right) - \frac{1}{2} \frac{m^2}{\hbar^2} \Psi\overline{\Psi} - \frac{1}{16\pi} F_{ab}F_{cd}g^{ac}g^{bd},$$

$e \rightarrow$ constant, $\overline{\Psi}$ is complex conjugate of Ψ.

Varying $\Psi, \overline{\Psi}, \text{ and } A_a$ independently, one obtains the following Euler–Lagrange equations

$$\Psi_{;ab}g^{ab} - \frac{m^2}{\hbar^2}\Psi + ieA_a g^{ab} \left(2\Psi_{;b} + ieA_b\Psi \right) + ieA_{a;b}g^{ab}\Psi = 0,$$

its conjugate, and

$$\frac{1}{4\pi} F_{ab;c}g^{bc} - ie\Psi \left(\overline{\Psi}_{;a} - ieA_a\overline{\Psi} \right) + ie\overline{\Psi} \left(\Psi_{;a} + ieA_a\Psi \right) = 0.$$

3. General Relativity

In general relativity, the action functional consists of two different entities namely, $S_G[g]$ from gravitational field g_{ab} and $S_M[\phi, g]$ from matter distribution. Further, the gravitational action consists of three different expressions such as the Hilbert term $S_H[g]$, a boundary term $S_B[g]$, and a nondynamical term S_0. The last term does not affect the equations of motion.

Hence,

$$S_G[g] = S_H[g] + S_B[g] - S_0,$$

where

$$S_H[g] = \frac{1}{16\pi} \int_W R\sqrt{-g}d^4x, \quad S_B[g] = \frac{1}{8\pi} \int_{\partial W} \epsilon k|h|^{\frac{1}{2}}d^3y, \quad S_0 = \frac{1}{8\pi} \int_{\partial W} \epsilon k_0|h|^{\frac{1}{2}}d^3y.$$

Here, W is the spacetime manifold bounded by a closed non-null hypersurface ∂W. R is the Ricci scalar in W, k is the trace of the extrinsic curvature of ∂W, $\epsilon = \pm 1$ depending on the time-like and space-like nature of ∂W and $h = | h_{ab} |$ where h_{ab} is induced metric on ∂W. x^α and y^a are the coordinates used in W and ∂W, respectively. k_0 is the extrinsic curvature of ∂W embedded in flat spacetime. We can write the whole action functional in the following compact form as

$$S[g;\phi] = S_G[g] + S_M[\phi, g] = \int_W \left(\frac{R}{16\pi} + L_m \right) \sqrt{-g}d^4x + \frac{1}{8\pi} \int_{\partial W} \epsilon(k - k_0)|h|^{\frac{1}{2}}d^3y.$$

The Einstein field equation $G_{\alpha\beta} = -8\pi T_{\alpha\beta}$ can be obtained by varying $S[g;\phi]$ with respect to $g_{\alpha\beta}$. Note that there is no variation of $g_{\alpha\beta}$ on ∂W, i.e. $[\delta g_{\alpha\beta}]_{\partial W} = 0$ and the induced metric h_{ab} is kept fixed during the variation.

Now, we will vary the Hilbert term, boundary term and matter distribution part with respect to $g_{\alpha\beta}$ to get Einstein field equation.

VARIATION OF THE HILBERT TERM

Variation of the Hilbert term with respect to $g_{\alpha\beta}$ yields

$$16\pi\delta S_H = \int_W \delta(g^{\alpha\beta}R_{\alpha\beta}\sqrt{-g})d^4x$$

$$= \int_W (R_{\alpha\beta}\sqrt{-g}\cdot\delta g^{\alpha\beta} + g^{\alpha\beta}\sqrt{-g}\delta R_{\alpha\beta} + R\delta\sqrt{-g})d^4x,$$

$$i.e., \quad 16\pi\delta S_H = \int_W (R_{\alpha\beta} - \frac{1}{2}Rg_{\alpha\beta})\delta g^{\alpha\beta}\sqrt{-g}d^4x + \int_W g^{\alpha\beta}\delta R_{\alpha\beta}\sqrt{-g}d^4x. \tag{B1}$$

The first gives the Einstein tensor, what we want for Einstein field equation.

Now we consider the second term.

[**Note:** For a given point P in spacetime, it is always possible to find a coordinate system $x^{\alpha 1}$ such that

$$g_{\alpha^1\beta^1}(P) = \eta_{\alpha^1\beta^1}, \quad \Gamma^{\alpha^1}{}_{\beta^1\gamma^1}(P) = 0.$$

($\eta_{\alpha^1\beta^1}$ is Minskowski metric)

Such a coordinate system will be called a local Lorentz frame at P or geodesic coordinate system]

To find the variation of the Ricci tensor $\delta R_{\alpha\beta}$, we notice that in a geodesic coordinate system, we have

$$\delta R_{\alpha\beta} = \delta[\Gamma^\mu{}_{\alpha\beta,\mu} - \Gamma^\mu{}_{\alpha\mu,\beta} + \Gamma^\sigma{}_{\alpha\beta}\Gamma^\mu{}_{\mu\sigma} - \Gamma^\sigma{}_{\alpha\mu}\Gamma^\mu{}_{\beta\sigma}]$$

$$= (\delta\Gamma^\mu{}_{\alpha\beta})_{,\mu} - (\delta\Gamma^\mu{}_{\alpha\mu})_{,\beta},$$

$$= (\delta\Gamma^\mu{}_{\alpha\beta})_{;\mu} - (\delta\Gamma^\mu{}_{\alpha\mu})_{;\beta}.$$

The above equation is, however, a tensorial equation. Hence, it must be valid in all coordinate systems and at all points of spacetime not only in the geodesic system. Now,

$$g^{\alpha\beta}\delta R_{\alpha\beta} = v^\mu{}_{;\mu}; \quad where \; v^\mu = g^{\alpha\beta}\delta\Gamma^\mu{}_{\alpha\beta} - g^{\alpha\mu}\delta\Gamma^\beta{}_{\alpha\beta}, \tag{B2}$$

is a contravariant vector. Now

$$\int_W g^{\alpha\beta}\delta R_{\alpha\beta}\sqrt{-g}d^4x = \int_W v^\mu{}_{;\mu}\sqrt{-g}d^4x$$

$$= \int_{\partial W} v^\mu d\Sigma_\mu = \int_{\partial W} \epsilon v^\mu \eta_\mu h^{\frac{1}{2}}d^3y,$$

where we have used the Gauss divergence theorem and η_μ is unit normal to ∂W with $\epsilon = \eta^\mu\eta_\mu = \pm 1$.

Now, we try to find $v^\mu\eta_\mu$ on ∂W. Here,

$$\left|\delta\Gamma^\mu{}_{\alpha\beta}\right|_{\partial W} = \frac{1}{2}g^{\mu\gamma}(\delta g_{\gamma\alpha,\beta} + \delta g_{\gamma\beta,\alpha} - \delta g_{\alpha\beta,\gamma}).$$

Putting this in (B2), we get

$$v_\mu = g^{\alpha\beta}[\delta g_{\mu\beta,\alpha} - \delta g_{\alpha\beta,\mu}],$$

so that

$$\left|\eta^\mu v_\mu\right|_{\partial W} = \eta^\mu(\epsilon\eta^\alpha\eta^\beta + h^{\alpha\beta})(\delta g_{\mu\beta,\alpha} - \delta g_{\alpha\beta,\mu})$$

$$= \eta^\mu h^{\alpha\beta}(\delta g_{\mu\beta,\alpha} - \delta g_{\alpha\beta,\mu}).$$

$$\left[g^{\alpha\beta} = \epsilon\eta^\alpha\eta^\beta + h^{\alpha\beta} \; ; \; h^{\alpha\beta} = h^{ab}e^\alpha_{(a)}e^\beta_{(b)}\right].$$

Since $\delta g_{\alpha\beta}$ is zero on ∂W, therefore, its tangential derivatives should vanish, i.e.,

$$\delta g_{\alpha\beta,\gamma} e^\gamma_{(c)} = 0.$$

This indicates that

$$h^{\alpha\beta}\delta g_{\mu\beta,\alpha} = 0.$$

Therefore, we get

$$\eta^\mu v_\mu \mid_{\partial W} = -h^{\alpha\beta}\delta g_{\alpha\beta,\mu}\eta^\mu. \tag{B3}$$

Note that this expression does not necessarily vanish as normal derivative of $\delta g_{\alpha\beta}$ does not mandatorily vanish on the hypersurface. Hence finally, we get

$$16\pi\delta S_H = \int_W G_{\alpha\beta}\delta g^{\alpha\beta}\sqrt{-g}d^4x - \int_{\partial W} \epsilon h^{\alpha\beta}\delta g_{\alpha\beta,\mu}\eta^\mu |h|^{\frac{1}{2}} d^3y. \tag{B4}$$

The extra second term in (B4) will be negated by the variation of the boundary term $S_B[g]$. That's why one needs to include a boundary term in the gravitational action.

VARIATION OF THE BOUNDARY TERM

One can notice that $S_B[g]$ contains two terms namely, h and k in which h is fixed and k will vary on ∂W. Here,

$$k = \eta^\alpha_{;\alpha} = g^{\alpha\beta}\eta_{\alpha;\beta},$$

$$= (\epsilon\eta^\alpha\eta^\beta + h^{\alpha\beta})\eta_{\alpha;\beta} = h^{\alpha\beta}\eta_{\alpha;\beta},$$

$$\left[as, \; \eta^\alpha\eta_{\alpha;\beta} = \frac{1}{2}(\eta^\alpha\eta_\alpha)_{;\beta} = 0\right]$$

$$= h^{\alpha\beta}[\eta_{\alpha;\beta} - \Gamma^\gamma_{\alpha\beta}\eta_\gamma].$$

Now,

$$\delta k = -h^{\alpha\beta}\delta\Gamma^{\gamma}_{\alpha\beta}\eta_{\gamma},$$

$$= -h^{\alpha\beta}\frac{1}{2}(\delta g_{\mu\alpha,\beta} + \delta g_{\mu\beta,\alpha} - \delta g_{\alpha\beta,\mu})\eta^{\mu},$$

$$= \frac{1}{2}h^{\alpha\beta}\delta g_{\alpha\beta,\mu}\eta^{\mu}.$$

[since tangential derivatives of $\delta g_{\alpha\beta}$ vanish on ∂W]

Hence, we get,

$$(16\pi)\delta S_B = \int_{\partial W} \epsilon h^{\alpha\beta}\delta g_{\alpha\beta,\mu}\eta^{\mu}(h)^{\frac{1}{2}}d^3y. \tag{B5}$$

This expression is canceled out with the second term on the right-hand side of (B4). For $\delta S_0 = 0$, the complete variation of the gravitational action yields the desired Einstein tensor

$$\delta S_G = \frac{1}{16\pi}\int_W G_{\alpha\beta}\delta g^{\alpha\beta}\sqrt{-g}d^4x. \tag{B6}$$

VARIATION OF MATTER ACTION

Variation of $S_\mu[\phi; g]$ gives

$$\delta S_M = \int_W \delta[L_m\sqrt{-g}]d^4x$$

$$= \int_W \left(\frac{\partial L_m}{\partial g^{\alpha\beta}}\delta g^{\alpha\beta}\sqrt{-g} + L_m\delta\sqrt{-g}\right)d^4x$$

or

$$\delta S_M = \int_W T_{\alpha\beta}\delta g^{\alpha\beta}\sqrt{-g}d^4x, \tag{B7}$$

where

$$T_{\alpha\beta} = 2\frac{\partial L_m}{\partial g^{\alpha\beta}} + L_m g_{\alpha\beta}, \tag{B8}$$

is known as stress-energy tensor. For $\delta S_0 = 0$, one obtains

$$\delta(S_G + S_M) = 0 \Rightarrow G_{\alpha\beta} = -8\pi T_{\alpha\beta}.$$

Let us consider the following Lagrangian

$$L = \frac{1}{2}(\psi_{,a} + ieA_a\psi)g^{ab}(\bar{\psi}_{,b} - ieA_b\bar{\psi}) - \frac{1}{2}\frac{m^2}{\hbar^2}\psi\bar{\psi} - \frac{1}{16\pi}F_{ab}F_{cd}g^{ac}g^{bd},$$

$$T_{ab} = \frac{1}{2}(\psi_{;a}\bar{\psi}_{;b} + \bar{\psi}_{;a}\psi_{;b})$$

$$+ \frac{1}{2}(-\psi_{;a}ie\,A_b\bar{\psi} + \bar{\psi}_{;b}ie\,A_a\psi + \bar{\psi}_{;a}ieA_b\psi - \psi_{;b}ieA_a\bar{\psi} + \frac{1}{4\pi}F_{ac}F_{bd}g^{cd} + e^2A_aA_b\psi\bar{\psi} + Lg_{ab}).$$

Note B1

Energy-momentum conservation, $T^{\alpha\beta}_{;\beta} = 0$, implies Euler–Lagrangian equation.

Now we search the role of the nondynamical term in the gravitational action functional,

$$S_0 = \frac{1}{8\pi}\int_{\partial W}\epsilon k_0 h^{\frac{1}{2}}d^3y.$$

Here, S_0 contains only h_{ab}, so that its variation with respect to $g_{\alpha\beta}$ will be zero and as a result, it has no contribution in the equations of motion. It contributes only to provide a numerical value in the gravitational action.

Let us consider the vacuum solution with $R = 0$. Then the gravitational action assumes the form (excepting the contribution of k_0 term)

$$S_G = \frac{1}{8\pi}\int_{\partial W}\epsilon k h^{\frac{1}{2}}d^3y. \tag{B9}$$

We evaluate this action for flat spacetime. Let us also assume that the hypersurface ∂M comprises two hypersurfaces, $t = $ constant and a large three cylinder at $r = R$.

The induced metric (on this cylinder) is

$$ds^2 = -dt^2 + R^2 d\Omega^2,$$

so that

$$|h|^{\frac{1}{2}} = R^2\sin\theta.$$

For this cylinder, one can easily calculate

$$k = n^\alpha_{;\alpha} = \frac{2}{R}.$$

[Here, we have used $n_\alpha = \partial_\alpha r$, so that $\epsilon = 1$]

Hence, equation (B9) yields

$$\int_{\partial M} \epsilon k h^{\frac{1}{2}} d^3 y = 8\pi R(t_2 - t_1).$$

The value of the integral may be infinite for infinitely large value of the spatial boundary R. Hence the value of gravitational action for flat spacetime is infinitely large though the spacetime manifold W is bounded by two hypersurfaces of constant time. Hence, the gravitational action is not suitable for asymptotically flat spacetime. The nondynamical term S_0 solves this as the difference $S_B - S_0$ is well defined in the limit $R \to \infty$.

Note B2

$S_g = 0$ for flat spacetime.

3+1 Decomposition

To describe the Hamiltonian formulation of a field theory in curved spacetime it is necessary to foliate W, a region in the spacetime, with a family of space-like hypersurfaces, $\sum$, in every "instant of time." This is the purpose of the breakup of spacetime into space and time, i.e., $3+1$ decomposition.

To express this decomposition one needs a scalar field $t(x^\alpha)$ such that the surface of constant time ($t = constant$) represents a family of nonintersecting space-like hypersurfaces $\sum_t$.

Let $n^\alpha \, \alpha \, \partial_\alpha t$, the unit normal to the hypersurfaces, be a future-directed time-like vector field. Now one can introduce new coordinates, y^a in each hypersurface $\sum_t$. Let us assume a congruence of curves γ intersecting the hypersurfaces $\sum_t$ (see Fig. 111).

[Let O be an open region in spacetime. A congruence in O is a family of curves such that through each point in O there passes one and only one curve from this family (the curves do not intersect)]

Let t^α be the tangent to the congruence satisfying

$$t^\alpha \partial_\alpha t = 1 \tag{i}$$

Let a specific congruence curve γ_p pass through a point P on $\sum_t$ and meet a point p' on $\sum'_t$, and then a point p'' on $\sum''_t$, etc. Now one can fix coordinates of p' and P'' by imposing

$$y^a(P'') = y^a(P') = y^a(P),$$

for a given $y^a(P)$ on $\sum_t$. Therefore, we can fix y^a for every member of the congruence. Thus, in this way we can describe a new coordinate system (t, y^a) in W.

Here spacetime is foliated in terms of space-like three-dimensional hypersurfaces $\sum_t$ and the spacetime manifold W is diffeomorphic to $R \times \sum_t$, where the manifold $\sum_t$ represents space and $t \in R$ represents time. One can note that the particular slicing of spacetime into instants of time is fully arbitrary.

It is obvious that spacetime coordinate x^μ must be some function of y^a and t, i.e., $x^\alpha = x^\alpha(t, y^a)$. Here,

$$t^\alpha = \left(\frac{\partial x^\alpha}{\partial t}\right)_{y^a} = \delta^\alpha_t \tag{ii}$$

Now, one can define the tangent vector on $\sum_t$ as

$$e^\alpha_a = \left(\frac{\partial x^\alpha}{\partial y^a}\right)_t = \delta^\alpha_a. \tag{iii}$$

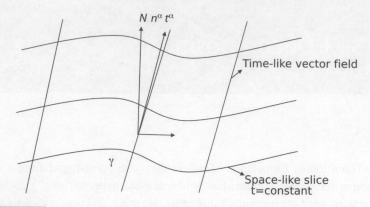

Figure 111 3+1 foliation of spacetime: Decomposition of t^α into lapse and shift.

Also Lie derivative of the tangent vector e_a^α is zero, i.e.,

$$L_t e_a^\alpha = 0.$$

The unit normal to the constraint hypersurface $\sum_t$ can be written as

$$n_\alpha = -N\partial_\alpha t, \tag{iv}$$

where the scalar function N is known as the **lapse function**. It helps to normalize n_α. Obviously,

$$n_\alpha e_a^\alpha = 0.$$

Note that it is not necessary that the curves γ do intersect $\sum_t$ orthogonally and as a result t^α is not parallel to n^α. The time-evolution vector field t^α can be decomposed to the spatial and normal parts as

$$t^\alpha = Nn^\alpha + N^a e_a^\alpha. \tag{v}$$

Here, the three vector N^a is known as the **shift vector**.

Since $x^\alpha = x^\alpha(t, y^\alpha)$, we have

$$
\begin{aligned}
dx^\alpha &= \left(\frac{\partial x^\alpha}{\partial t}\right) dt + \left(\frac{\partial x^\alpha}{\partial y^a}\right) dy^a \\
&= t^\alpha dt + e_a^\alpha dy^a \\
&= (Ndt)n^\alpha + (dy^a + N^a dt)e_a^\alpha
\end{aligned}
$$

The line element, $ds^2 = g_{\alpha\beta}dx^\alpha dx^\beta$, then can be found as

$$ds^2 = g_{\alpha\beta}[(Ndt)n^\alpha + (dy^a + N^a dt)e_a^\alpha]\,[(Ndt)n^\beta + (dy^b + N^b dt)e_b^\beta]$$

$$= N^2 dt^2 g_{\alpha\beta}n^\alpha n^\beta + g_{\alpha\beta}e_a^\alpha e_b^\beta (dy^a + N^a dt)(dy^b + N^b dt)$$

$$= N^2 dt^2 n_\alpha n^\alpha + h_{ab}(dy^a + N^a dt)(dy^b + N^b dt)$$

or

$$ds^2 = -N^2 dt^2 + h_{ab}(dy^a + N^a dt)(dy^b + N^b dt), \tag{vi}$$

where

$$h_{ab} = g_{\alpha\beta}e_a^\alpha e_b^\beta,$$

is the induced metric on $\sum_t$.

$[n^\alpha n_\alpha = -1$ as n^α is perpendicular to the space like hypersurface]

This indicates that the spacetime geometry is described not by a single metric but by the induced metric h_{ab} together with deformations of neighboring slices containing the lapse function N and the shift vector N^a.

Now one can find the determinant of the metric $g_{\mu\nu}$ in terms of determinant of the induced metric h_{ab} and lapse function N. We know from matrix theory

$$g^{tt} = \frac{cofactor\,(g_{tt})}{g} = \frac{h}{g}. \tag{vii}$$

However using (iv), one can have

$$g^{tt} = g^{\alpha\beta}t_{,\alpha}t_{,\beta} = g^{\alpha\beta}n_\beta n_\alpha N^{-2} = n^\alpha n_\alpha N^{-2} = -N^{-2}. \tag{viii}$$

Using (vii) and (viii), one find the required relation as

$$\sqrt{-g} = N\sqrt{h}. \tag{ix}$$

For 3+1 decomposition of spacetime equations (v), (vi), and (ix) are the fundamental results.

Bibliography

H. Atwater, *General Relativity* (Pergamon Press, Oxford, UK, 1974).

K. D. Abhyankar, *Astrophysics* (Universities Press, Hyderabad, India, 2001).

N. Banerji and A. Banerjee, *General Relativity and Cosmology* (Elsevier India, Gurgaon, Haryana, India, 2007).

G. Barton, *Introduction to the Relativity Principle* (John Wiley and Sons Ltd, New Jersey, USA, 1999).

P. G. Bergmann, *Introduction to the Theory of Relativity* (Prentice Hall of India, New Delhi, India, 1992).

S. K. Bose, *An Introduction to General Relativity* (John Wiley and Sons, New Jersey, USA, 1980).

M. Carmeli, *Cosmological Special Relativity* (World Scientific, Toh Tuck Link, Singapore, 2002).

M. Carmeli, *Classical Felds: General Relativity and Gauge Theory* (John Wiley and Sons, New Jersey, USA, 1982).

S. Carroll, *Spacetime and Geometry: An Introduction to General Relativity* (Pearson, Cambridge, UK, 2003).

S. Chandrasekhar, *Mathematical Theory of Black Holes* (Springer, New York City, USA, 1994).

F. deFelice and C. J. S. Clarke, *Relativity on Curved Manifolds* (Cambridge, Cambridge, UK, 1992).

A. Das and A. DeBenedictis, *The General Theory of Relativity* (Springer, New York City, USA, 2012).

R. D'Inverno, *Introducing Einstein's Relativity* (Clarendon Press, Gloucestershire ,England, 1992).

S. S. De and F. Rahaman, *Finsler Geometry of Hadrons and Lyra Geometry: Cosmological Aspects* (Lambert, Mauritius, 2012).

A. Eddington, *Space Time and Gravitation: An outline of the general relativity theory* (Cambridge University Press, Oxford, UK, 1920).

R. Feynman, *Lectures on Gravitation* (Westview Press, Colorado, USA, 2002).

S. Gibilisco, *Understanding Einstein's Theories of Relativity* (Dover Publication, New York City, USA, 1983).

H. Goldstein, *Classical Mechanics* (Narosa, Delhi, India, 1998).

D. J. Griffiths, *Electrodynamics* (Wiley Eastern, Delhi, India, 1978).

J. B. Griffiths, *Exact Space-Times in Einsteins General Relativity* (Cambridge University Press, Cambridge, UK, 2009).

O. Gron, *Einstein's General Theory of Relativity* (Springer, New York City, USA, 2007).

J. B. Hartle, *Gravity: An Introduction to Einstein's General Relativity* (Pearson, New York City, USA, 2014).

S. Hawking and G. F. R. Ellis, *The Large Scale Structure of Space-Time* (Cambridge, Cambridge, UK, 1975).

Hubble, Edwin "A relation between distance and radial velocity among extragalactic nebulae". *PNAS*. **15**(3): 168–173 (Washington, DC 20001, USA, 1929).

J. N. Islam, *An Introduction to Mathematical Cosmology* (Cambridge University Press, Cambridge, UK, 1992).

P. S. Joshi, *Global Aspects in Gravitation and Cosmology. International Series of Monographs on Physics* (Clarendon Press, Gloucestershire, England, 1993).

J. D. Jackson, *Electrodynamics* (Wiley Eastern, Delhi, India, 1978).

J. B. Kogut, *Introduction to Relativity* (Academic Press, Massachusetts, USA, 2000).

L. D. Landau and E. M. Lifshitz, *Classical Theory of Fields* (Pergamon Press, Oxford, UK, 1975).

E. A. Lord, *Tensor Relativity and Cosmology* (Tata McGraw-Hill, New York City, USA, 1976).

M. Ludvigsen, *General Relativity: A Geometric Approach* (Cambridge University Press, Cambridge, UK, 1999).

C. W. Misner, K. Throne and J. Wheeler, *Gravitation* (Freeman, New York City, USA, 1973).

C. Moller, *The Theory of Relativity* (Oxford University Press, Oxford, UK, 1972).

J. V. Narlikar, *General Relativity and Cosmology* (Macmillan, New York City, USA, 1979).

J. V. Narlikar, *An Introduction to Cosmology* (Cambridge University Press, Cambridge, UK, 2002).

E. T. Newman and A. I. Janis, Note on the Kerr spinning-particle metric. *Journal of Mathematical Physics*, **6** (American Institute of Physics, Maryland, USA, 1965), 915.

E. T. Newman, E. Couch, K. Chinnapared, et al., Metric of a rotating, charged mass. *Journal of Mathematical Physics*, **6** (American Institute of Physics, Maryland, USA, 1965), 918.

E. T. Newman, Complex coordinate transformations and the Schwarzschild-Kerr metrics. *Journal of Mathematical Physics,* **14** (American Institute of Physics, Maryland, USA, 1973), 774.

E. P. Ney, *Electromagnetism and Relativity* (John Weatherhill, Cork, Ireland, 1965).

Olbers, Heinrich Wilhelm Matthias, Encyclopædia Britannica (11th ed.). Cambridge University Press, UK, Cambridge, UK.

R. K. Pathria, *The Theory of Relativity* (Dover Publications, New York City, USA, 1974).

W. Pauli, *Theory of Relativity* (Dover Publications, New York City, USA, 1958).

Perlmutter, S., *et al.*, *ApJ*, **517**, 565 (University of Chicago, USA, 1999).

E. Poisson, *A Relativist's Toolkit* (Cambridge University Press, Cambridge, UK, 2004).

S. Prakash, *Relativistic Mechanics* (Pragati Prakashan, Meerut, Uttar Pradesh, India, 2000).

S. P. Puri, *General Theory of Relativity* (Pearson, New York City, USA, 2013).

F. Rahaman, *The Special Theory of Relativity-A Mathematical Approach* (Springer, New York City, USA, 2014).

A. K. Raychaudhuri, S. Banerjee and A. Banerjee, *General Relativity, Astrophysics, and Cosmology* (Springer, New York City, USA, 1992).

Riess, A. G., *et al.*, *AJ*, **116**, 1009 (University of Chicago, USA, 1998).

W. Rindler, *Relativity* (Clarendon Press, Gloucestershire, England, 2002).

H. P. Robertson, *Kinematics and world structure II", Astrophysics Journal*, **83**: 187–201, (University of Chicago, USA, 1936).

W. G. V. Rosser, *An Introduction to the Theory of Relativity* (Butterworths, Oxford, UK, 1964).

B. A. Schutz, *First Coursein General Relativity* (Cambridge University Press, Cambridge, UK, 2009).

I. I. Shapiro, *et al.*, *Physical Review Letters*, **26**, 1132 (American Physical Society, Maryland, USA, 1971).

R. Sharipov, *Classical Electrodynamics and Theory of Relativity* (Samizdat Press, Toronto, Ontario, Canada, 2003).

H. Stephani, *General Relativity: An Introduction to the Theory of the Gravitational Field* (Cambridge University Press, Cambridge, UK, 1985).

H. Stephani, D. Kramer, M. MacCallum, C. Hoenselaers and E. Herit,a *Exact Solutions of Einstein's Field Equations* (Cambridge, Cambridge, UK, 2009).

M. M. Schiffer, Kerr geometry as complexified Schwarzschild geometry. *Journal of Mathematical Physics*, **14** (American Institute of Physics, Maryland, USA, 1973), 52.

J. L. Synge and B. A. Griffith, *Classical Mechanics* (McGraw Hill, New York City, USA, 1949).

E. Taylor and J. Wheeler, *Spacetime Physics* (Freeman, New York City, USA, 1992).

R. Tolman, *Relativity, Thermodynamics and Cosmology* (Dover, New York City, USA, 1934).

P. Tourrence, *Relativity and Gravitation* (Cambridge University Press, Cambridge, USA, 1992).

M. Visser, *Lorentzian Wormholes: From Einstein to Hawking* (AIP, California, USA, 1997).

M. Visser, The Kerr spacetime: a brief introduction. Online Archive arXiv:0706.0622v3 [gr-qc] (2008).

R. M. Wald, *General Relativity* (The University of Chicago Press, Chicago, USA, 1984).

A. G. Walker, "On Milne s theory of world-structure", *Proceedings of the London Mathematical Society, Series 2*, **42**(1): 90–127 (London, UK, 1937).

S. Weinberg, *Gravitation and Cosmology* (Wiley, New Jersey, USA, 1972).

Index